D1228594

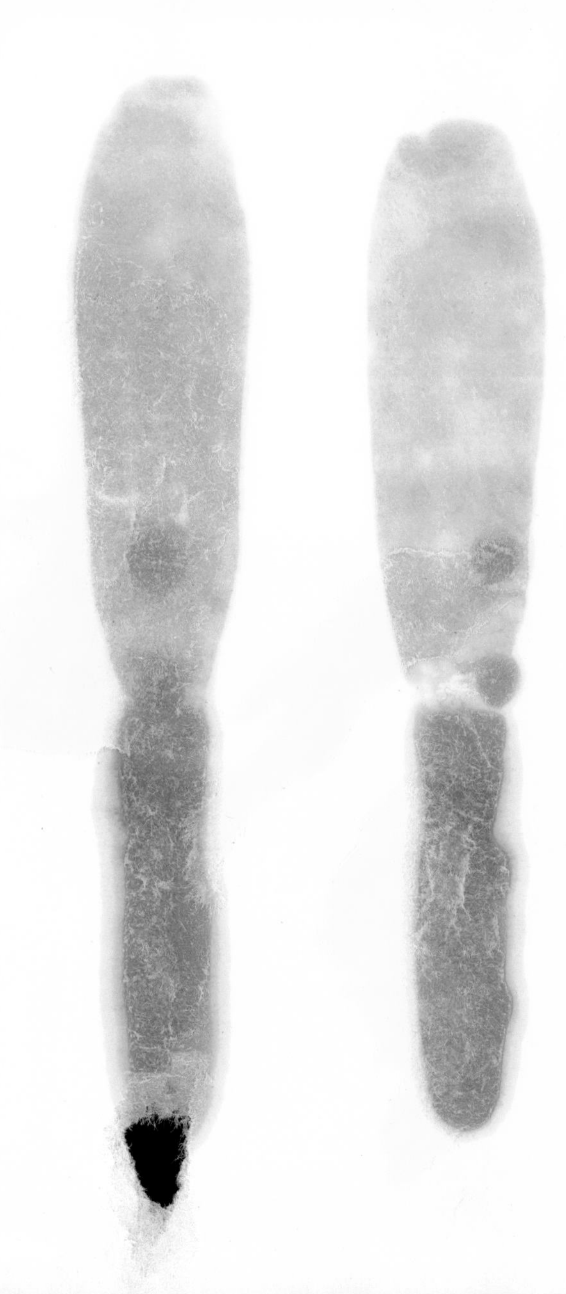

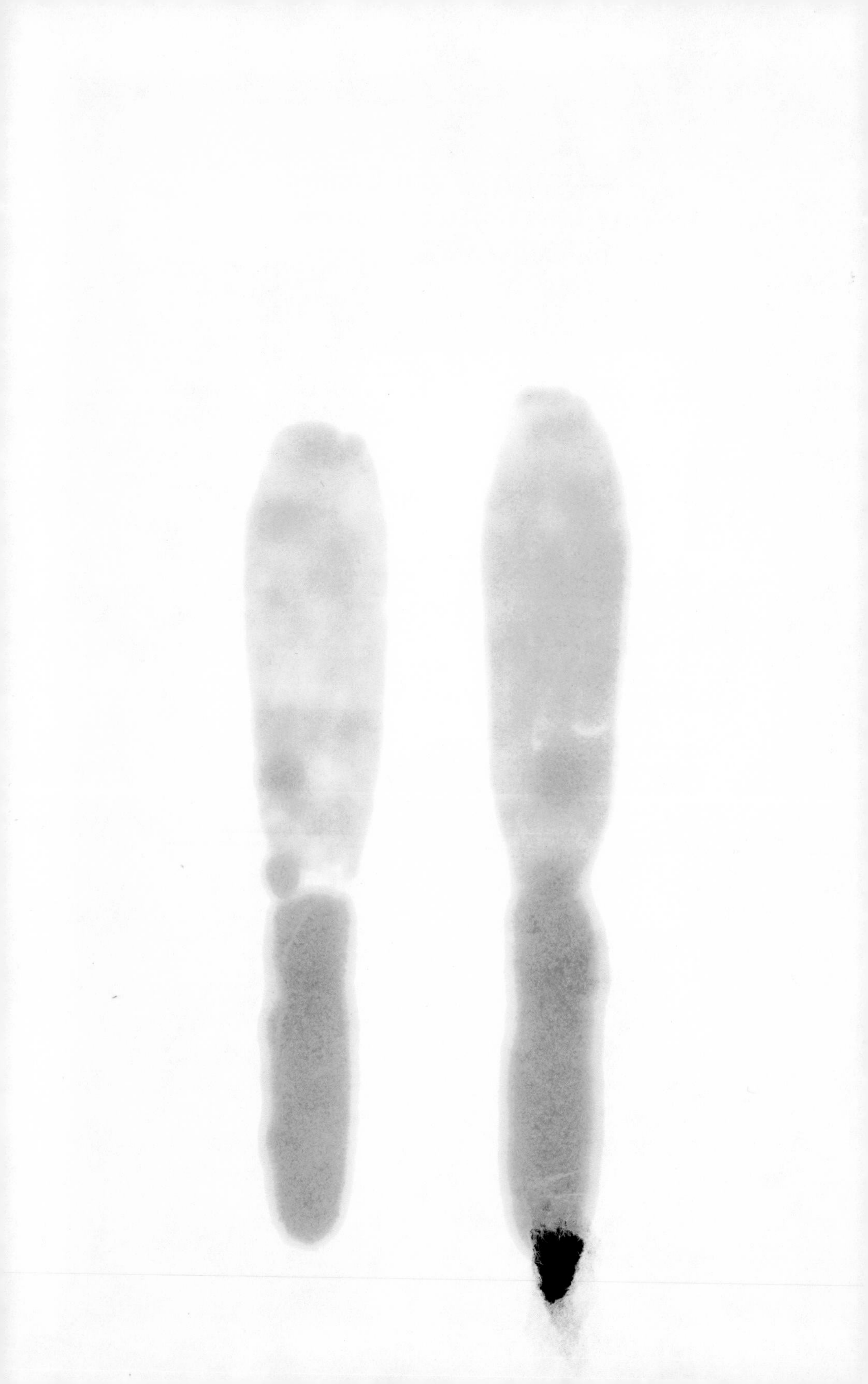

The Analysis and Control of Less Desirable Flavors in Foods and Beverages

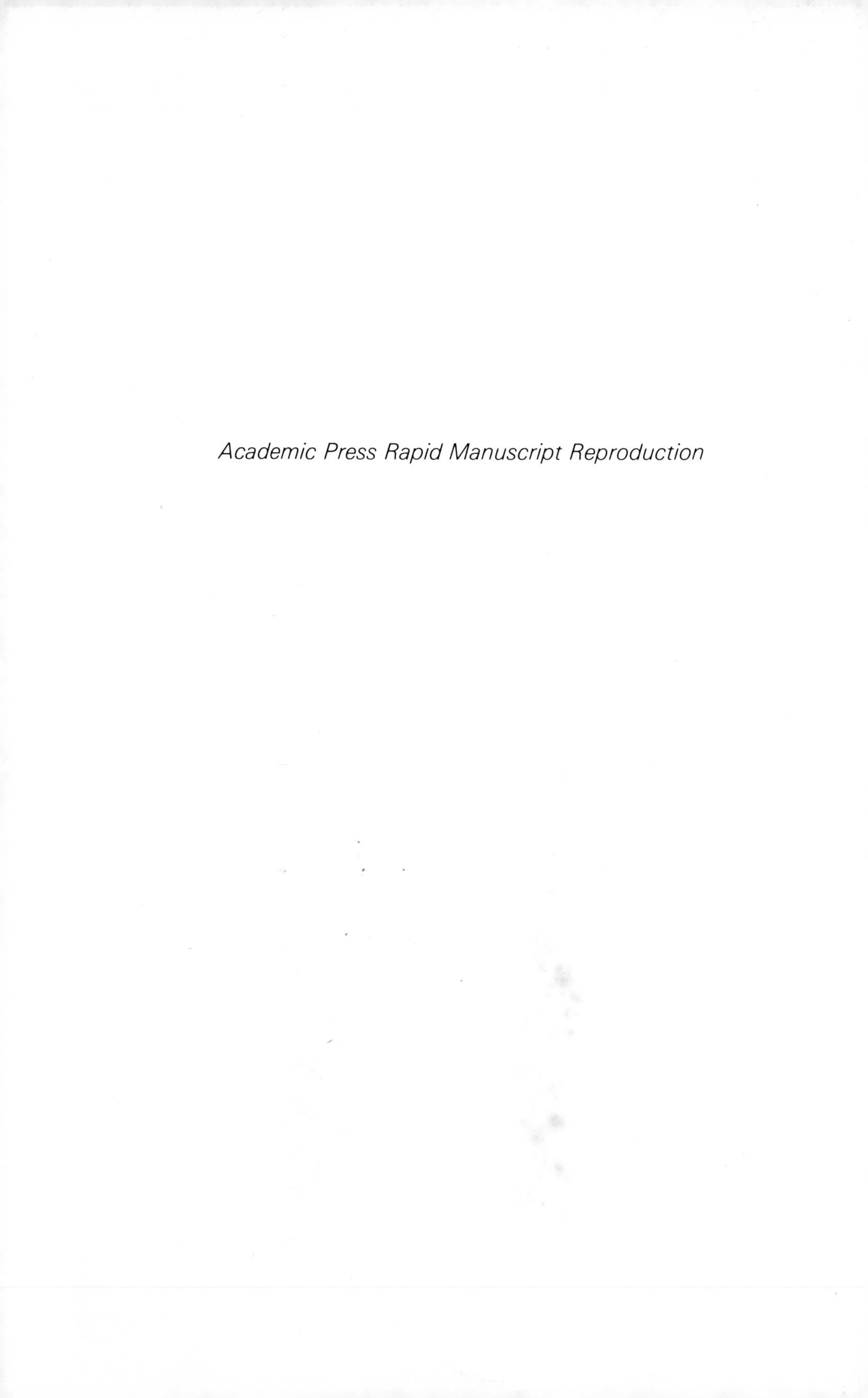

Academic Press Rapid Manuscript Reproduction

The Analysis and Control of Less Desirable Flavors in Foods and Beverages

Edited by

George Charalambous

ACADEMIC PRESS 1980

A Subsidiary of Harcourt Brace Jovanovich, Publishers

New York London Toronto Sydney San Francisco

ACADEMIC PRESS, INC.
111 Fifth Avenue, New York, New York 10003

United Kingdom Edition published by
ACADEMIC PRESS, INC. (LONDON) LTD.
24/28 Oval Road, London NW1 7DX

Library of Congress Cataloging in Publication Data

The Analysis and control of less-desirable flavors
in foods and beverages.

Proceedings of a symposium held at the Second
Chemical Congress of the North American Continent
in Las Vegas in Aug. 1980.

1. Food—Analysis—Congresses. 2. Flavor—
Congresses. I. Charalambous, George, DATE
II. Chemical Congress of the North American
Continent, 2d, Las Vegas, 1980.
TP372.5.A5 664'.07 80-22291
ISBN 0-12-169065-2

PRINTED IN THE UNITED STATES OF AMERICA

80 81 82 83 9 8 7 6 5 4 3 2 1

Contents

Contributors

Numbers in parentheses indicate the pages on which authors' contributions begin.

A. Aitoku (241), Ogawa & Co., Ltd., 6-32-9 Akabanemishi, Kita-Ku, Tokyo, Japan

M. A. Amerine (319), Department of Viticulture and Enology, University of California, Davis, California 95616; and Wine Institute, San Francisco, California 94108

S. Arai (1, 133), Department of Agricultural Chemistry, The University of Tokyo, Bunkyo-ku, Tokyo, Japan

D. Bahri (293), Technische Universität Berlin, D-1000 Berlin 65, West Germany

M. E. Bailey (31), Food Science and Nutrition Department, University of Missouri, Columbia, Missouri 65211

J. A. Burnette (17), Department of Food Science and Technology, Virginia Polytechnic Institute and State University, Blacksburg, Virginia 24061

S. Damodaran (95), Institute of Food Science, Cornell University, Ithaca, New York 14853

H. P. Dupuy (31), V-Labs, Covington, Louisiana

G. J. Flick (17), Department of Food Science and Technology, Virginia Polytechnic Institute and State University, Blacksburg, Virginia 24061

R. L. Hagy (71), Grain Processing Corporation, Muscatine, Iowa 52761

J. Hayashi (241), Ogawa & Co., Ltd., 6-32-9 Akabanemishi, Kita-Ku, Tokyo, Japan

W. G. Jennings (3), Department of Food Science and Technology, University of California, Davis, California 95616

Y. Kamiya (241), Ogawa & Co., Ltd., 6-32-9 Akabanemishi, Kita-Ku, Tokyo, Japan

J. E. Kinsella (95), Institute of Food Science, Cornell University, Ithaca, New York 14853

D. L. Kiser (71), Grain Processing Corporation, Muscatine, Iowa 52761

M. Kossa (293), Technische Universität Berlin, D-1000 Berlin 65, West Germany

M. G. Legendre (17, 31), Science and Education Administration, Agricultural Research, Southern Regional Research Center, United States Department of Agriculture, New Orleans, Louisiana 70179

S. D. Leonard (149), Water Quality Division, San Francisco Water Department, Millbrae, California 94030

M. R. McDaniel (267), Department of Foods and Nutrition, University of Manitoba, Winnipeg, Manitoba, Canada R3T 2N2

S. Mihara (241), Ogawa & Co., Ltd., 6-32-9 Akabanemishi, Kita-Ku, Tokyo, Japan

G. J. Moskowitz (53), Dairyland Food Laboratories, Inc., Waukesha, Wisconsin 53187

S. Nagy (171), Florida Department of Citrus, University of Florida, AREC, Lake Alfred, Florida 33850

O. Nishimura (241), Ogawa & Co., Ltd., 6-32-9 Akabanemishi, Kita-Ku, Tokyo, Japan

R. L. Olson (71), Grain Processing Corporation, Muscatine, Iowa 52761

R. L. Ory (17), Science and Education Administration, Agricultural Research, Southern Regional Research Center, United States Department of Agriculture, New Orleans, Louisiana 70179

R. Rouseff (171), Florida Department of Citrus, University of Florida, AREC, Lake Alfred, Florida 33850

T. Shibamoto (241), Department of Environmental Toxicology, University of California, Davis, California 95616

W. F. Shipe (201), Institute of Food Science, Cornell University, Ithaca, New York 14853

A. J. St. Angelo (17), Science and Education Administration, Agricultural Research, Southern Regional Research Center, United States Department of Agriculture, New Orleans, Louisiana 70179

H. W. Tracy (149), Water Quality Division, San Francisco Water Department, Millbrae, California 94030

R. Tressl (293), Technische Universität Berlin, D-1000 Berlin 65, West Germany

Preface

The flavor of foods and beverages is a matter of perennial importance to growers, processors, fabricated food and beverage manufacturers, brewers, wine makers, distillers, and, ultimately, the consumer. Nutrition, naturalness, appearance, cost, packaging—all of the several inherent or acquired desirable characteristics of a food or beverage—remain incomplete in the absence of a unique asset, an acceptable flavor.

Conversely, the presence of less desirable or downright off-flavors will mar the acceptability of any food or beverage. Such undesirable tastes and aromas may arise in a great variety of ways, through spoilage, staling, heating and other processing steps, packaging.

This book is the proceedings of a symposium on the analysis and control or prevention of less desirable flavors in foods and beverages. The symposium was held in August 1980 at the Second Chemical Congress of the North American Continent under the auspices of the Agricultural and Food Chemistry Division of American Chemical Society. In keeping with the character of this venue, contributions to this symposium are from Canada, Europe, Japan, and the United States, from both academia and industry.

The scope of the symposium is comprehensive. The stage is set by an authoritative discussion of very recent progress made on scattered fronts concerned with improved methods of glass capillary gas chromatographic columns and other devices for better separation of volatile components. There follow expert accounts of up-to-date research and technological developments leading to the analysis and control or prevention of less-desirable taste and aroma factors in: fish and shellfish; meat, cheese, soy proteins, corn syrup, water, citrus juices, milk (including the human variety), and beer. The chapter on wine is a "first" in that it embodies a unique compilation, source material, and discussion of the vocabulary used to describe abnormal appearance, odor, taste, and tactile sensations in wine. This contribution is the distillation of a lifetime's expertise in this context. All chapters conclude with very comprehensive bibliographies.

Paraphrasing Murphy's law, if something abnormal can occur, it will. It is therefore to be hoped that this book—which concerns itself with the pathology (Webster: deviation from assumed normal state) of certain foods and beverages, will prove useful to students, chemists, food technologists, and executives in their continuing efforts to analyze, categorize, avert, and control less desirable flavors in foods and beverages.

Best thanks are due to the symposium participants, Professor Soichi Arai of the University of Tokyo for contributing the Introduction, and to the publishers for their guidance and assistance.

INTRODUCTION

Soichi Arai

Department of Agricultural Chemistry
University of Tokyo
Bunkyo-ku, Tokyo 113, Japan

Man survives only by consuming other organisms as his food. All food organisms go through their own life cycle until harvested. The preharvest process consists of these features in chronological order: genesis, growth, maturation and reproduction. A variety of environmental factors affect such a life process in whole or in part. These include ambient temperature, moisture, light, composition of the atmosphere, nutrient concentration, etc. Invading microorganisms and even some pollutants also may be included. Failure to control properly such factors will eventually have an undesirable effect on the quality of food organisms at the time of harvesting.

Most food organisms after harvest receive artificial treatments for processing and preservation, prior to being consumed. A main purpose of the treatments is to lengthen the time from harvest (H) to consumption (C). The longer the time HC, the farther the food can be transported from its site of harvest to the consumer. This relation is particularly important in the efficient feeding of people living in highly sophisticated society (1). Modern science and technology have developed a high level of methods for processing and preservation of food organisms in the postharvest stage. Food processing in the forms of peeling, grinding, extracting, mixing, cooking, solubilizing, the use of enzymes and of microorganisms in fermentation, etc., and also food preservation in the forms of drying, heating, sterilizing, irradiating, packaging, refrigerating, freezing, etc. are even a major part of food industry in economically developed areas of the world. If for any reason, any one of the several necessary treatments fails to function properly, it will undesirably affect the quality of food at the time of its consumption.

Flavor is a most important index of food quality and can sensitively reflect the total career of a food organism

ISBN 0-12-169065-2

through its pre- and postharvest process. Both odor and taste substances induce the sensation, flavor. Food texture also may often act as a factor responsible for this sensation (2). Flavor thus plays a central part in the quality of our food, even determining its acceptability in most cases. It is extremely important to determine a way of controlling less-desirable flavors in foods and beverages, in order to increase their acceptability for our consumption.

Several chapters in the present volume deal with this aspect in detail. Examples are given concerning how to control less-desirable flavors occurring in a number of food products of animal origin (fish and shellfish, cow's and human milk, meat, etc.) and of plant origin (citrus juices, beer, wines, corn syrup, plant proteins and their hydrolysates, etc.). A unique paper is presented on the taste and odor of public water supplies which are used commonly for food processing and manufacture.

The process control for minimizing less-desirable flavors requires in the first place the detection, measurement, separation and identification of responsible factors. Recent progress in instrumental analysis has satisfied this requirement to a large measure. Gas chromatography, while restricted to volatile compounds, is the most widely-used technique in the world today. High performance liquid chromatography also has become an indispensable method especially for the separation of compounds of lower volatility. As two chapters emphasize, further methodological development in the field of flavor research would permit more and more profound analysis of less-desirable flavors and proper control of the chemical reactions by which these flavors are produced.

All the topics discussed in this volume contribute greatly to the knowledge of why and how undesirable reactions take place in food organisms during the period from genesis to harvest and in their products during the period from harvest to consumption. The importance of improving and maximizing the quality of foods and beverages by means of analysis and control of their less-desirable flavors is stressed. We cannot lose sight of the fact that flavor is a most sensitive index of the quality of our food, often influencing its total value in a definite manner.

REFERENCES

1. Deatherage, F. E., "Food for Life", p. 247. Plenum Press, New York and London, (1975).
2. Schutte, L., *in* "Phenolic, Sulfur, and Nitrogen Compounds in Food Flavors" (G. Charalambous and I. Katz, eds.), p. 101. American Chemical Society, Washington, D.C., (1976).

RECENT ADVANCES IN THE SEPARATION OF VOLATILE COMPONENTS

Walter Jennings

Department of Food Science & Technology
University of California
Davis, California

I. INTRODUCTION

Studies concerned with flavor--regardless of the degree of response elicited and whether it is attraction or repulsion--must at some point in time become concerned with determining which of the constituent compounds elicit that particular response. While this information will allow the analyst to detect (and hopefully to quantitate) those compounds, frequently he wants to progress a bit further; he'd like to establish their chemical identities, and elucidate the reaction mechanisms by which they are produced and altered. In some environments we can justify these efforts on a "pursuit of knowledge" argument; but of course this information can also have very practical ramifications. It can suggest causative factors--and/or control mechanisms, i.e. avoidance of light, elimination of oxygen, adjustment of pH, or removal of a reactant.

In most cases it simply is not possible to accomplish any of these goals--detection, measurement, characterization, or elucidation of biological pathways--unless the culprit compound(s) is isolated from other compounds in the mixture. The final results, of course, will also be influenced by the fidelity of our sampling procedures and the individual stabilities of the compounds toward our analytical system, but these cans of worms I prefer to leave unopened for this presentation. Permit me instead to concentrate my efforts on that step which is at least equally important and which probably is most frequently the limiting step: the separation of this complex mixture.

ISBN 0-12-169065-2

Those separation techniques employed most widely are chromatographic, and the chromatographic methods consuetudinary to the study of flavor compounds are gas chromatography (GC) and high performance (or, as most practice it, high pressure) liquid chromatography (HPLC). The former of course is restricted to volatile components and compounds that can be converted to volatile derivatives, while the latter is primarily, but not exclusively, applied to the non-volatile constituents. I'd like to set the stage for this series of papers by reviewing a few recent developments in the area of volatile component separation.

GC is still today the single most widely-used technique, not only in flavor studies, but in all forms of analysis (1). Several factors have contributed to this wide and general acceptance: A high degree of individual component separation can be achieved, analysis times are reasonably short, and sensitivities are usually high. It is a bit surprising that the flavor chemist, who was very active in the development of packed and large-bore capillary GC, has been relatively passive in the newer developments of glass capillary GC which, when compared with packed column analyses, can yield gains in separation efficiencies (e.g. 2), analysis times (e.g. 3) and sensitivities that are measurable in orders of magnitude (4). But even with these more powerful systems, the complexities of the mixtures which concern the flavor chemist are usually so great that it is probably safe to conclude that a complete separation is rarely achieved. This dilemma is also of concern to other analytical chemists, and I'd like to concentrate this paper on reviewing some of the progress that has been made on scattered fronts concerned with improved methods of gas chromatographic separation.

II. TWO DIMENSIONAL GC

Bertsch (5) suggested that this much-abused term be restricted to systems that fit one of two criteria: systems containing two columns of different selectivity, used in a manner that permits the assignment of retention indices, or systems permitting the selective transfer of a portion of a chromatographic run to a second column of different selectivity. Such methods have been developed and explored by many investigators (e.g. 5-8). It is immediately apparent that two-(or multi-) dimensional GC requires redirecting of the carrier gas stream, and this poses certain problems. Early efforts in this direction involved packed columns, whose relatively large carrier gas flows made a variety of valves and connectors tolerable. When these devices were used with the more efficient open tubular capillary columns,

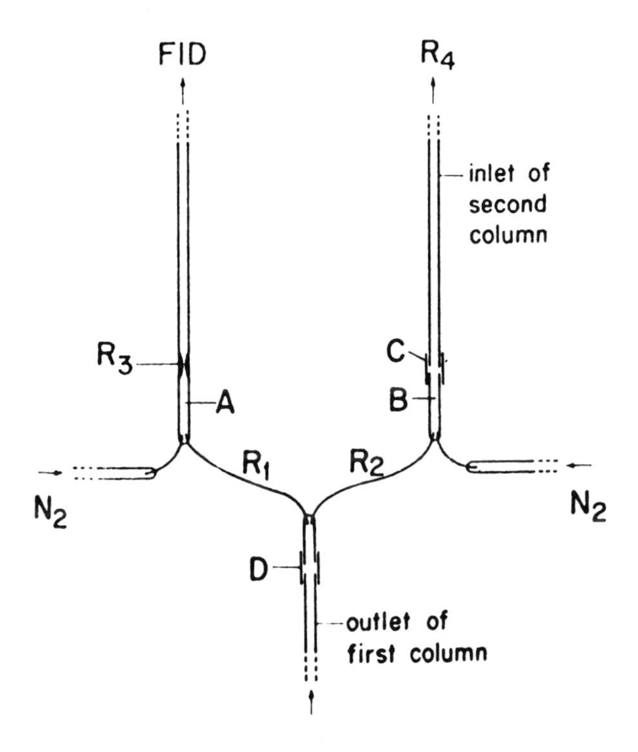

Fig. 1. Use of a Deans' switch to direct the effluent of a glass capillary column to either a flame ionization detector, or to the inlet of a second column. Switching is achieved by adjustment of the relative pressures in the left and right nitrogen streams, relative to that of the column outlet. After Bertsch (5).

band broadening in the valve and its connecting tubing, solute absorption in the valve lubricant, and adsorption on elastomeric and/or metallic components invariably resulted in the loss of an unacceptable amount of system efficiency. Deans (9) introduced a novel method of pressure switching which solved most of these problems; a simple Deans' switch application is shown in Fig. 1, and Fig. 2 shows a more elaborate system (5). Fig. 3 illustrates a poorly resolved section of a chromatogram from one glass capillary switched to a second glass capillary where full resolution was

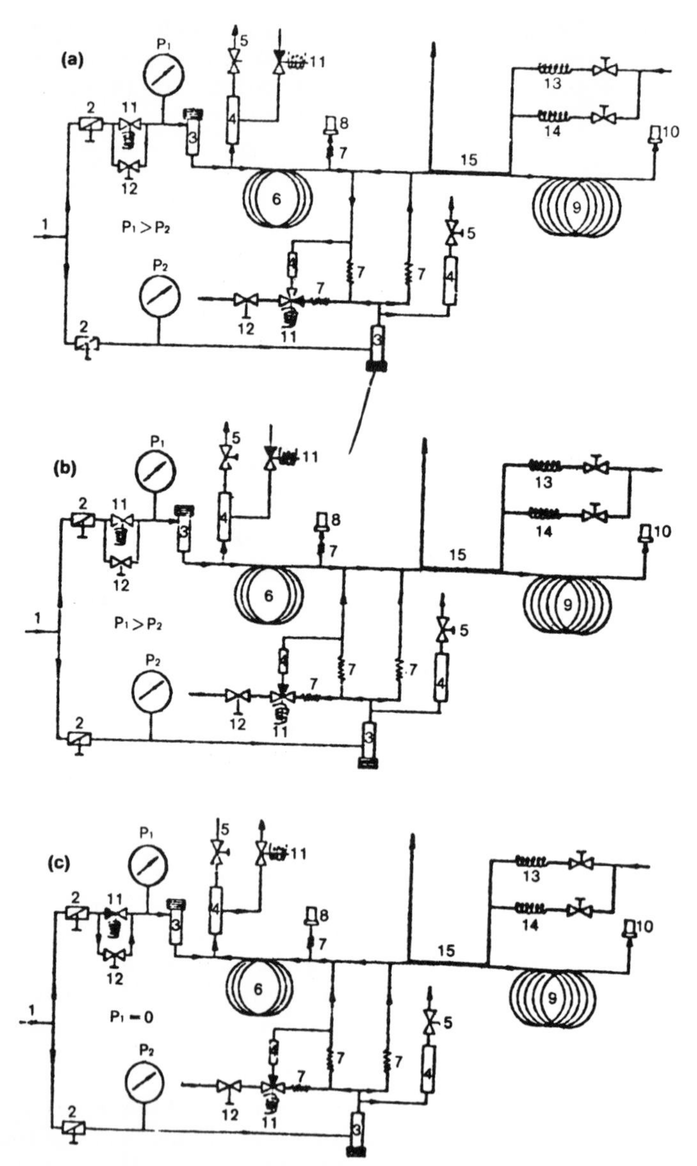

Fig. 2. Automated system with coupled columns and intermediate trapping. 1 = carrier; 2 = pressure control; 3 = injector; 4 = filter; 5 = split; 6 = precolumn; 7 = throttles; 8 = control FID; 9 = main column; 10 = FID; 11= solenoid valves; 12 = needle valves; 13 = cooled coil; 14 = heated coil; 15 = trap. Status: (a) precolumn eluate vented; (b) precolumn eluate to trap; (c) backflush of slower components from precolumn. After Schomburg et al.(10).

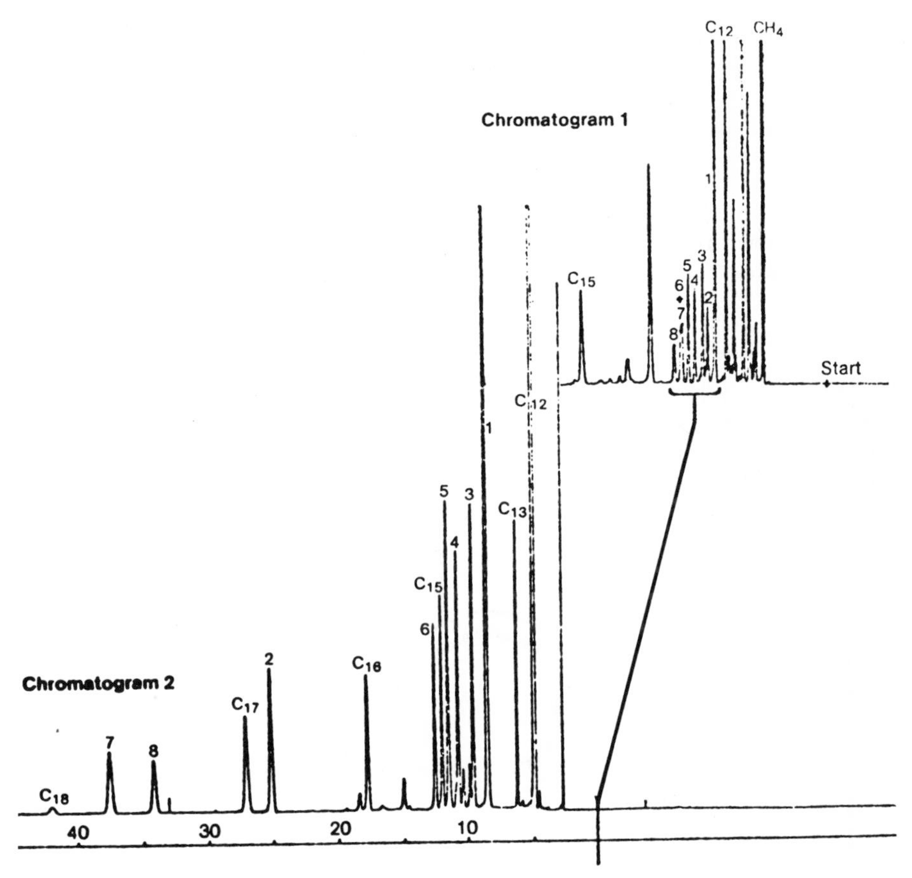

Fig. 3. Isothermal dual column chromatography with glass capillary columns. (1), 20 m x 0.25 mm OV-101 column at 160°; sample: perfume oil, CH_4, C_8, C_{12} and C_{15}. (2), 35 m x 0.25 mm OS 138 column at 160°; cut of chromatogram #1 after trapping and reinjection. After Schomburg et al. (10).

achieved (10). Because the Deans' switch requires careful balancing of the gas streams, its use in a programmed mode can lead to pressure imbalances and inadvertent flow stream switching.

Miller et al. (11) described a mechanical rotary valve and associated connections whose internal volumes were sufficiently small to match the flow volumes commensurate with glass capillary GC. Because gas flows in the two columns are entirely independent, the system can be operated either isothermally, or in a programmed mode. Fig. 4 illustrates the system, and Fig. 5 shows a classical example of two-dimensional GC (11). These methods seem to offer considerable promise and are worthy of additional effort.

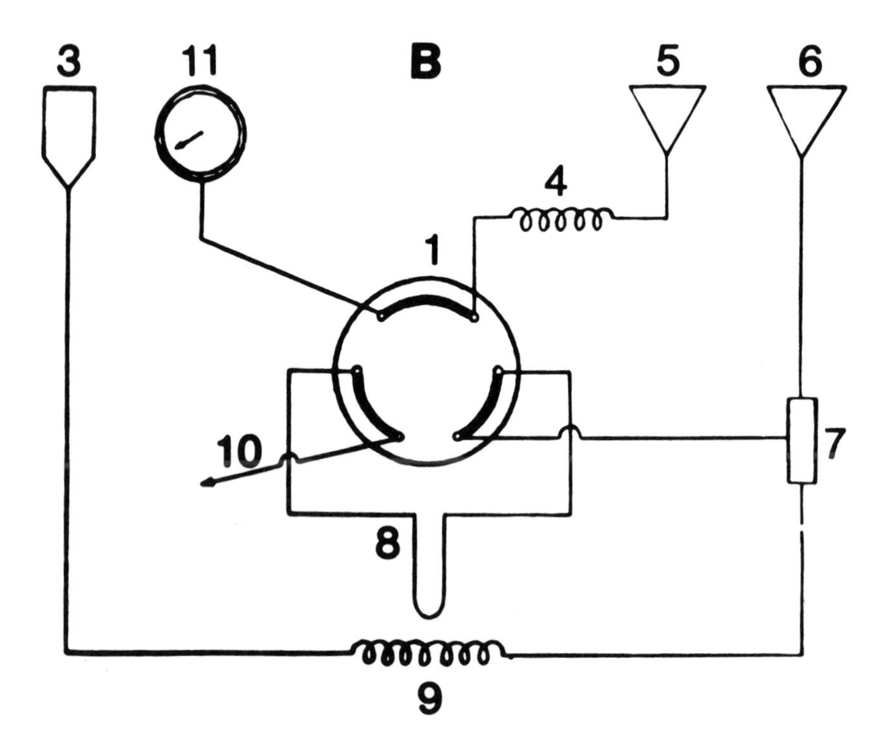

Fig. 4. Schematic of two-dimensional GC with dual glass capillary columns and a mechanical flow switching valve. 1= valve; 3 = inlet; 4 = analytical column; 5 = B FID; 6 = A FID; 7 = splitter; 8 = trap; 9 = precolumn; 10 = vent. After Miller et al. (11).

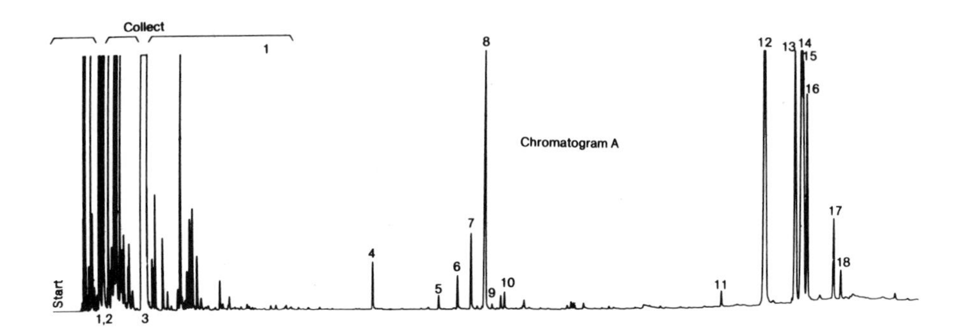

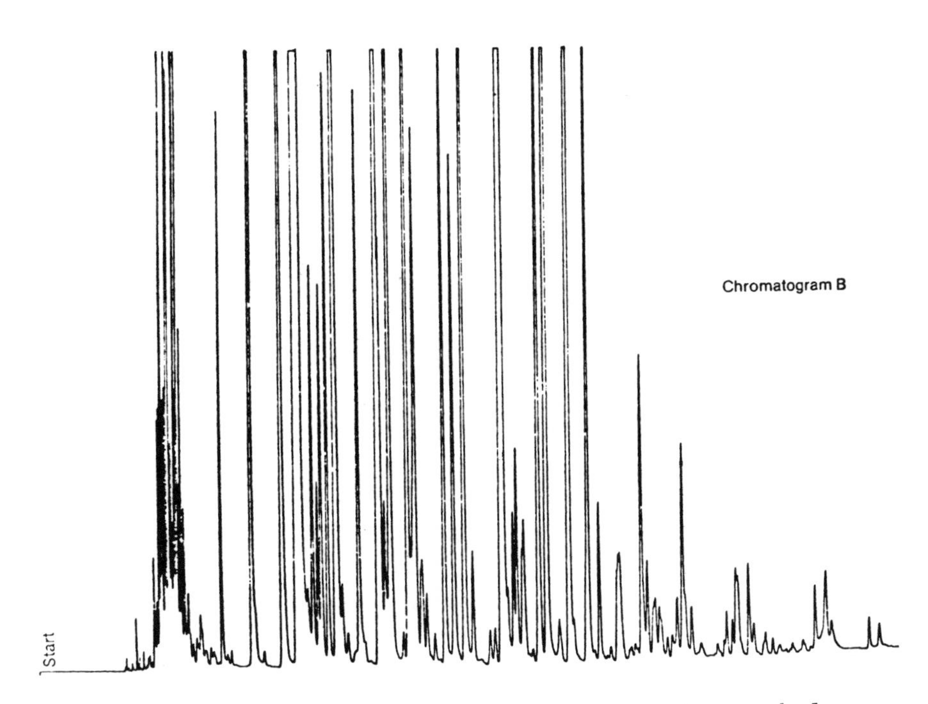

Fig. 5. Heart cutting of alkyl benzenes from a hydrocarbon feedstock of wide volatility range. Chromatogram A, precolumn separation on a 60 m x 0.25 mm SP 2100 glass capillary 100° 10 min and programmed at 2°/min to 280°, final hold 30 min. Chromatogram B, alkyl benzene fraction shunted to a 60 m x 0.25 mm capillary coated with SP 1000; 100° 12 min, 2°/min to 220°. After Miller et al. (11).

III. SECAT GC

Pretorius et al. reported that slight changes in the temperature of one or both of two sequentially coupled columns containing dissimilar liquid phases could exercise a dramatic effect on the overall relative retentions of solutes, and that it was even possible to change elution orders (Fig. 6) (12). Although their experiments utilized

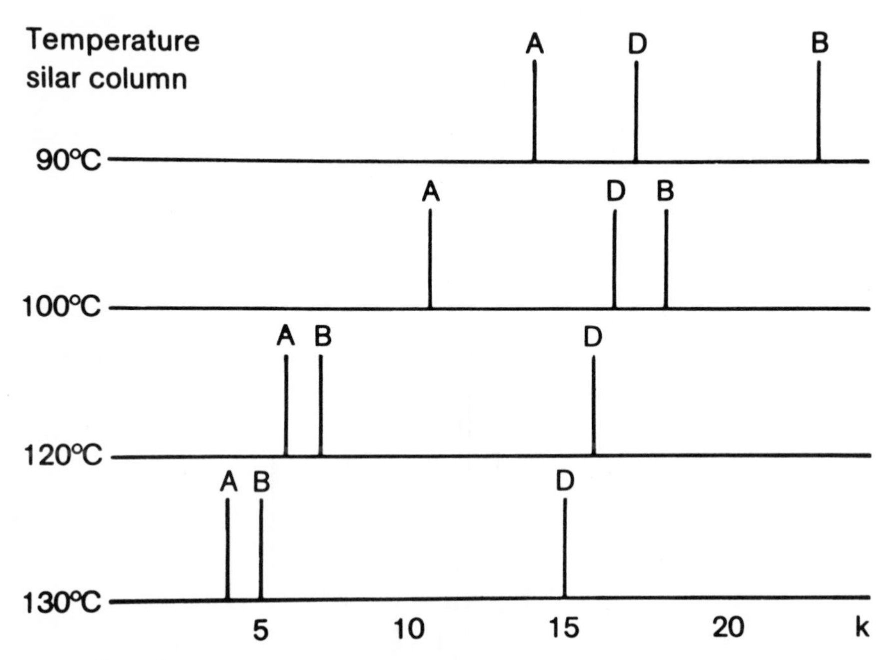

Fig. 6. Effect of temperature in SECAT chromatography. Column 1, OV 101 held at 80°; column 2, Silar 5 at temperatures as shown. A = acetonitrile; B = benzyl alcohol; D = dodecane. After Pretorius (12).

packed columns, they went on to suggest that the conclusions, relative to the effect on relative retentions, of changing individually the temperature of one of two serially coupled columns, were also applicable to open tubular systems. Kaiser used micro-packed columns to explore this continuous two-column two-temperature system (13). A schematic diagram of their apparatus is shown in Fig. 7, and Fig. 8

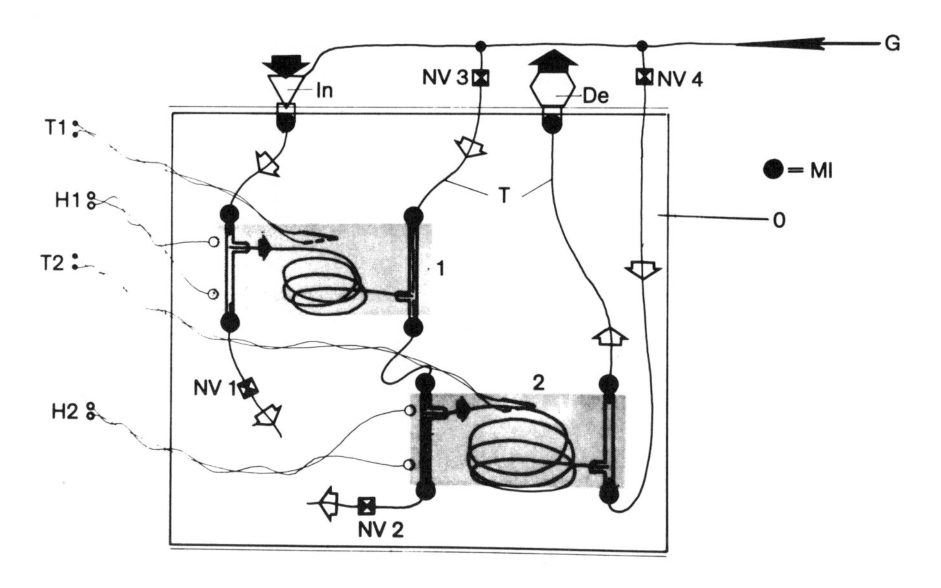

Fig. 7. Schematic of a SECAT installation. 1,2 = capillary column cassettes, OV 101 and CW HP; In = injector; NV 1 = needle valve restriction on splitter outlet; De = FID with make-up via needle valve 4; T = connecting tubing; T1, T2 = temperature sensors; H1, H2 = cassette heaters; G = carrier gas. After Kaiser and Rieder (14).

illustrates its application to a model system. Proposals for automating the system, which they termed SECAT chromatography, have also been advanced (14).

IV. MULTIPLE-PASS GC

It is well established that the optimum practical gas velocity (15) varies inversely with column length (e.g. 16). As a consequence, shorter columns exhibit greater resolution per unit time; indeed, the sole limitation of the short column is that it has limited powers of separation (17). It is possible to realize both the higher OPGV and large plate numbers by recycling to achieve multiple passes through a short column. Over 2,000,000 theoretical plates have been developed in slightly less than 16 minutes in a recycle unit operated at its optimum average linear carrier gas velocity; at OPGV, the unit achieved in excess of 3,600 theoretical plates/sec (18). These are exciting developments, but the

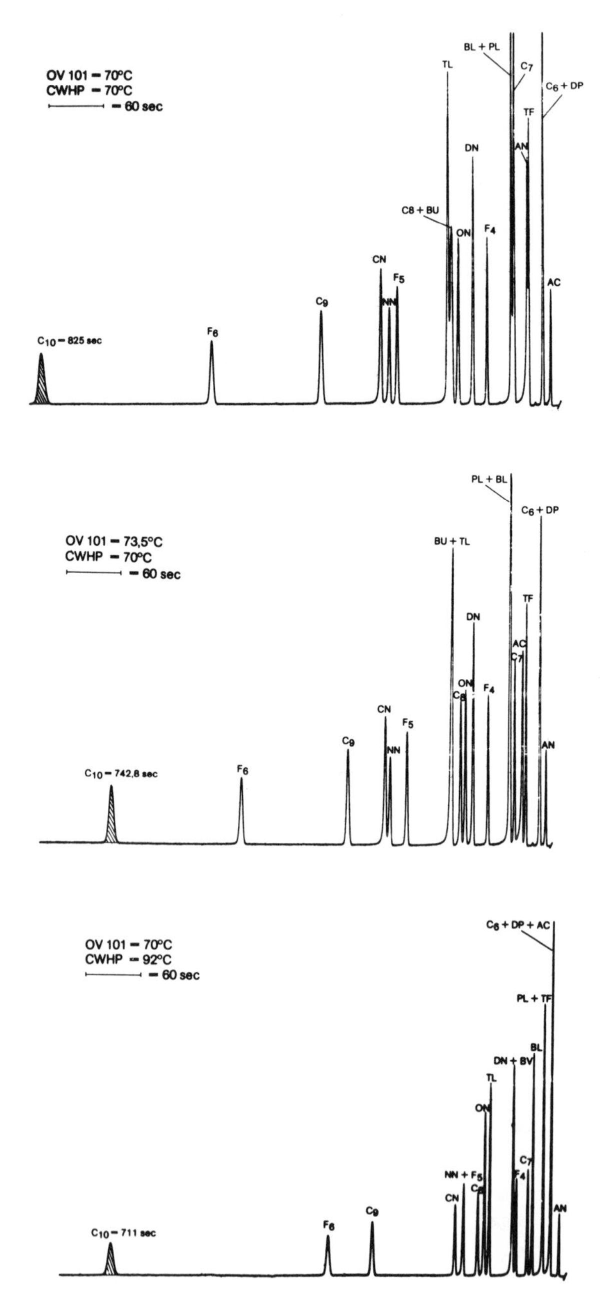

Fig. 8. SECAT chromatograms. 1 = both columns 70°; 2 = OV 101 at 73.5°, CW HP at 70°; 3 = OV 101 at 70°, CW HP at 92°. After Kaiser and Rieder (14).

other advantages that can be realized from multiple-pass GC are even more exciting.

A. Column Length as an Operational Parameter

Although the samples that concern the flavor chemist are usually complex, many of the constituents can be separated on a relatively low-resolution system. On those occasions where these more easily-separated compounds are the ones of interest, there exists, from a pragmatic point of view, scant reason to dedicate additional time and equipment to a more complete separation of the other constituents in which the analyst has little interest. Especially where time is also a factor, we wish to use the shortest possible column possessing the highest possible separation efficiency (4). Fig. 9 shows a practical way of achieving this goal. A standard high-efficiency column with sufficient length to

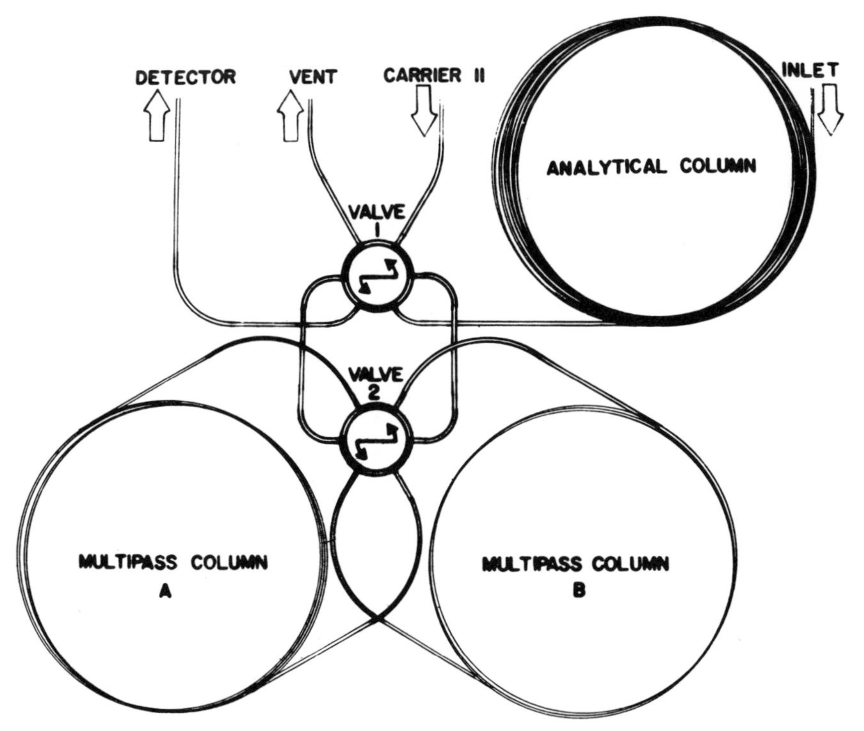

Fig. 9. Schematic representation of a recycle unit in combination with a standard analytical column. After Jennings et al. (17).

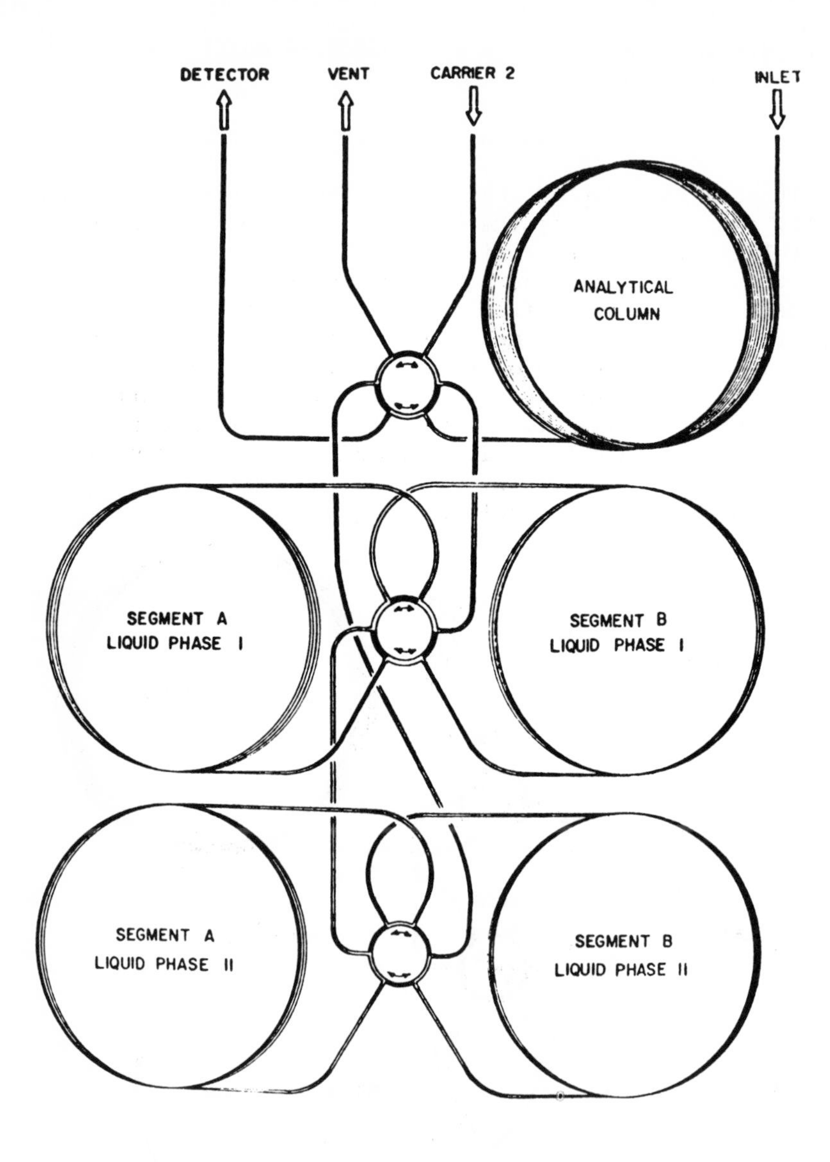

Fig. 10. Schematic representation of two recycle units, containing different liquid phases, used in conjunction with an analytical column. See text for details. After Jennings et al. (18).

barely achieve the separation of the least-demanding samples is used as the analytical column, which is operated in a normal manner. Restricted portions of more-demanding samples can be shunted to the multiple-pass system and subjected to recycling to achieve a much higher degree of separation. Limitations imposed by the range of partition ratios (18) and prospects for refocussing the partially-separated components to permit continued recycling (4) have been discussed elsewhere.

B. Column Polarity as an Operational Parameter

Laub and Purnell (19, 20) suggested a method of "window diagramming" to establish binary mixtures of liquid phases that would achieve the largest possible relative retentions of all of the components of a given mixture. The method involved preliminary separation of that mixture on several packed columns containing different relative proportions of those two liquid phases; from these data, the ideal mixture for the analytical column could be calculated. Liquid phase mixtures can also be achieved by preparing series-coupled columns, each containing a single liquid phase, and varying the lengths of the column segments to correspond to the correct proportion of that liquid phase (21). Figure 10 illustrates a simpler approach involving two recycle units. Data for the window diagram can be collected by ratios of the times spent in the two liquid phases; by adjusting these times, a fraction can then be subjected to any desired "length of column" containing the required "binary mixture."

Much more work will be required before the full potential of these devices can be realized. As mentioned above, progress is being made on techniques for periodic refocussing of the recycling components, and on the development of an in-line detector that can be used to trigger pulses for the recycle switch and focussing device. Adjustment of the recycle ratio and recycle time will then permit the analyst to control column polarity and column length from the front panel, and subject the sample to that combination of these variables necessary for any desired degree of resolution.

REFERENCES

1. "The Analytical Instrument Industry 1979", published by Centcom, Ltd., Advertising Management for Am. Chem. Soc. publications (1979) p. 35.
2. Grob, K. and Grob, G. HRC&CC 2, 109 (1979).
3. Rooney, T. A., Altmayer, L. H., Freeman, R. R. and Zerenner, E. H. Am. Lab. 11 (No. 2), 81 (1979).

4. Jennings, W. "Gas Chromatography with Glass Capillary Columns", Academic Press, New York, London, 2nd ed. (1980).
5. Bertsch, W. HRC&CC 1; 85, 187, 289 (1978).
6. Rijks, A. J. Doctoral thesis, Tech. Univ. Eindhoven, The Netherlands 1973.
7. Schomburg, G., Husmann, H. and Weeke, F. J. Chromatogr. 112, 205 (1975).
8. Jennings, W. G., Wyllie, S. G. and Alves, S. Chromatographia 10, 426 (1977).
9. Deans, D. R. Chromatographia 1, 18 (1968).
10. Schomburg, G., Husmann, H. and Weeke, F. J. Chromatogr. 112, 205 (1975).
11. Miller, R. J., Stearns, S. D. and Freeman, R. R. HRC&CC 2, 55 (1979).
12. Pretorius, V., Smuts, T. W. and Mancrieff, J. HRC&CC 1, 200 (1978).
13. Kaiser, R. E. and Rieder, R. I. HRC&CC 1, 201 (1978).
14. Kaiser, R. E. and Rieder, R. I. HRC&CC 2, 416 (1979).
15. Scott, R.P.W. and Hazeldean, G.S.F. in "Gas Chromatography 1960", R.P.W. Scott, (ed) Butterworths, London, 1960 p. 144.
16. Yabumoto, K. and VandenHuevel, W.J.A. J. Chromatogr. 140, 197 (1977).
17. Jennings, W., Settlage, J. A. and Miller, R. J. HRC&CC 2, 441 (1979).
18. Jennings, W., Settlage, J. A., Miller, R. J. and Raabe, O. G. J. Chromatogr. (in press).
19. Laub, R. J. and Purnell, J. H. Anal. Chem. 48, 799 (1976).
20. Laub, R. J. and Purnell, J. H. Anal. Chem. 48, 1720 (1976).
21. Littlewood, A. B. "Gas Chromatography", Academic Press, New York, London 2nd ed. (1970) p. 250.

ANALYSIS AND CONTROL OF LESS DESIRABLE FLAVORS IN FISH AND SHELLFISH

George J. Flick, Jr.
Janis A. Burnette

Department of Food Science and Technology
Virginia Polytechnic Institute and State University
Blacksburg, Virginia

Michael G. Legendre
Allen J. St Angelo
Robert L. Ory

Southern Regional Research Center[1]
U. S. Department of Agriculture
New Orleans, Louisiana

I. INTRODUCTION

The acceptance of food is based on many factors. Of primary importance are those attributes related to texture, flavor, odor, and appearance. Two of these, flavor and odor, are significantly related to the consumption of fresh and further processed fish and shellfish. Many of the factors that affect fish and shellfish flavor and/or odor are related to environmental factors, composition, metabolic processes of the animal, primary and secondary processing methods, storage conditions, sanitation, packaging materials and the presence of specific microbiological flora. However, with all the controls or manipulations available to the industry to prevent or limit undesirable sensory changes, the analysis and control of less desirable flavor is of immediate and serious concern. The development of off or undesirable flavors usually does not present a public health hazard but the economic aspects can be of some consequence.

[1]*One of the facilities of the U.S. Department of Agriculture, Science and Education Administration.*

ISBN 0-12-169065-2

II. HISTORY

A. Fish and Shellfish With an Undesirable Flesh Chemical Composition

V. P. Bykov (1974) has reported that there are some fish with adequate protein contents but have a high degree of substances endowing them with a disagreeable flavor. For example, the high urea content of sharks and skates imparts a bitter taste. The meat of alewife (Pomolobus mediocris), sea clams "mahogany" (Arctica islandica), and gold-lined grunt (Bathystoma aurolineatum) has a sharp iodine taste. Sea catfish (Tachysurus filiceps) have been reported as having a bad-egg odor. The dark meat of tuna cannot always be used because of the high content of volatile bases. To eliminate or reduce the undesirability of such natural defects, it is necessary to apply various processing techniques. Some of these will be discussed in greater detail in this paper.

It is also known that undesirable flavors may develop post-harvest because of specific chemical or biochemical compositions. Most of these are dependent on enzyme, microbiological, or oxidative processes. Many of the changes caused by these procedures are based on the formation of volatile bases, carbonyls, dicarbonyl compounds, sulfides, and mercaptans.

B. Fish and Shellfish Contaminated Through Environmental Factors

The effects of organisms as being responsible for specific types of odors and flavors in fresh water has been reported by Silvey (1966) in Texas, Sipma et al., (1972) in Holland, Tocher and Ackman (1966) in Canada, Medsker et al., (1968) in California, Dor (1974) in Israel, and Kikuchi and Mimura (1972) in Japan.

A muddy-earthy flavor in rainbow trout (Salmo gairdneri), channel catfish (Ictalurus punctatus), and carp (Cyprinns carpio) has been associated with the presence of species of actinomycetes and blue green algae in water environments (Thaysen, 1936; Thaysen and Pentelow, 1936; Gerber and Lechevalier, 1965; Safferman et al., 1967; Aschner et al., 1967; Lovell, 1972).

It is important to realize that the off-flavor producing algae and actinomycetes have been found to produce the noxious odors or flavors under one set of conditions but not other conditions (Leventer and Eren, 1969; Lovell, 1971). In channel catfish the undesirable flavor was shown to be absorbed by the fish through the digestive system and the gill membranes (Lovell and Sackey, 1973).

The culturing of rainbow trout in prairie pothole lakes in Central Canada has been impaired by the occurrence of this muddy-earthy flavor (Irendale and Shaykewich, 1973). Off-flavor has also been reported as a common occurrence in carp ponds in China, Japan, and Europe (Lovell et al., 1975).

The flavor of fish is also known to be dependent on local environmental conditions, as well as the presence of organisms. Vale et al., (1970) reported a kerosene-like flavor in mullet due to pollution and Mann (1969) has reviewed much of the early literature on fish taints due to phenolic and mineral oil contamination. Off-flavor due to feed in salmon has been reported by Motohiro (1962) and by Ackman and associates (Sipos and Ackman, 1964; Ackman et al., 1966; Ackman et al., 1967) in cod. Mann (1969) mentioned that farmed carp developed undesirable flavors when fed a diet of maiz or barley. Baeder et al., (1945) stated that wild brook trout were preferred to the cultured variety although both fish were judged to be acceptable.

C. Compounds Responsible for Undesirable Flavor in Fish and Shellfish

It has now been established that geosmin is probably the major cause of the muddy flavor (Yurkowski and Tabachek, 1974) reported in fish. Various authors have related the fish flavor to extracellular products of actinomycetes (Thaysen, 1936; Thaysen and Pentelow, 1936; Ilncykyj, 1951; West, 1953), some of which are: geosmin, 2-exo-hydroxy-2-methylbornane, 2-methylisoborneol, and/or mucidone (Dougherty et al., 1966; Gerber and Lechevalier, 1965; Gerber 1967, 1968, 1969; Kikuchi and Mimura, 1972; Medsker et al., 1968; Rosen et al., 1968, 1970; Yurkowski and Tabachek, 1974). The known or proposed structures of these compounds are contained in Table I. There are other reports relating the muddy flavors to particular species of blue-green algae (Cornelius and Bandt, 1933; Aschner et al., 1967; Lovell and Sackey, 1973) which also produce geosmin (Safferman et al., 1967; Medsker et al., 1968).

Table I. Reported Chemical Structures of Odoriferous Metabolites

Name	Proposed Structure(s)	Reference
Geosmin	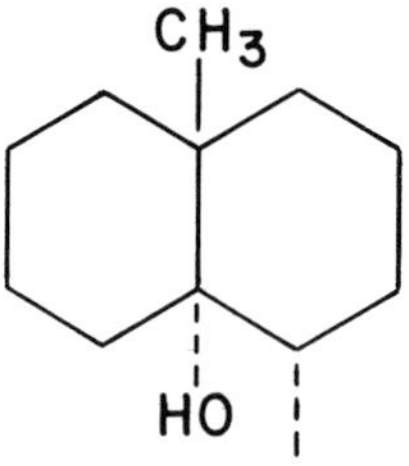 CH_3, HO	Kikuchi and Mimura 1972
Mucidone	I* 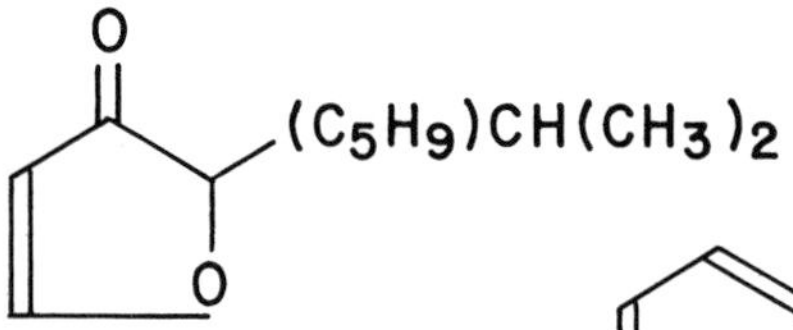 $(C_5H_9)CH(CH_3)_2$, O, O; II $CH_2CH(CH_3)_2$, C_2H_5, O, O 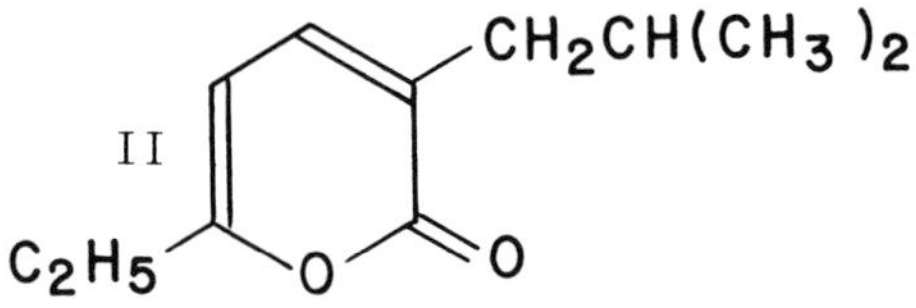	Sipma *et al.*, 1972
2-methylisoborneol	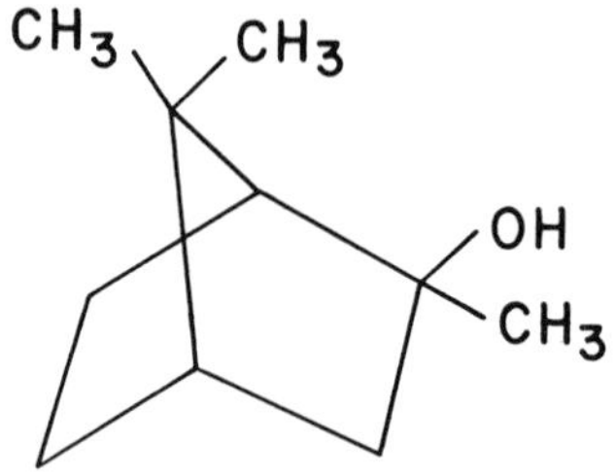 CH_3, CH_3, OH, CH_3	Gerber 1969
2-exo-hydroxy-2-methylbornane	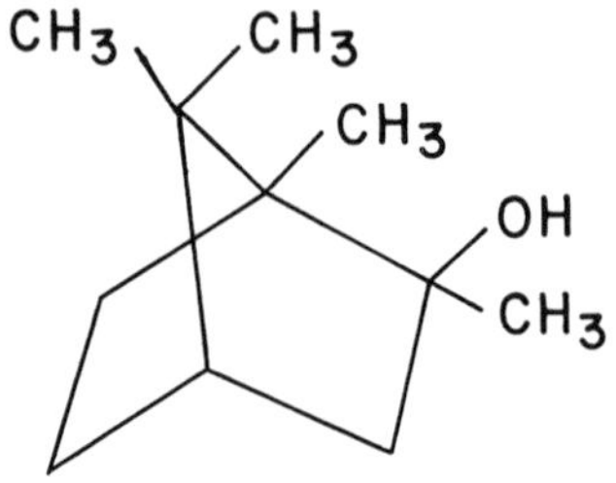 CH_3, CH_3, CH_3, OH, CH_3	Medsker *et al.*, 1968

*Most Probable Structure.

D. Detection and Quantitation of Muddy Flavor in Fish

Informal evaluations have been reported by various authors in the literature. Hedonic scales were used with a consumer panel by Iredale and Shaykewich (1973) with smoked rainbow trout. Paired comparison tests have been successfully utilized (Iredale and Rigby, 1972; Iredale and York, 1976). Linear scales were also employed to measure the intensity of the flavor (Lovell and Sackey, 1973; Iredale and Shaykewich, 1973) and with a flavor profile method using a descriptive vocabulary developed for the study (Maligalig *et al.*, 1973).

III. APPROACHES TO CONTROL OR ELIMINATE OFF-FLAVORS AND OFF-ODORS

A. Control of Growing Environment

Control of the growing environment is dependent on economics and current science or technology. This aspect is most difficult since cultured systems promote the growth of organisms which produce undesirable flavors. Practices suggested by Lovell (1976) include:

1. Minimizing feed waste by using good feeds and feeding practices. Growth of geosmin-producing microorganisms is stimulated by unconsumed and unabsorbed organic (actinomycetes) and inorganic (algae) nutrients.
2. Exchange of water through the culture system will remove unused nutrients and minimize the growth potential for off-flavor causing microorganisms. It will also reduce concentration of microorganisms and off-flavor compounds in the water.
3. Increasing the turbidity or muddiness of pond water through mechanical agitation or with bottom-feeding fish will suppress phytoplankton growth. There is also support for the idea that the suspended clay particles act as adsorbants for the off-flavor compounds.

B. Control or Removal of the Undesirable Flavor or Odor by Chemical Treatment of Fish Ponds

Important considerations are the long-term effects on the culturing environment and the economic considerations. One acceptable method is the use of chemicals to control algae growth. Copper sulfate crystals have been used; however, there are reports indicating that this practice may have questionable results. McKee and Wolf (1963) showed trout to have little tolerance to the chemical concentrations of 0.14 mg/liter being lethal. Moyle (1949) reported that an accumulation of precipitated forms of copper in lake mud could affect the growth of bottom organisms, which could be important in the food chain. There is also the possibility of increasing the algae blooms to copper sulphate. A commercial herbicide, Simizine, has been approved for use in food fish ponds.

Meade (1975) found that adding salt to water in a closed system raceway, to achieve a salinity of 10 parts per 1,000, several days before harvesting solved the off-flavor problem. The salt destroyed the microorganism in the system.

C. Removal of the Undesirable Flavor or Odor by Depuration

The most used practice is depurating the undesirable odor/flavor by changing the environment and holding for a definite period of time. Usually the flavor will be acceptable after 5 (geosmin concentration of 1.1 μg/100 g of flesh) to 14 (0.3 μg/100 g of flesh) days depending on the temperature and the intensity of the off-flavor.

Since the undesirable flavors/odors are usually short lived, many producers will delay harvesting the fish until flavor improve. This practice is accompained with pond water exchange and a feeding program. This may be impractical however, in cold climates where ponds may freeze (Yurkowski and Tabachek, 1974; Lovell and Sackey, 1973; Maligalig _et al_., 1973; Aschner _et al_., 1967; Thaysen and Pentelow, 1936).

Lovell (1976) held channel catfish in a 77°F (25°C) tank having a distinct earthy-musty flavor for 14 days. The fish were then placed in a clean water aquarium and after 3, 6, 10, and 15 days the fish were removed and evaluated for flavor by four experienced judges. Fish remaining for 3 days in clean water had significantly ($P<.05$) improved flavor. After 10 days, the flavor was not significant ($P<.05$) from that of control fish.

Iredale and York (1976) determined the length of time required to purge undesirable flavor taints from pond cultured rainbow trout transferred to two different clear water environments. Sensory data from trained judges show that this required 5 days for fish transferred to a rapidly changing purified artificial water environment and 16 days for fish transferred to a relatively static natural water environment to reduce this taint to or below threshold levels of recognition. This practice presents minimal difficulty in cultured situations; however, in a natural growing environment, the problems of efficient and effective harvesting and transplanting are significant.

Shellfish (oysters, clams, and mussels) obtained in growing areas producing off-flavored products can be improved through relaying or depuration. Relaying is the practice of moving the tainted shellfish from one growing area to another. Depuration is the process of placing the shellstock in tanks that have: closed systems of recirculated natural or artificial seawater; or open circulated seawater systems. The amount of time the shellfish remain relayed or in a depurated system depend on the flavor taint and water temperature.

D. Removal of the Undesirable Flavor or Odor by Primary or Secondary Processing

Processing of rainbow trout (Iredale and Shaykewich, 1973) and catfish (Lovell, 1972) to eliminate or minimize muddy-earthy off-flavors by smoking for 5 to 6.5 hours with simultaneous cooking has been shown to be acceptable to consumers.

Iredale and Shaykewich (1973) reported that steam precooking of fillet strips of rainbow trout and subsequently adding vegetable oil (fish packed in cans; steam cooked at 210°F for 10 min.; drained; 25 ml corn oil + 1% salt added) before canning eliminated the earthy-musty flavor. The use of ribotide and citric acid alone or together was not effective in masking muddy taste; in fact, these treatments resulted in a higher intensity of muddiness than the control sample.

E. Unit Operations Affecting Undesirable Odors and Flavors

The influence of skin as an important carrier of odors was credited to Obata *et al*., (1950) and Bramstedt (1957). However, work by Yamanishi *et al*., (1956) proved to be in conflict with the earlier hypothesis. A report by Baldwin *et al*., (1962) showed that undesirable flavor or aroma could not be sufficiently altered by the presence of skin or bone during baking, or by cooking in hot fat.

Sharks, skates, and rays have as much as 2.0 to 2.5 percent urea in their blood, while teleost fishes have only 0.01 to 0.03 percent (Smith, 1953). The deterioration of the urea into ammonia and the reduction of trimethylamine oxide, both immediate results of autolytic and microbial action, are responsible for the pungent shark odor (Tsachiza *et al*., 1951). According to Warfel and Clague (1950) the simplest way to neutralize any residual ammonia is to presoak the fillets in a citric acid solution prior to cooking or freezing. Arundale and Herborg (1971) suggest gutting and heading and cutting off the tail to accelerate bleeding. Subsequently, the fish should be thoroughly washed and stored in ice. They recommend processing into frozen fillets, smoked frozen steaks, dried salted fillets, and smoked-salmon substitute. The fish are then acceptable and often considered as highly desirable.

When air packed, soft shell clam (*Mya arenaria*) meats were irradiated and/or heated, the concentration of carbonyl compounds increase immediately. Storage at 33 - 35°F cause a gradual increase in the volatile carbonyl concentrations until about the 20th day. If the meats were placed in vacuum-packed containers, the effects of storage were minimized (Gadbois *et al.*, 1967).

The lack of characteristic flavor in seafood should also be considered as an undesirable flavor attribute. Spinelli and Miyauchi (1968) irradiated commercially filleted vacuum packaged fish and placed them in storage at 33 - 35°F for 2 weeks. When 1 μm of IMP (inosine monophosphate) was added to the fillets which contained no IMP at the end of the storage period, a significant ($P<.05$) difference was noted and they were preferred over those containing no IMP.

Yanagisawa and Masayama (1960) received a patent for the improvement of whale or fish meat. Whale meat cut 2 cm thick is immersed for 30 min. in 0.1 - 3.0 percent calcium mesotartrate and washed with water. The blood corpuscles in the meat are removed by the osmotic pressure of the chemical and the unsaturated compounds in the meat are deodorized by combination with the chemical.

IV. CONCLUSION

The control of less desirable flavor in fish and shellfish is of importance to the fresh and saltwater fish and shellfish industry. Unfortunately, the success that has been reported is somewhat disappointing. This situation is the result of several factors:

A. The Nature of the Off-Flavors and Odors.

Most of the undesirable flavors are caused by autolytic, microorganisms, and oxidation processes. The end products of the complex reactions are usually highly obnoxious even in trace quantities (Table II). Attempts to remove or lessen their effect usually results in the production of an unacceptable product. Most processing procedures (as smoking, canning, and additiyes) used to minimize or eliminate undesirable taints only result in intensifying them.

B. Value of the Fish and Shellfish.

With few exceptions, the market value of fish (both volume and unit price) at the wholesale level is not competitive with other protein foods. Consequently, most firms would not commit financial support to improve undesirable quality since their product may be uncompetitive in the market place. Shellfish products on the other hand are higher priced, but their volume is low when compared to the total volume of fish and seafood. The laboratory requirements and unit processing operations to effectively and efficiently handle the various undesirable taints become economically unfeasable compared to the relatively small volume of product.

C. Production of the Fish and Shellfish.

Fish and shellfish resources are for the most part, an uncontrolled food resource. In fact, it is one of the last remaining "wild" foods that is of dietary importance. Because of this, large quantities of fish and shellfish are not harvested with the same undesirable taint. Since the resource is spread over a wide geographical area, the taste of the products may vary but this variation has generally been accepted by the consumer. Only extreme off-flavors and odors are rejected by the typical less-discriminating consumer.

D. Volume of the Fish and Shellfish Production.

It is important to note that the United States and other nations are seafood importing countries. Approximately two-thirds to three-fourths of the U.S. fish and shellfish are obtained through imports. Therefore, if undesirable products are received, importing firms will usually reject the shipment. Consequently, the burden of quality assurance for many countries' products are left to the producer.

E. Historical Consumption of Fish and Shellfish.

It has been long postulated that tainted fish and shellfish are usually acceptable to the consumer. In many countries and communities, fish and shellfish constitute a small portion of the diet and the consumers are not able to adequately distinguish quality attributes. This is not surprising since fish and shellfish utilization practices do not equal those of other food products.

The utilization of marine bioproducts are receiving substantial attention on a national and international level. Within the immediate future, more information on the presence and control of undesirable flavors and odors should be available. The future appears encouraging to all and exciting to many.

Table II. Some volatile compounds identified in stored seafood.

methyl mercaptan
dimethyl sulfide
acetone
diacetyl
2-butanone
trimethylamine
dimethylamine
dimethyl disulfide
pyridine
benzaldehyde

ACKNOWLEDGMENTS

The authors wish to express their appreciation to Jean B. Brewer for her invaluable help in preparing the first drafts of this manuscript and to Joyce Smoot for her cooperation in the typing of the final copy. Thanks are also extended to Sharon Chiang for preparation of the illustrations.

REFERENCES

1. Ackman, R. G., Dale, J., and Hingley, J., J. Fish Res. Bd. Can. 23:487-497 (1966).
2. Ackman, R. G., Hingley, J., and May, A. W., J. Fish Res. Bd. Can. 24:457-461 (1967).
3. Arundale, J., and Herborg, L., UNDP/FAO Caribbean Fishery Development Project, report SF/CAR/REG 189 M 18, 23 p. (1971).

4. Aschner, M., Laventer, Ch., and Chorin-Kirsch, I., Bamidgeh. 19:23-25 (1967).
5. Baeder, H. A., Tack, P. I., Hazzard, A. S., Trans. Am. Fish. Soc. 75:181-185 (1945).
6. Baldwin, R. E., Strong, D. H., and Torrie, J. H., Food Technol. 16:115-118 (1962).
7. Bramstedt, F., Fisheries Research Board Can., Transl. Ser. No. 235 (1957).
8. Bykov, V. P., in "Fishery Products", (R. Kreuzer, ed.), p. 154. Fishing News (Books) Ltd. Surrey, England, 1974.
9. Cornelius, W. O., and Bandt, H. J., A. Fisch. Hilfswiss. 31:675-686 (1933).
10. Dor, I., Hydrobiologia. 44(2-3):255-264 (1974).
11. Dougherty, J. D., Campbell, R. D., Morris, R. L., Science. 152:1372 (1966).
12. Gadbois, D. F., Mendelsohn, J. M., and Ronsivalli, L. J., J. Food Sci. 32:511-515 (1967).
13. Gerber, N. N., Biotechnol. Bioeng. 9:321-327 (1967).
14. Gerber, N. N., Tetrahedron Lett. 25:2971-2974 (1968).
15. Gerber, N. N., J. Antibiot. (Tokyo). 22:508-509 (1969).
16. Gerber, N. N., and Lechevalier, H. A., Appl. Microbiology. 13:935-938 (1965).
17. Ilnyckyj, S., M. Sc. Thesis. Univ. Saskatchewan, Saskatoon, Sask. 85 p. 1951.
18. Iredale, D. G., and Rigby, D., J. Fish. Res. Bd. Can. 29:1365-1366 (1972).
19. Iredale, D. G., Shaykewich, K. J., J. Fish. Res. Bd. Can. 30:1235-1239 (1973).
20. Iredale, D. G., and York, R. K., J. Fish. Res. Bd. Can. 33:160-166 (1976).
21. Kikuchi, T., and Mimura, T., Yakugaku Zasshi. 92(5):652-653 (1972).
22. Leventer, H., and Eren, J., in "Development in water quality research". Proc. Jersulem International Conf. on Water Quality p. 19-37 (1969).
23. Lovell, R. T., Proc. Ass. South. Agr. Workers 67th Annual Meeting p. 102. 1971.
24. Lovell, R. T., Trans. Am. Fish. Soc., 103:775-777 (1972).
25. Lovell, R. T., Proc. First Ann. Trop. and Subtrop. Fish. Tech. Conference p. 467. 1976.
26. Lovell, R. T., and Sackey, L. A., Trans. Am. Fish. Soc. 4:774-777 (1973).
27. Lovell, R. T., Smitherman, R. O., and Shell, E. W., in "New Protein Foods" Progress and Prospects in Fish Farming. Academic Press, New York, 1975.
28. Maligalig, L. L., Caul, J. F., and Tiemeier, O. W., Food Prod. Dev. 7:86-92 (1973).
29. Mann, H., Mittel. 71(12):1021-1024 (1969).

30. McKee, J. E., and Wolf, H. E., in "Water Quality Criteria" 2nd ed. Calif. State Water Quality Control Board Publ. 3-A:548 p. 1963.
31. Meade, T. L., Fish Culture in Closed-System Raceways. Reported at the Midwest Fish Disease Workshop, June 12-13, Carbondale, IL. 1975.
32. Medsker, L. L., Jewkins, D., and Thomas, J. F., Env. Sci. Tech. 3:476-477 (1968).
33. Motohiro, T., Mem. Fac. Fish. Hokkaido Univ. 10(1):1-65 (1962).
34. Moyle, J. B., in "Limnological aspects of water supply and waste disposal". Am. Assoc. Advanc. Sci., Washington, D.C. p. 79-87. 1949.
35. Obata, Y., Yamanishi, T., and Ishida, M., Bull. Jap. Soc. Sci. Fish. 15:551-553 (1950).
36. Rosen, A. A., Mashni, C. I., and Safferman, R. S., Water Treat. Exam. 19:106-119 (1970).
37. Rosen, A. A., Safferman, R. S., Mashni, C. I., and Romano, A. H., Appl. Microbiology. 16:178-179 (1968).
38. Safferman, R. S., Rosen, A. A., Mashni, C. I., and Morris, M. E., Env. Sci. Tech. 1:429-430 (1967).
39. Silvey, J. K. G., J. Am. Water Works Assoc. 58(6):706-715 (1966).
40. Sipma, G., van der Wal, B., and Kettenes, D. K., Tetrahedron Lett. 41:4159-4160 (1972).
41. Sipos, J. C., and Ackman, R. G., J. Fish Res. Bd. Can. 21:423-425 (1964).
42. Smith, H. W., in "From Fish to Philosopher", Little, Brown and Co., Boston, 1953.
43. Spinelli, J., and Miyauchi, D., Fd. Technol. (Champaign, IL). 22:781-783 (1968).
44. Thaysen, A. C., Ann. Appl. Biol. 23:99-104 (1936).
45. Thaysen, A. C., and Pentelow, F. T. K., Ann. Appl. Biol. 23:105-109 (1936).
46. Tocher, C. S., and Ackman, R. G., Can. J. Biochem. 44:519-522 (1966).
47. Tsachiza, Y., Takahashi, I., and Yoshida, S., Tohoku J. Agr. Res. 2:119-126 (1951).
48. Vale, G. L., Sidhu, G. S., Montgomery, W. A., and Johnson, A. R., J. Sci. Food Agric. 21:429-432 (1970).
49. Warfel, H. E., and Clague, J. E., U.S. Wildl. Serv. Res. Rep. 15:19 p. 1950.
50. West, M. O., M.Sc. Thesis. Univ. Saskatchewan, Saskatoon, Sask. 90 p. 1953.
51. Yamanishi, T., Yamashita, S., Yamazaki, A., and Tokue, Y., Bull. Jap. Soc. Sci. Fish. 22:480-485 (1956).
52. Yanagisawa, F., and Masayama, Y., Japanese patent 18,415 Dec. 20, 1960.

53. Yurkowski, M., and Tabachek, J. L., J. Fish. Res. Bd. Can. 31:1851-1858 (1974).

UNDESIRABLE MEAT FLAVOR AND ITS CONTROL

Milton E. Bailey

Department of Food Science and Nutrition
University of Missouri
Columbia, Missouri

Harold P. Dupuy

V-Labs
Covington, Louisiana

Michael G. Legendre

U.S.D.A., S.E.A. Southern Regional Research Center
New Orleans, Louisiana

I. INTRODUCTION

Meat is the most highly desirable and sought after food product throughout the world and many types of muscle foods are enjoyed in different countries. Meat flavor is perhaps the most important criterion of quality and many investigators have studied the chemistry of this important attribute. Meat products are excellent sources of protein, lipids and other nutrients, but consumption is largely a matter of flavor satisfaction. If judged undesirable in flavor, these important nutrient sources are not consumed.

Most measures of flavor desirability are subjective because acceptance of foods varies considerably among different cultures. Food flavors considered objectionable to consumers in one country might be very acceptable to those in a different location. It would therefore be impossible to agree on

ISBN 0-12-169065-2

universal food flavor standards particularly for meat (1).

Much emphasis has been placed on objective analyses of flavor volatiles in meat, and flavor chemistry methodology is important as a tool for enhancing the palatability of many non-muscle proteins through study of ingredients which make them taste like meat products. The chemistry of undesirable meat flavor is of equal importance since chemical ingredients responsible for these flavors might be removed from products once their identity is known.

Domesticated animals used for meat can utilize many materials in the environment for feed and in many instances do not compete with man for land that could be used for growing other foods. Consumption of these different feeds is an important factor because of production of flavors considered undesirable by many consumers.

Other factors responsible for undesirable flavor of meat are: oxidation of meat lipids, animal species, animal sex and processing and storage procedures. Three comprehensive reviews discussing some of these factors have been published recently (2,3,4). In this discussion, major attention is given to the influence of feed and autoxidation on undesirable flavor of meat and other factors will be discussed briefly.

II. ANIMAL FEED

A. Beef

Much contradictory data are published concerning the influence of feed on the flavor of beef. Most of these data indicate that palatability characteristics of beef can be affected by nutritional regimen. Beef from cattle fed forage is usually reported to be less desirable in flavor than beef from grain-finished cattle (5-15). This type of undesirable flavor is associated with fat.

A project underway at the University of Missouri (16) includes a study of volatile compounds in fat from forage and grain-fed beef as they relate to sensory characteristics of steaks and roasts. Following a wintering phase of grazing on pasture and a diet supplemented with hay, corn and soybean meal, steers were assigned to three nutritional regimens as follows: (1) fescue (_Festuca Arundinacea_) pasture for six months, (2) fescue pasture for 3 months followed by _ad libitum_ corn grain while on fescue pasture for approximately 5 months and (3) fescue pasture for six months and then _ad libitum_ corn grain and protein supplement in dry lot for approximately three months. Steaks and roasts from these animals were evaluated for flavor and overall acceptability and these results compared to data

TABLE I. Mean Sensory Panel Flavor Scores of Loin Steaks and Round Roasts as Influenced by Nutritional Regimen

Nutritional treatment	Flavor scores[a]	
	Steaks	Roasts
Fescue pasture	5.07[b]	5.21[b]
Grain *ad libitum* on fescue pasture	6.21[c]	6.20[c]
Fescue pasture followed by *ad libitum* grain in dry lot	6.48[d]	6.17[c]

[a]Range of scores: 1, extremely undesirable to 8, extremely desirable. Number of animals represented by each value was 27.

[b,c,d]Mean values bearing different superscripts within the same column are significantly ($P<.05$) different.

for volatile compounds from loin subcutaneous fat. The volatiles were separated and analyzed by the direct sampling GLC-MS procedure of Dupuy et al. (17) as modified by Legendre et al. (18).

Mean sensory panel flavor scores of loin steaks and round roasts as influenced by nutritional regimen are given in Table I.

Steaks and roasts from animals fed grain were judged significantly more acceptable than those from animals on fescue pasture. Steaks from animals fed grain ad libitum in dry lot were judged more desirable than those from animals in the other two treatments. Feeding animals grain while grazing on fescue or feeding animals on grain in dry lot a short period following pasturing on fescue improved the flavor of meat from these animals.

GLC-MS profiles and identities of volatile compounds from fat of animals on the three nutritional regimens are presented in Figure 1. Fat from animals on fescue produced the most total volatiles which are believed related to undesirable flavor. A peak appearing at MS-scan #1100 appeared to be associated with undesirable flavor since it was very large in fat from animals on fescue and small in animals fed grain. This peak contains several volatile compounds which cause identification to be difficult. Octadecane, δ-decalactone, diethylphthalate and δ-dodecalactone were identified by the mass spectra data. Octadecane and δ-decalactone have retention times similar to compounds that produce grassy and oily odors as measured by Watanabe and Sato (19). Other compounds appearing at higher concentrations in fat from grass-fed animals that appear to be associated with grassy or oily odors include: C_5-C_{10} aldehydes (peaks - 13, 17, 19, 21 and 23) and 2,4-decadienal (peak 28).

Hydrocarbons are apparently derived from decarboxylation and splitting of saturated fatty acids and some hydrocarbons are found in leaves of many grasses (20). The lipids from fescue contain a high percentage of linolenic acid (20) which becomes saturated due to the hydrogenation activity of rumen microorganisms. Beef cattle fed forage diets have been found to contain more saturated fatty acids than those fed grain (21, 22).

γ- and δ-lactones can be formed by oxidizing unsaturated fatty acids or from dehydration of hydroxy acids formed by δ-oxidation of saturated fatty acids (23). Watanabe and Sato (19) studied the conversion of saturated fatty acids, aldehydes and alcohols into γ- and δ-lactones in meat and other foods. Fatty aldehydes make up a large fraction of volatiles from beef fat as shown later in this discussion and these are possible precursors for saturated and unsaturated lactones.

In other instances where undesirable flavor of beef could

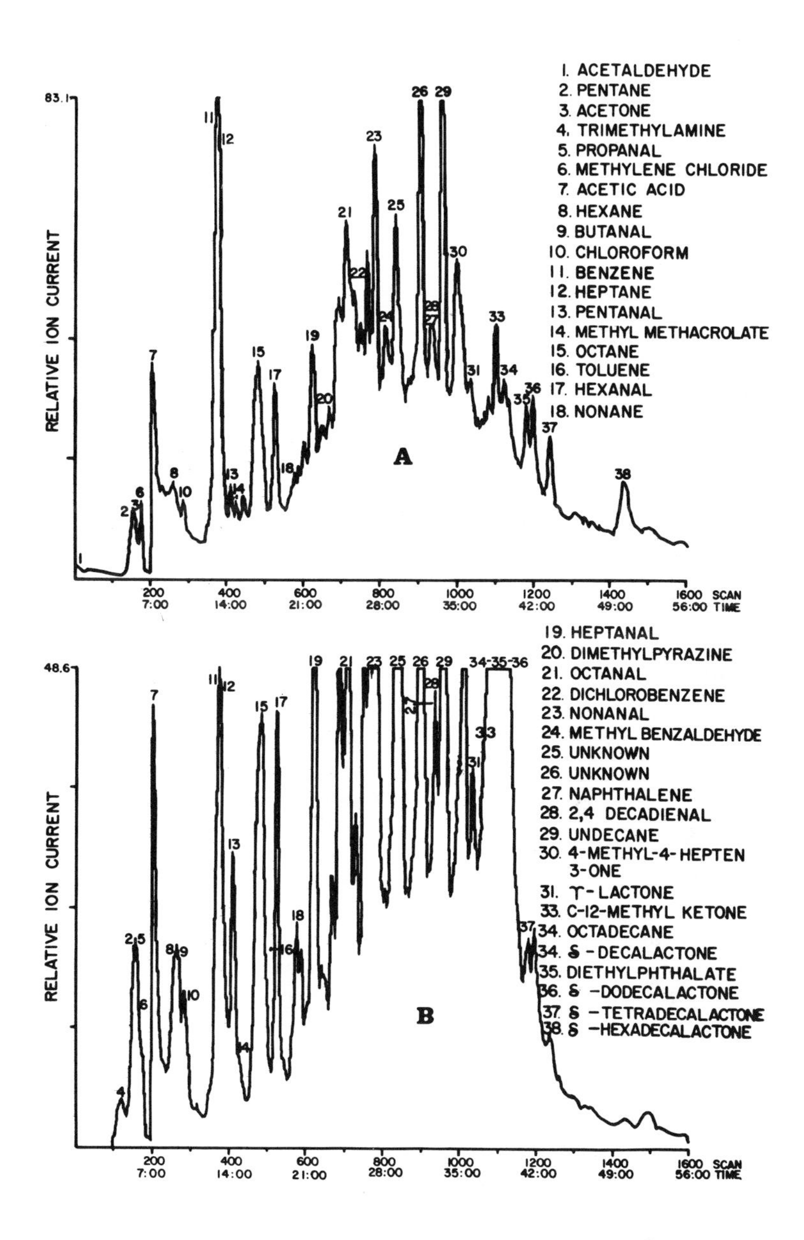

FIGURE 1. Direct sampling (18) ion-current chromatograms of volatile compounds from beef subcutaneous fat. A. Fat from corn-fed animal. B. Fat from fescue-fed animal.

be attributed to forage diets, the off-flavor problem was corrected by supplemental feeding on grain for short periods of time (24-26). It must be emphasized that forage-fed beef is quite acceptable in many countries where the consumer is accustomed to meat from this type of animal (2).

B. Lamb

Lamb is a good animal model for studying the influence of animal feed on meat flavor. There have been many studies relating lamb flavor to type of pasture, but most of the off-flavors have been characterized subjectively and not chemically. Results of some of these studies were discussed in detail by Patterson (2) and Reineccius (4). Numerous compounds present in forages may survive metabolic conditions of the rumen and are deposited in the carcass fat. These may constitute normally accepted flavor or they may have undesirable sensory properties.

Shorland *et al.* (27) and others have concluded that compounds in forage are rapidly deposited in animal tissues and can influence flavor acceptability of cooked lamb. These workers (27) found that fat from lambs fed white clover had more undesirable flavor than that from lambs pastured on perennial rye grass. Indications of off-flavor production have been found in lamb on other forages such as lucerne (28) rape and oats (29) and *glycine wishtii* (30).

More recently Park and coworkers published data that demonstrated that lambs fed "protected" lipid supplements produced meat that was less desirable in flavor than that from conventional pasture-fed lambs. In one of several papers, Park *et al.* (31) found that lambs fed a diet of formaldehyde-protected sunflower seed-casein or sunflower oil-casein produced meat with an "oily" flavor due to the presence of deca-2,4-dienal after cooking. A "sweet" flavor was also found in meat from lambs fed the protected sunflower seed-casein which was attributed to the presence of increased quantities of 4-hydroxy-dodec-*cis*-6-enoic acid lactone [dihydro-5(2[Z]-octenyl)2(3H)-furanone] (32). Presumptive evidence was obtained that the lactone was generated from a monohydroxydodecenoic acid triglyceride ester.

C. Non-Lipid Soluble Components

Although many off-flavors of meat due to feeding are associated with volatiles from fat, others may be due to components soluble in the non-lipid fraction. These include organic sulfides and disulfides that may have their origin in wild onions or similar weeds (2).

III. LIPID OXIDATION

The most common cause of undesirable meat flavor is lipid oxidation. Some oxidative reactions occur spontaneously, but others are catalyzed by enzymes and other catalysts. Most of these reactions result in the development of undesirable odors and flavors, but lipid oxidation contributes to desirable flavors as well.

Prior to 1950, the type of lipid oxidation most readily recognized in meat was autoxidation of adipose tissue lipids such as that occurring during long term frozen-storage. These oxidative changes resulted in "rancidity" the nature and mechanisms of which have been reviewed by several authors (33-38).

A more important aspect of meat lipid deterioration is the catalytic oxidation of muscle phospholipids by metal ions (39) and by hematin compounds. Undesirable flavor resulting from these reactions is called "stale", "old" or "warmed over" flavor. Many theoretical aspects of lipid oxidation by hematin compounds have been published (40), but pioneering work concerning the flavor attributes of meat resulting from catalytic oxidation of muscle lipids has been published by Watts and her students (41-47).

Various aspects of lipid oxidation and "warmed-over" flavor were recently reviewed by Greene (46), by Sato and Herring (48) and by Pearson *et al.* (49). The latter review is very detailed and informative so only selected areas relative to this problem need be mentioned here.

"Warmed-over" flavor is caused by rapid catalytic oxidation of unsaturated fatty acids in cooked meat by iron and its heme complexes (50). The undesirable flavor results from increased amounts of aldehydes, alcohols, furans, hydrocarbons and other low molecular weight volatile compounds (51).

The 2-thiobarbituric acid (TBA) test is often used to measure oxidative deterioration of meat lipids and results from this test parallel flavor deterioration of some meat samples (43). The TBA test, however, has never been adequately standardized as a measure of meat acceptability and it is not always reliable as a measure of undesirable meat flavor due to the interaction of malonaldehyde and closely related compounds with other food constituents such as amino acids, proteins, glycogen and other food ingredients (52). It is particularly unreliable in frozen foods and for certain cured meat products. Other undesirable aspects of the TBA test utilizing heat and acid were discussed by Tarladgis *et al.* (53).

Another method of measuring volatile compounds responsible for undesirable flavors of meat is gas-liquid chromatography (GLC) or gas-liquid chromatography-mass spectroscopy (GLC-MS).

The low molecular weight volatiles of cooked fresh meat increased rapidly following cookery as demonstrated by gas-liquid chromatograms of beef roasted at 163°C to an internal temperature of 71°C and stored at 4°C (Figure 2). The names of compounds identified by high resolution mass spectrometry and the percentage total area for each compound are listed in Table II.

The volatiles were extracted from a lean meat homogenate in a Likens and Nickerson extractor with diethyl ether. Identification of volatiles was based on retention times, mass spectral comparison with spectra of authentic compounds and by high resolution mass spectral data.

There were little, if any, qualitative differences in volatiles produced during storage at 4°C for 3 days, but there were quantitative differences. The chromatograms all represent samples of similar amounts of lean roast beef, so peak size is an indication of concentration of volatiles in the cooked meat.

Generally, the short-chain compounds increased in concentration during storage while the less volatile components decreased in concentration. Compounds increasing most appreciably were n-hexanal and 2-pentyl furan while long chain (C_{16}-C_{18}) aldehydes decreased in concentration. The former compounds are very important constituents in "warmed-over" flavor since they are associated with fatty acid oxidation and undesirable flavor of many lipid-containing foods (36,37).

Increased formation of volatiles during storage of cooked meat is demonstrated most readily by chromatograms in Figure 3. These data were obtained using the direct sampling GLC-MS procedure of Dupuy et al. (17) as modified by Legendre et al. (18). Analysis time is shortened considerably using this method compared to conventional GLC-MS methods above for study of cooked roast beef and it is more informative than the TBA test since selective individual compounds can be quantitative. Data obtained from freshly cooked Boston butt compared to that from samples stored for 1 day at 4°C following cookery indicate that n-hexanal and 2-pentyl furan are excellent indices of oxidative change. Other aldehydes also increased in concentration during 1 day storage.

Nitric oxide upon reacting with the heme porphyrins myoglobin and hemoglobin forms a heat-stable complex with iron which otherwise becomes a very active phospholipid-oxidation catalyst when protein of these pigments become denatured by heating. The flavor-preserving quality of nitric oxide hemochromogen with its less active iron compared to heat-denatured myoglobin is the most important reason for adding nitrite as a curing ingredient as previously discussed by Bailey and Swain (51) and by Westerberg (54).

Heat-processed nitrite-cured meat has a desirable flavor during refrigerator storage for several months, whereas fresh meat that has been processed similarly has undesirable flavor

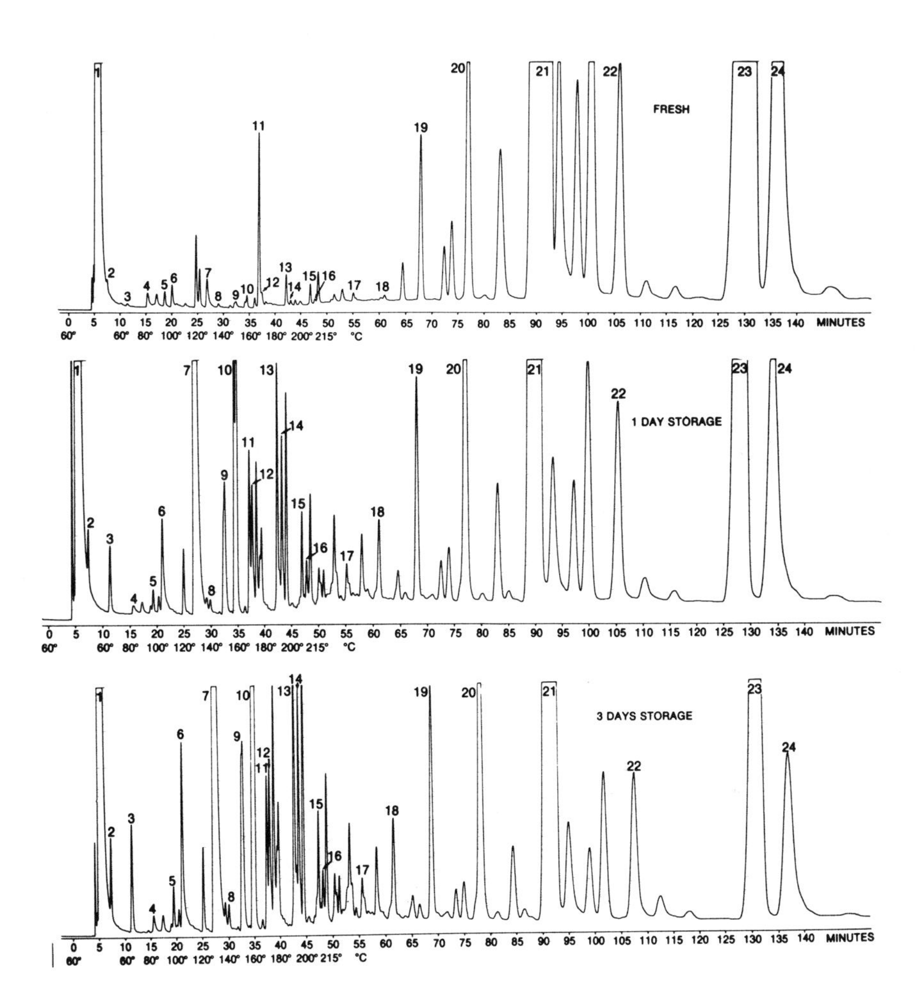

FIGURE 2. Chromatograms of ether-extracted volatiles from roast beef. Samples were extracted immediately following roasting (top), 1 day after roasting (middle) and 3 days after roasting (bottom). See Table II for peak identification.

TABLE II. Relative Concentration[a] of Volatile Compounds from Roast Beef During 4°C-Storage

Peak no.[b]	Compound	Days of storage 0	1	3
		Area %	Area %	Area %
2	n-Heptane	tr	0.05	0.14
3	n-Octane	tr	0.15	0.33
4	Ethyl acetate	0.03	0.02	0.07
5	2-Ethylfuran	tr	0.05	0.14
6	n-Pentanal	tr	0.29	0.76
7	n-Hexanal	0.06	7.06	14.42
8	2-n-Butylfuran	tr	0.04	0.09
9	n-Heptanal and 2-heptanone	0.02	0.50	1.03
10	2-n-Pentylfuran	0.03	1.30	2.89
11	3-Hydroxy-2-butanone	0.25	0.34	0.56
12	n-Octanal	0.02	0.32	0.78
13	n-Nonanal	0.05	0.53	0.90
14	n-Tetradecane	0.01	0.39	0.69
15	2-Decanone	0.03	0.19	0.38
16	Benzaldehyde	0.01	0.11	0.16
17	n-Heptadecane	0.01	0.13	0.20
18	n-Tridecanal	0.01	0.31	0.54
19	n-Tetradecanal	0.44	0.88	1.29
20	n-Pentadecanal	1.23	2.55	3.35
21	n-Hexadecanal	65.35	53.42	43.35
22	n-Heptadecanal	1.42	1.84	1.82
23	n-Octadecanal	18.57	16.18	11.91
24	2-Octadecenal	5.45	4.48	3.35

[a]Area relative to total area of volatiles.

[b]Peak nos. correspond to peak nos. in Figure 2.

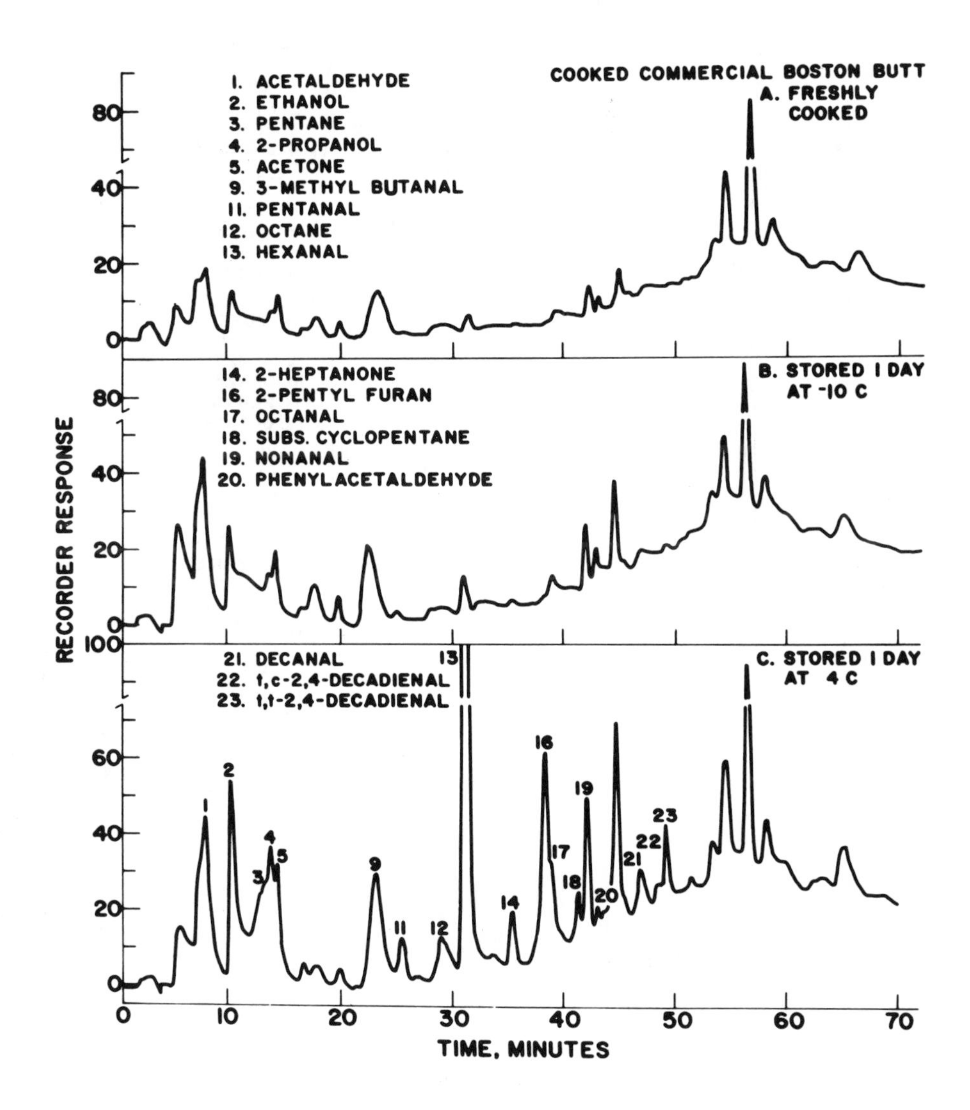

FIGURE 3. Direct sampling (17,18) GLC-MS profiles of volatiles from fresh Boston Butt.

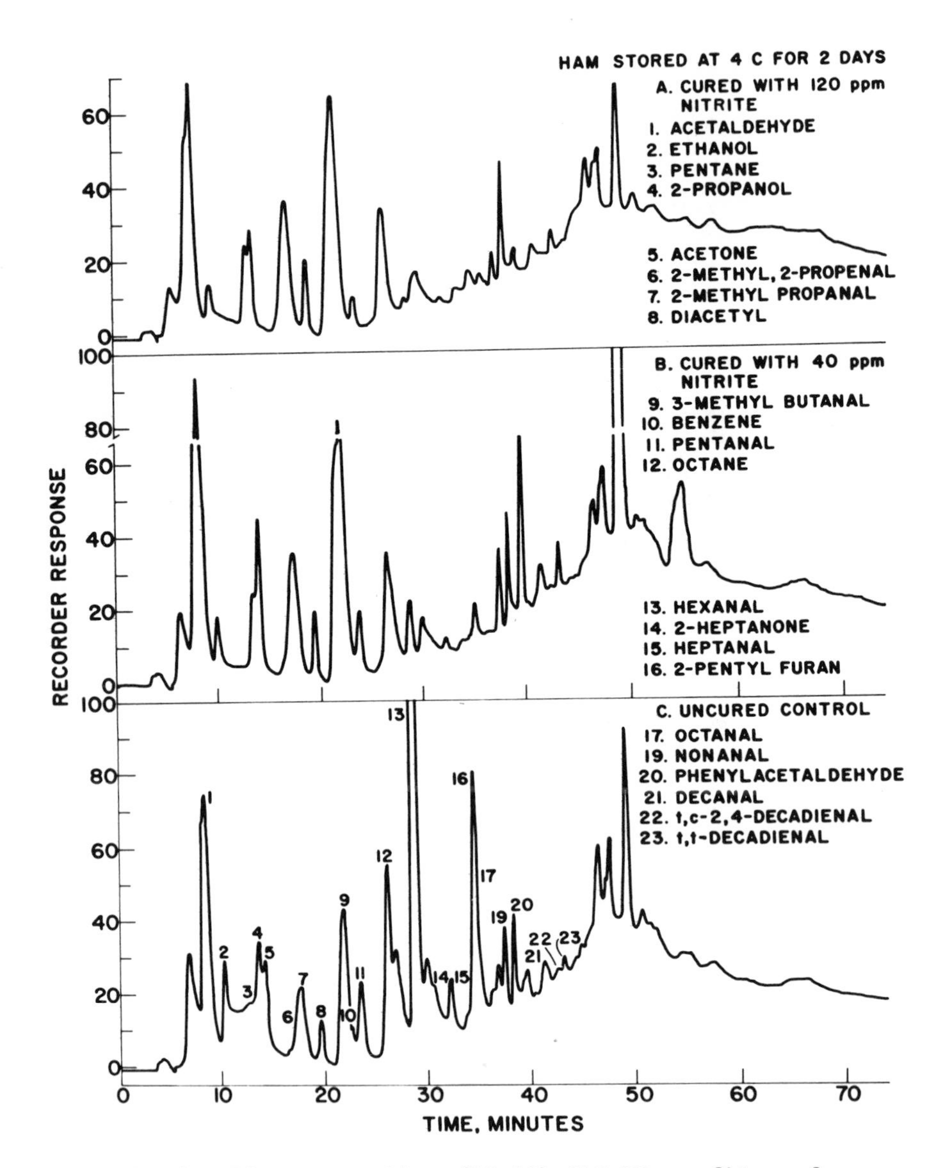

FIGURE 4. Direct sampling (17,18) GLC-MS profiles of volatiles from ham cured with (A and B) and without (C) sodium nitrite.

shortly following cookery. Nitrite also retards formation of low molecular weight aldehydes (C_6-C_{12}) that might be expected to form from oxidation of oleic and linoleic acids. These volatiles were studied by Bailey and Swain (51) using classic GLC-MS. The concentration level of low molecular weight aldehydes was a good indication of lipid oxidation.

Direct sampling GLC-MS (18) was used as a rapid method of studying the influence of nitrite on volatile organic compounds formed during storage of uncured and nitrite-cured ham and pepperoni (Figures 4 and 5, respectively).

Uncured ham stored for 2 days at 4°C yielded greater amounts of octane, hexanal, 2-pentyl furan and octanal than ham cured with either 40 or 120 ppm sodium nitrite (Figure 4). Hexanal and 2-pentyl furan would undoubtedly be useful for measuring the degree of oxidized flavor of this product. A similar study was made of bacon and the amount of hexanal produced during storage paralleled undesirable flavor of bacon cured without nitrite (55).

The direct sampling method was found useful for measuring the influence of nitrite on volatiles produced during heating of pepperoni (Figure 5). The major difference in the chromatograms is the increase during storage in the content of hexanal and 2-pentyl furan in the pepperoni prepared without nitrite. Several other volatiles such as octane, substituted furans and 2,4-decadienal were also more concentrated in the pepperoni cured without nitrite after storage for 1 week at 4°C.

GLC utilizing direct sampling has tremendous potential as a quality control procedure during processing and storage of meat products. It presently is widely accepted and is in use in the quality control of many other lipid-containing food products.

The most efficient additive used for preventing "warmed-over" flavor is sodium nitrite and despite its implication in nitrosamine formation, it is used primarily for this purpose in meat curing. No other additive has been found that can replace nitrite for this purpose. Other methods of retarding "warmed-over" flavor have been suggested and discussed (48,49), many of which were previously mentioned by Watts (43,44).

IV. ANIMAL SPECIES

Hornstein (56) was the first to emphasize that constituents responsible for differences in flavor of meat from various species were located in fat and not in muscle tissue. Hornstein and Crowe (57,58) examined volatiles from beef, pork and lamb

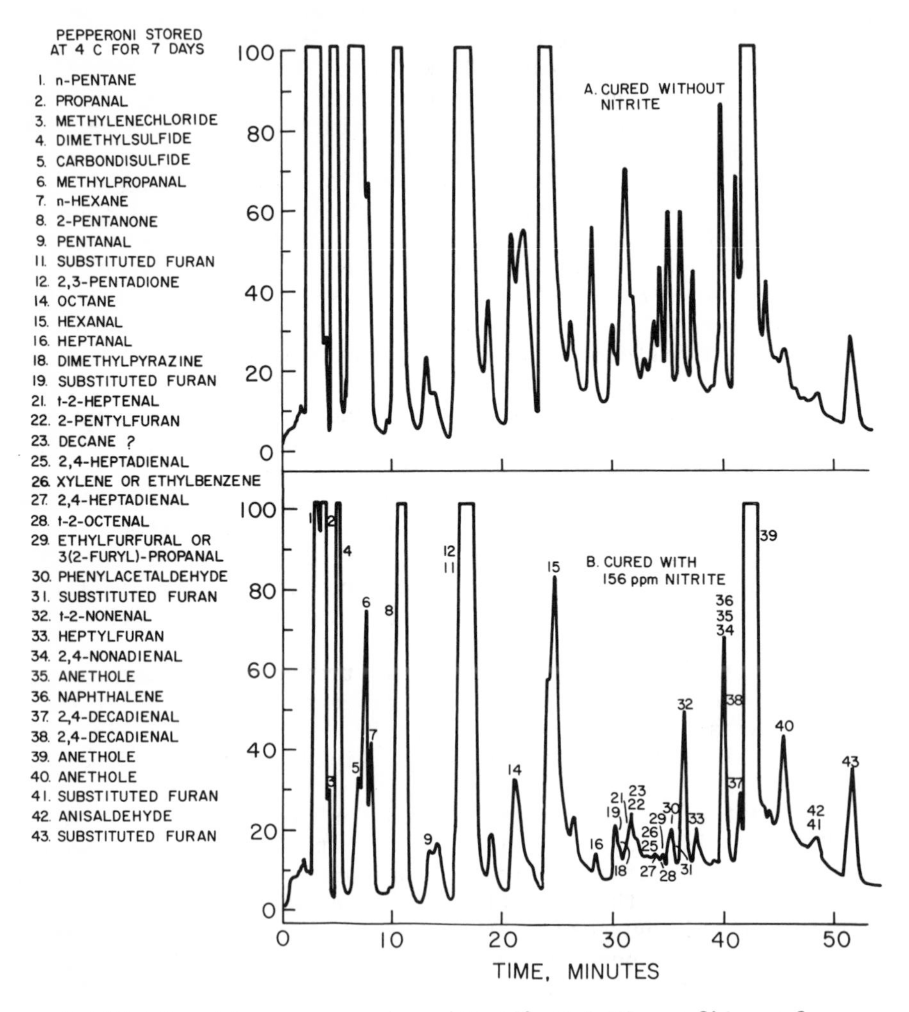

FIGURE 5. Direct sampling (17,18) GLC-MS profiles of volatiles from pepperoni cured with and without sodium nitrite.

fat and found that heating produced characteristic odors associated with the individual species. Fewer carbonyl compounds were found in lamb than in beef or pork because the fatty acids were more highly saturated. Pork had the greatest percent polyunsaturated fatty acids and produced the most volatile carbonyl compounds.

Hornstein and Crowe (58) concluded that a high proportion of "mutton aroma" is carbonyl and highly polar in nature and C_6-C_{18} saturated aldehydes were identified as contributing to mutton aroma. More recently, branched chain and unsaturated fatty acids having 8 to 10 carbon atoms have been identified as responsible for "mutton aroma" or "SOO" (59,60). These acids are associated with the sweaty-sour odor note from heated mutton. 4-Methyloctanoic acid was considered primarily responsible for the undesirable flavor. Other medium chain branched-saturated and unsaturated fatty acids were identified which were considered important to the flavor of lamb and mutton. The species effect on the flavor of mutton and lamb is probably expressed through their metabolism of lipid feeds, but the above authors have preliminary results that indicate that the level of medium-length, branched-chain fatty acids can be affected by animal diet. This is a possible method of modifying flavor of meat from this species.

A recent patent (61) claims that mutton flavor is improved by adding asparagine, glutamine, alanine, or glycine to the meat prior to cooking. Amino acids could be effective by reacting with undesirable volatile acids or carbonyls to reduce their volatility and diminish their influence on flavor.

The characteristic flavor of fresh pork may be associated with the presence of steroids in fatty tissue and these are believed responsible for "sex odor" in this species.

V. ANIMAL SEX

The predominant undesirable meat flavor resulting from animal sex is that recognized as "sex odor" in the uncastrated male pig (boar). This odor is recognized by a large percentage of consumers when fat from this animal is heated and the distinctive odor has been described as "urine-like." The occurrence and chemistry of this aspect of undesirable flavor has been reviewed by Pearson *et al.* (62) and by Sink (3,63).

"Sex odor" occurs in about 75% of boars and about 5% in barrows, gilts and sows. Sink (64) postulated that the odor was caused by C_{19}-16ene steroids and Patterson (65) identified the sex odor as 5 α-androst-16-ene-3-one. Male sex hormones appear to be precursors for formation of this steroid. Metabolism of C_{19}-16-ene steroids from pregnenolone in boar testis has been discussed by Gower (66.)

Diethylstilbesterol implants during late stages of growth decrease the objectionable aroma in some animals (67). Boar meat can best be used in products that are normally consumed cold such as many types of cured meats where the undesirable odor appears to be minimized (62). Naturally occurring amines such as certain amino acids might also alleviate this problem as well.

VI. PROCESSING AND STORAGE

Processing procedures resulting in undesirable meat flavors have been discussed in detail by Peineccius (4) and by Lawrie (68). Processing procedures and storage conditions causing most severe undesirable flavors include irradiation sterilization, heating at high temperature (retorting), microbial growth during storage and environmental contamination.

A. Irradiation

The undesirable odor developing when meat products are sterilized by irradiation has been characterized as being "sulfury", "wet dog", "goaty", or "burnt" (69-71). This problem has been studied for over thirty years and the undesirable flavor caused by this processing procedure remains one of the limiting features of using this method in commercial practice.

Undesirable flavor is caused by free radical interaction and oxidation of natural constituents of meat. Compounds responsible include carbonyls, sulfur compounds, alcohols and hydrocarbons (72-74).

Many procedures have been recommended to decrease the severity of off-odors formed during meat irradiation. Lower irradiation dosages, removal of oxygen prior to irradiation, use of spices and antioxidants and low temperature processing would appear to be the most promising (69).

B. High Temperature Processing

High temperature processing such as retorting causes chemical change of most meat constituents. Enormous amounts of ammonia, carbon dioxide and hydrogen sulfide are produced by amino acid and protein degradation. These gases interact with other products in the closed system resulting in formation of numerous volatile compounds that may be responsible for desirable or undesirable flavor of meat. Over 500 compounds have been identified by heating meat ingredients most of which are formed from lipid oxidation and degradation (75). Constituents responsible for "retort flavor" include sulfur compounds, carbonyls, alcohols, and substituted furans (76).

In some canning procedures hydrogen sulfide from meat protein can react with mesityloxide (from acetone in lacquer) to form 4-methyl-4-mercaptopenta-2-one which produces a "catty" odor (77).

Undesirable flavors resulting from retorting can be diminished by reducing heating time and sterilizing smaller samples at higher temperatures and by addition of certain amino and other organic acids (76).

C. Microbial Growth

Odors produced by microorganisms during meat storage are usually objectionable. The nature of the volatiles formed will depend on the type of microorganisms growing and also upon the storage environment. Prepackaged cured meat products and deep-seated portions of the carcass can be sources of growth of anaerobic organisms which produce putrefactive enzymes (68). *Proteus inconstans* causes "cabbage odor" in sliced vacuum packaged bacon due to formation of methane diol (78). Other types of compounds produced by microbiological enzyme degradation include amines, sulfur compounds and increased total volatiles (4).

D. Environmental Contamination

Frequently meat contains undesirable flavorants due to contamination of the animal or its carcass during processing and storage. Phenolic substances from animal dips have been identified and other disinfectants such as methyl bromide have been detected (79). Other environmental contaminants such as insecticides, bactericides, industrial solvents and related fat-soluble substances can penetrate carcass tissues and cause off-odors.

The undesirable flavor properties relative to microbiological growth and environmental contamination can be controlled by using good production and manufacturing practices, including good sanitation, good temperature control and good management.

VI. SUMMARY

The influence of animal feed, meat lipid oxidation, animal species, animal sex and processing and storage procedures on undesirable meat flavor is discussed. Meat from beef pastured on fescue has less desirable flavor than that from animals fed corn. Volatiles responsible for this undesirable flavor include octadecane, diethylphthalate, δ-dodecalactone and C_5-C_{10} aldehydes.

Meat from lambs consuming white clover, rape, oats and other forages has less desirable flavor than that of animals not consuming these feeds. Volatiles identified from feeding formaldehyde-protected sunflower seed-casein which causes undesirable flavor in lambs are trans, trans-2,4-decadienal and 4-hydroxydodec-cis-6-enoic acid lactone.

These problems can be diminished by finishing cattle for short periods on corn or grazing sheep on certain grasses as ryegrass 3 to 4 weeks prior to slaughter.

The most important type of undesirable flavor in meat is that due to oxidation. "Warmed-over" flavor is caused by the catalytic oxidation of phospholipid fatty acids by iron from heat-denatured hematin compounds in muscle. The most frequently used additive for preventing "warmed-over" flavor is sodium nitrite. Ingredients added to remove this flavor in fresh meat are: polyphosphates, sodium ascorbate, tocopherols, flavonoids and various Maillard-type browning products.

"Mutton aroma" is the most frequently recognized undesirable flavor caused by animal species. It may be caused by feed ingredients for lambs and mature sheep or may result from intrinsic lipid metabolism by these animals. Low molecular weight carbonyls (C_8-C_{10}) and branched chain fatty acids (4-methyloctanoic acid) have been identified as contributing to this flavor.

The most common undesirable flavor due to sex differences is caused by boar "sex odor." The volatile responsible for this characteristic is 5α-androst-16-ene-3-one and perhaps other C_{19}-16-ene steroids. Curing, hormone implants and meat chilling might make boar meat products more consumer acceptable.

Irradiation, high temperature processing, microbial growth and certain environmental contaminants also cause undesirable flavors in meats. Irradiation flavor can be retarded by using lower irradiation dosages or by irradiating samples at below-freezing temperatures in vacuum. The other processing and storage related off-flavors can be controlled by using good production and manufacturing practices including good sanitation and temperature control.

REFERENCES

1. Pagborn, R. M., Food Technol. 29:34 (1975).
2. Patterson, R. L. S., in "Meat" (D. J. A. Cole and R. A. Lawrie, eds.), p. 359. Butterworths, London, 1975.
3. Sink, J. D., J. Food Sci. 44:1 (1979).
4. Reineccius, G. A., J. Food Sci. 44:12 (1979).
5. Wanderstock, J. J. and Miller, J. C., Food Research 13:291 (1948).
6. Meyer, G., Thomas, J., Buckley, R. and Cole, J. W., Food Technol. 14:4 (1960).
7. Dube, G., Bramblett, V. D., Howard, R. D., Hamler, B. E., Johnson, R. B., Harrington, R. B. and Judge, M. D., J. Food Sci. 36:147 (1971).
8. Bond, J., Hooven, N. W., Jr., Warwick, E. J., Hiner, R. L. and Richardson, G. V., J. Anim. Sci. 34:1046 (1972).
9. Purchas, R. W. and Davies, L., Australian J. Agr. Res. 25:183 (1974).
10. Bidner, T. D., Proc. Reciprocal Meat Conference 28:301 (1975).
11. Kroft, D. H., Allen, D. M. and Thouvenelle, G. J., 67th Am. Soc. Anim. Sci. (1975) (Abstract).
12. Bowling, R. A., Dutson, T. R., Carpenter, A. L., Smith, G. C. and Oliver, W. M., J. Anim. Sci. 42:254 (1976) (Abstract).
13. Moody, W. G., Proc. Reciprocal Meat Conference 29:128 (1976).
14. Smith, G. M., Crouse, J. D., Mandigo, R. W. and Neer, K. L., J. Anim. Sci. 45:236 (1977).
15. Bowling, R. A., Riggs, J. K., Smith, G. C., Carpenter, A. L., Reddish, R. L. and Butler, O. D., J. Anim. Sci. 46:333 (1978).
16. Hedrick, H. B., Bailey, M. E., Krouse, N. J., Dupuy, H. P. and Legendre, M. G., Proc. 26th European Meeting of Meat Research Workers, U.S.A. (1980).
17. Dupuy, H. P., Brown, M. L., Legendre, M. G., Wadsworth, J. I. and Rayner, E. T., in "Lipids as a Source of Flavor" (M. K. Supran,ed.), p. 60. American Chemical Society, Washington, D. C., 1978.
18. Legendre, M. G., Fisher, G. S., Schuller, W. H., Dupuy, H. P. and Rayner, E. T., J. Am. Oil Chem. Soc. 56:552 (1979).
19. Watanabe, K., and Sato, Y., Agr. Biol. Chem. 35:756 (1971).
20. Hawke, J. C., in "Chemistry and Biochemistry of Herbage" Vol. I (G. W. Butler and R. W. Bailey, eds.), p. 213, Academic Press, London, 1973.

21. Westerling, D. B. and Hedrick, H. B., J. Anim. Sci. 48:1343 (1979).
22. Rumsey, T. S., Oltjen, R. R., Bovard, K. P. and Proide, B. M., J. Anim. Sci. 35:1069 (1972).
23. Demick, P. S., Walker, N. and Patton, S., Biochem. J. 111:395 (1969).
24. Reagan, J. O., Carpenter, J. A., Bauer, F. T. and Lowrey, R. S., J. Anim. Sci. 45:716 (1977).
25. Harrison, A. R., Smith, M. E., Allen, D. M., Hunt, M. G., Kastner, C. L. and Kroft, D. H., J. Anim. Sci. 47:383 (1978).
26. Skelly, G. C., Edwards, R. L., Wardlow, F. B. and Torrence, A. K., J. Anim. Sci. 47:1102 (1978).
27. Shorland, F. B., Szochanska, Z., Moy, M., Barton, R. A. and Rae, A. L., J. Sci. Fd. Agric. 21:1 (1970).
28. Park, R. J., Corbett, J. L. and Furnival, E. P., J. Agr. Sci. (Camb.) 78:47 (1972).
29. Park, R. J., Spurway, R. A. and Wheeler, J. L., J. Agr. Sci. (Camb.) 78:53 (1972).
30. Park, R. J. and Minson, D. J., J. Agr. Sci. (Camb.) 79:473 (1972).
31. Park, R. J., Ford, A. L. and Ratcliff, D., J. Food Sci. 41:633 (1976).
32. Park, R. J., Murray, K. E. and Stanley, G., Chem. and Ind. (London),p. 380 (1974).
33. Watts, B. M., Adv. Food Res. 5:1 (1954).
34. Lea, C. H., in "Lipids and Their Oxidation" (H. W. Schultz, E. A. Day and R. O. Sinnhuber, eds.), p. 3. AVI Publishing Co., Westport, CT., 1962.
35. Frankel, E. N., in "Lipids and Their Oxidation" (H. W. Schultz, E. A. Day and R. O. Sinnhuber, eds.), p. 51. AVI Publishing Co., Westport, CT., 1962.
36. Kenney, M., in "Lipids and Their Oxidation" (H. W. Schultz, E. A. Day and R. O. Sinnhuber, eds.), p. 79. AVI Publishing Co., Westport, CT., 1962.
37. Dugan, L., Jr., in "Food Chemistry" (O. R. Fennema, ed.), p. 169. Marcel Dekker, New York, 1976.
38. Lillard, D. A., in "Lipids as a Source of Flavor" (M. K. Supran, ed.), p. 68. American Chemical Society, Washington, D. C., 1978.
39. Ingold, K. U., in "Lipids and Their Oxidation" (H. W. Schultz, E. A. Day and R. O. Sinnhuber, eds.), p. 93. AVI Publishing Co., Westport, CT., 1962.
40. Tappel, A. L., in "Lipids and Their Oxidation" (H. W. Schultz, E. A. Day and R. O. Sinnhuber, eds.), p. 122. AVI Publishing Co., Westport, CT., 1962.
41. Tims, M. J. and Watts, B. M., Food Technol. 12:240 (1958).
42. Younathan, M. T. and Watts, B. M., Food Res. 24:728 (1959).

43. Watts, B. M., in Proc."Flavor Chemistry Symposium", p. 83. Campbell Soup Co., Camden, N. J., 1961.
44. Watts, B. M., in "Lipids and Their Oxidation" (H. W. Schultz, E. A. Day and R. O. Sinnhuber, eds.), p. 202. AVI Publishing Co., Westport, CT., 1962.
45. Kendrick, J. and Watts, B. M., Lipids 4:454 (1969).
46. Greene, B. E., J. Food Sci. 34:100 (1969).
47. Liu, H., J. Food Sci. 35:590 (1970).
48. Sato, K. and Herring, H. K., Proc. Reciprocal Meat Conference 26:64 (1973).
49. Pearson, A. M., Love, J. D. and Shorland, F. B., Adv. Food Res. 23:1 (1977).
50. Liu, H. and Watts, B. M., J. Food Sci. 35:596 (1970).
51. Bailey, M. E. and Swain, J. W., Proc. Meat Ind. Res. Conf., p. 29, 1973.
52. Kwon, T. and Menzel, D. B., J. Food Sci. 30:808 (1965).
53. Tarladgis, B. G., Pearson, A. M. and Dugan, L. R., Jr., J. Sci. Fd. Agric. 15:602 (1964).
54. Westerberg, D. O., Proc. Reciprocal Meat Conference 26:64 (1973).
55. Ihekoronye, A. I., Effect of Nitrite on Flavor and Nitrosamine Content of Bacon. M. S. Thesis, Univ. of Missouri (1978).
56. Hornstein, I., in "The Chemistry and Physiology of Flavors" (H. W. Schultz, E. A. Day and L. M. Libbey, eds.), p. 228. AVI Publishing Co., Westport, CT., 1967.
57. Hornstein, I. and Crowe, P. F., J. Agric. Food Chem. 8:494 (1960).
58. Hornstein, I. and Crowe, P. F., J. Agric. Food Chem. 11:147 (1963).
59. Wong, E., Food Technol. New Zealand. January, p. 13, 1975.
60. Wong, E., Nixon, L. N. and Johnson, C. B., J. Agric. Food Chem. 23:495 (1975).
61. Japanese Examined Patent. "Mutton Flavor Improvement." No. 5406623 (1979).
62. Pearson, A. M., Thompson, R. H. and Price, J. F., Proc. Meat Ind. Res. Conf., p. 145, 1969.
63. Sink, J. D., J. Amer. Oil Chem. Soc. 50:470 (1973).
64. Sink, J. D., J. Theoret. Biol. 17:174 (1967).
65. Patterson, R. L. S., J. Sci. Food Agric. 19:31 (1968).
66. Gower, D. B., J. Steroid Biochem. 3:45 (1972).
67. Teague, H. S., Plimpton, R. F., Cahill, V. R., Grifo, A. P., Jr. and Kunkle, L. E., J. Anim. Sci. 23:332 (1964).
68. Lawrie, R. A., "Meat Science", 3rd Ed., p. 362. Pergamon Press, Oxford, 1979.
69. Huber, W., Brasch, A. and Waly, A., Food Technol. 7:109 (1953).

70. Batzer, O. F., Sliwinski, L. C., Pih, K., Fox, J. B., Jr., Doty, D. M., Pearson, A. M. and Spooner, M. E., Food Technol. 13:501 (1959).
71. Batzer, O. F. and Doty, D. M., J. Agr. Food Chem. 3:64 (1955).
72. Merritt, C., Jr., "Food Irradiation: Proceedings of Symposium, Karlsruhe, Germany." International Atomic Energy Agency (Vienna), 1966.
73. Wick, E. L., Murry, E., Mizutani, J. and Koshika, M., "Radiation Preservation of Foods." American Chemical Society, Washington, D. C., 1967.
74. Angelini, P., Merritt, C., Jr., Mendelsohn, J. M. and King, F. J., J. Food Sci. 40:197 (1975).
75. Ching, J., Volatile Flavor Compounds From Beef and Beef Constituents. Ph.D. Dissertation, Univ. of Missouri (1979).
76. Persson, T. and von Sydow, E., J. Food Sci. 38:377 (1973).
77. Aylward, F., Coleman, G. and Harsman, D. R., Chem. Ind. 1563 (1967).
78. Gardner, G. A. and Patterson, R. L. S., J. Appl. Bact. 29:263 (1975).
79. Rhodes, D. N., Patterson, R. L. S., Puckey, D. J., Heuser, S. G., Wainman, H. E., Chakrabarti, B. and Allen, E. N. W., J. Sci. Fd. Agric. 26:1375 (1975).

FLAVOR DEVELOPMENT IN CHEESE

G.J. Moskowitz

Dairyland Food Laboratories, Inc.
Waukesha, Wisconsin

The origins of cheesemaking are lost in antiquity. References both in the bible and in the records of ancient Egypt, indicate that the art of cheesemaking was well advanced by the time man began to record history. It is conceivable that nomadic tribesmen accidentally discovered the art while trying to store milk in animal stomachs. Over the centuries, the art has been refined to where today, hundreds of cheese varieties are available (Davis, J.G., 1965).

I. CHEESE RIPENING

As a first step in cheesemaking, bacteria, referred to as starter organisms, are added to the milk prior to coagulation. These organisms ferment lactose (cows' milk contains approximately 4.8% lactose) to lactic acid. Rennet is then added to produce the curd, the curd is cut and the whey is separated. The curd is treated to effect whey syneresis and curd knitting, placed in forms and allowed to age. The proper rate and quantity of acid development is essential throughout the process as the pH environment at each step significantly affects the final outcome.

ISBN 0-12-169065-2

α, β and k caseins exist suspended in milk in a micellar structure supported by k casein. Rennet hydrolyzes the k casein which destablizes the micelle allowing the α and β casein to interact with soluble Ca^{+2} ion. The result is a precipitation of a cross linked and bonded fiberous protein network termed dicalcium paracaseinate that traps water, sugar, salt and fat (Kosikowski, 1978). This complex is called curd. During the ripening process, the fat and protein are partially hydrolyzed to produce precursor compounds which, together with sugar, are converted to flavor components of the particular cheese variety. The type of flavor components, and therefore the type of cheese produced, is determined by the manufacturing procedure and the choice of microorganisms.

These complex reactions can be described by a series of equations depicting the major chemical changes that occur.

$$\text{Dicalcium Paracaseinate} \xrightarrow[\text{Acid}]{\text{Enzymes}} \text{Monocalcium Paracaseinate}$$

$$\text{Monocalcium Paracaseinate} \xrightarrow[\text{Acid}]{\text{Enzymes}} \text{Paracasein}$$

$$\text{Paracasein} \xrightarrow[\text{Enzymes}]{} \text{Flavor Precursors}$$

$$\text{Triglyceride} \xrightarrow[\text{Lipase}]{} \text{Diglyceride \& Fatty Acid}$$

$$\text{Lactose} \xrightarrow[\text{Enzymes}]{} \text{Lactic Acid \& Other Metabolites}$$

A. The Role of Microorganisms in Cheese Flavor Development

Cheese flavor development is thought to occur through a complex series of biochemical reactions requiring bacterial enzymes. It is therefore necessary that these intracellular enzymes come into contact with the paracaseinates and other substrates.

Fine structural analysis of 5 month old Cheddar cheese indicated that the microorganisms were found clustered around the fat globules. Some of their cell walls had been disrupted resulting in the formation of both intact and ruptured protoplasts. These ruptured cells had leaked their cell contents into the body of the cheese thus providing contact between the bacterial enzymes and the cheese substrate (Umemoto et al., 1978).

Studies by Law and Sharpe (1975) showed that the decrease in cell viability in Cheddar cheese after 2-4 months, correlated with the detection of Cheddar flavor (Table 1). Further studies showed that the release of the intracellular enzyme dipeptidase into the cheese also correlated with the decrease in cell viability (Law and Sharpe, 1977). Since some of these intracellular enzymes are probably unstable, the gradual loss of viability could provide a continuing replacement of required enzymes over time.

II. IMPORTANT COMPOUNDS IN CHEESE FLAVOR

Kosikowski and Mocquot (1958) postulated the component balance theory principally for hard cheeses such as Cheddar. This theory states that flavor is not due to any single component, but is the result of a synergistic blend of various compounds in proper proportion. The definition of

TABLE 1. Starter Viable Counts in Cheddar Cheese During Maturation[a]

Age of Cheese (Months)	0	1	2	4
Viable Count x 10^6	2000	10-70	0.9-8	0.07-0.8

[a]Law and Sharpe (1975).

this complex blend continues to elude the flavor chemist but some basic information is available for several cheese types.

A. Italian Cheese Flavor

The characteristic flavor of Italian cheese is primarily due to low molecular weight free fatty acids. Approximately 12% of the fatty acids of milk are low molecular weight (C_4 - C_{12}). This lipid composition provides a unique substrate for lipolytic enzymes with a degree of specificity for these low molecular weight fatty acids.

Farnham (1950) found that the characteristic flavor of Italian cheese was produced by an enzyme system prepared from the pharyngeal region of suckling calves, kids and lambs. These enzymes are relatively specific for short chain fatty acids as compared to pancreatic lipase (Table 2).

The three pregastric esterases derived from calf, kid and lamb, also have their own specificity for each particular low molecular weight fatty acid. Therefore, milk fat hydrolyzed by these enzymes contains various ratios of C_4, C_6, C_8 and C_{10} fatty

TABLE 2. Fatty Acid Specificity of Pregastric Esterases[a]

Enzyme	Butyric	Caproic	Caprylic	Capric
Calf Oral	36.7	8.9	4.8	10.7
Kid Oral	44.4	15.2	7.6	12.3
Lamb Oral	48.1	8.6	14.2	9.3
Pancreatic Lipase	8.4	2.1	Trace	Trace

[a]Harper (1957).

acids. Lamb enzyme produces a higher proportion of caprylic acid than either calf or kid enzyme, while kid produces a higher proportion of caproic acid than calf or lamb. These ratios are responsible for the characteristic flavors of Romano, Parmesan and Provolone cheeses. Lamb enzyme produces a peccorino flavor found in Provolone cheese and kid produces a goaty or piccante flavor, while calf produces a sweet buttery flavor. Various combinations of these enzymes are used to produce a desired balanced flavor.

Additional properties of the pregastric esterases are outlined in Figures 1 and 2. The pregastric esterases have a pH optimum of 5.5 (Fig. 1) and are stable over the pH range of 5-10 (Fig. 2). They are most stable in the pH range

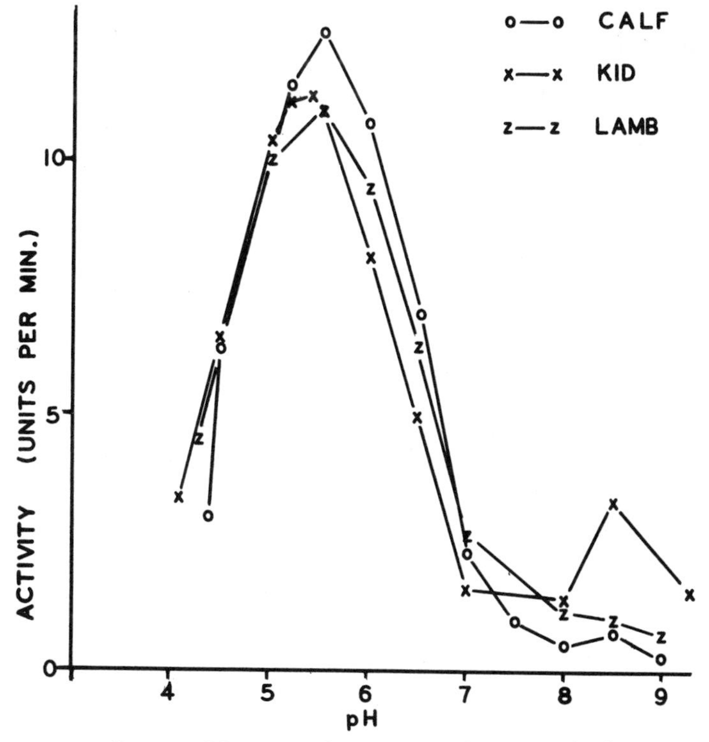

FIGURE 1. Effect of pH on the activity of pregastric esterases[a].

[a]Richardson and Nelson (1967).

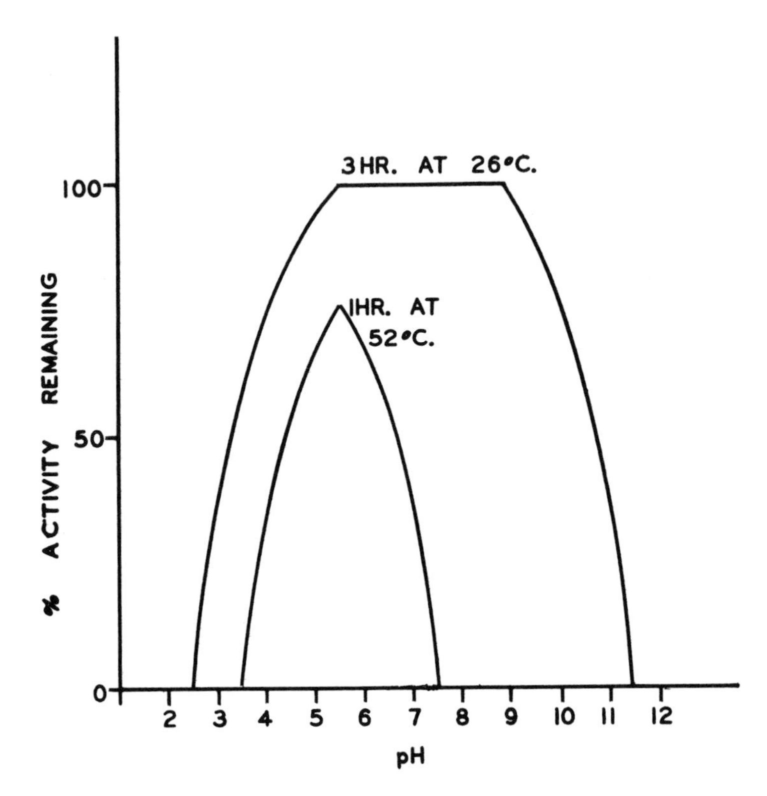

FIGURE 2. Effect of pH on the stability of calf pregastric esterase[a].

[a]Richardson and Nelson (1967).

of 5.5-6.0. These properties provide the highest enzyme efficiency at the pH of cheese ripening (Richardson and Nelson, 1967).

Other enzymes reported to produce good cheese flavors include the lipase from *Mucor miehei* (Huang and Dooley, 1976), *Aspergillus niger* (Harper, 1957) and *Candida cylindracea* (Sood and Kosikowski, 1979).

B. Cheddar Cheese Flavor

Cheddar flavor is a chemically complex system which is only partially dependent on the volatile fraction. Salts, peptides and amino acids previously considered either precursor products or background flavor agents were shown by McGugan to be the major contributors to flavor intensity. Fat and the insoluble residue were also important in flavor perception (McGugan et al., 1979). Table 3 outlines some of the volatile compounds considered important in Cheddar cheese flavor.

Manning (1978a) prepared Cheddar cheese in a microbiologically controlled environment using gluconolactone to adjust the pH of the milk. Cheese was aged up to 1 year and sampled at 2 month intervals.

Headspace volatiles that correspondingly increased with increasing Cheddar flavor included 2-pentanone, methanethiol, acetone and methanol. Methanethiol was considered the characterizing compound since the Cheddar cheese aroma was lost when it was removed from the volatile fraction (Manning, 1978b).

Other compounds deemed important include diacetyl, H_2S, butanone, dimethylsulfide, ethanol, ketones derived from amino acids by deamination, lactones, phenols and pyrazines (McGugan, 1975).

TABLE 3. Some Important Compounds Identified as Components of Cheddar Cheese Flavor

Free Fatty Acids
Methanethiol
Methanol
Dimethyl Sulfide
Diacetyl
Acetone
Butanone
2-Pentanone
Lactic Acid
Acetic Acid

1. Development of Cheddar Flavor. The development of Cheddar flavor is, in part, enzymatic and is dependent on the presence of microorganisms such as Streptococcus cremoris. Ubiquitous organisms apparently also play a role in the development of good quality Cheddar flavor and some of them are capable of producing methanethiol (Law and Sharpe, 1978).

The exact role of these microorganisms is unclear. Cheese made asceptically in the absence of starter and therefore bacterial enzymes, failed to produce either flavor or detectable levels of methanethiol (Manning, 1978b). Cheese flavor also failed to develop in cheese made by chemical acidification of the milk and the addition of microbial enzymes from lysozyme treated cells, yet cell lysis was detected by the increase of DNA and intracellular dipeptidase levels in the cheese. Furthermore, cell free extracts of the starter organisms did not convert L-methionine to methanethiol (Law et al., 1976). Based on these observations, Law and Sharpe (1978) postulated that methanethiol, and therefore presumably other flavor compounds, are produced non enzymatically. The role of the starter was postulated to be twofold, to provide precursor compounds such as L-methionine by protein hydrolysis and to produce a reducing environment (redox potential -150 to -200 mV) in the cheese to prevent oxidation.

C. Swiss Cheese Flavor

Swiss cheese flavor is characterized as nutty and sweet and is the result of the fermentation of Streptococcus thermophilus, Lactobacillus bulgaricus and Propionibacterium shermanii. P. shermanii produces CO_2 gas by converting lactic acid to acetic acid + proprionic acid + CO_2. It is this gas evolution that produces the characteristic eyes or holes. Interestingly, L. bulgaricus and S. thermophilus are the organisms of choice for the manufacture of Romano cheese, yet when combined with P. shermanii and the proper manufacturing procedure, the result is Swiss cheese.

1. Compounds Important in Swiss Cheese Flavor. Biede and Hammond (1979a,b) studied Swiss cheese flavor by fractionation of the cheese into water and oil soluble fractions. The sweet flavor, concentrated in the water soluble, non volatile fraction, correlated with the presence of amino acids, including proline, and small peptides. The sweet flavor perception required the presence of Ca^{+2} and Mg^{+2} ions. The nutty flavor correlated with the oil soluble portion and appeared to be due to free fatty acids. In addition, pyrazines, carbonyls, lactones and phenols were also detected. Table 4 lists several compounds considered important in Swiss cheese flavor.

D. Mold Ripened Cheeses

The typical flavor of mold ripened cheese is a result of the metabolic activity of particular mold species. Examples of mold ripened cheeses are Blue cheese, prepared with <u>Penicillium roqueforti</u>, Gorgonzola and Stilton cheese. Basically, these are manufactured by the introduction of mold spores into the milk or curd and the development of the typical flavor through vegetative cell outgrowth of the spores within the body of the cheese.

Camembert and Brie are examples of surface ripened cheeses. Curds are prepared in the usual manner, shaped and the surface inoculated with the particular mold. The flavor is developed by surface fermentation.

TABLE 4. Some Important Compounds Identified as Components of Swiss Cheese Flavor

Acetic Acid	Amino Acids and Small Peptides
Proprionic Acid	Dimethylsulfide
Butyric Acid	Isobutyric Acid
Diacetyl	Isovaleric Acid
Proline	Short Chain Thioesters

1. Blue cheese Flavor Development. Molds are potent producers of lipolytic and proteolytic enzymes. In the manufacture of Blue cheese _P. roqueforti_ produces a lipase that hydrolyzes the milk fat to glycerol esters plus free fatty acids. These fatty acids are then metabolized via the β-oxidative pathway to β-ketoacyl-CoA which is further hydrolyzed and decarboxylated to form methyl ketones with one carbon atom less than the precursor fatty acid (Figure 3).

These methyl ketones and secondary alcohols of 5-11 carbon atoms are the key flavor components of the cheese, the most abundant of which are 2-heptanone and 2-nonanone. In addition, free fatty acids are also important in Blue cheese flavor.

Proteases break down the curd, modifying the texture and producing amino acids which contribute to the background flavor and undoubtedly produce precursors for further metabolism (Moskowitz, 1979).

A natural Blue cheese flavor has been developed by Dairyland Food Laboratories, Inc. based on the fermentation of a milk fat substrate by _P. roqueforti._ The flavor components of the fermented milk substrate are similar to those found in Blue cheese (Watts and Nelson, 1963).

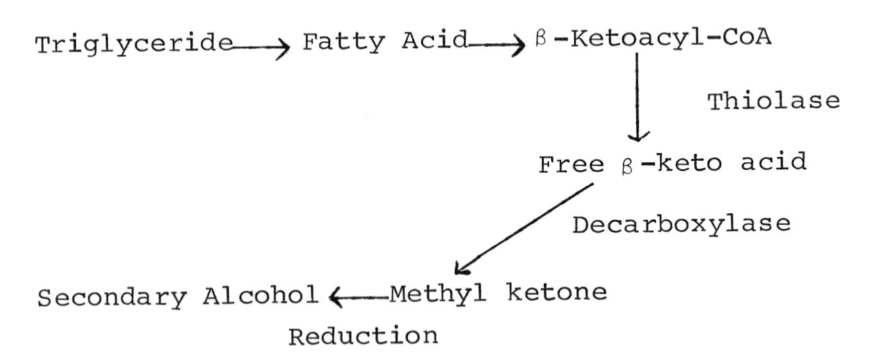

FIGURE 3. Production of methyl ketones and secondary alcohols from free fatty acids by _P. roqueforti_.

E. Accelerated Cheese Flavor Development

The time of ripening varies from several days for Mozzarella cheese to over 12 months for aged Cheddar. Longer ripening times add to the cost of the product. It is therefore desirable to develop systems that will enhance flavor development and decrease storage costs.

1. Enzyme Modified Cheese. Fatty acids and butterfat play a major role in the flavor perception of cheese. Low fat cheeses prepared from low fat milk or skim milk do not possess the intensity or typical flavor of full fat cheese. The role of lipid has not been fully defined although two aspects may be important. The triglycerides are precursors of free fatty acids and they may also serve as a carrier for fat soluble flavor components.

Enzyme modified cheese, the major flavor component of processed cheese, is prepared, in part, by lipolytic treatment of young cheese. The resultant fatty acids produced provide an intensified cheese flavor characteristic of the starting material. When added to a low flavor, young cheese, the intensity of the flavor is synergistically increased. Thus, free fatty acids, while not the principal characteristic flavor component of Cheddar or Swiss, nevertheless play a major role in flavor perception and intensity.

Table 5 demonstrates the significant increase in fatty acids produced by enzyme modification. In these examples, butyric acid levels have been elevated 10 - 100-fold, while palmitic acid has been elevated 8 - 50-fold. The degree of lipolysis is determined by the intensity of flavor desired and the application of the product.

The relative ratios are compared in Table 6. The similarity of the ratios for Swiss and Cheddar indicate the potentiating and modifying effect free fatty acids have on cheese flavor.

TABLE 5. Free Fatty Acid Composition of Young Swiss and Cheddar Cheese and Enzyme Modified Cheese

	Moles/100 g dry weight			
Fatty Acid	60 Day Swiss	EMC Swiss	60-90 Day Cheddar	EMC Cheddar
Acetic	75.0	76.7	1.3	63.3
Proprionic	105.4	89.2	---	---
Butyric	6.6	69.0	1.5	169.1
Caprioc	1.7	12.1	0.7	34.0
Caprylic	---	11.1	---	21.1
Capric	2.6	21.8	0.5	52.1
Lauric	2.0	18.8	0.7	51.6
Myristic	6.1	44.7	2.3	137.0
Palmitic	15.2	124.2	5.7	285.5
Stearic	0.5	48.4	1.6	85.7
Oleic	11.7	101.4	5.0	210.3
Linoleic	2.1	5.4	0.8	17.3
Linolenic	1.4	7.2	0.3	20.1

Enzyme modified cheeses include Cheddar, Swiss, Italian, Cream, Edam, Blue and Brick. Various flavor profiles are developed with combinations of the above. For example, a particular Cheddar profile can be developed using Cheddar, Blue and Swiss.

2. Accelerated Ripening by Fermentation. The approach of using an excess of cells to accelerate flavor development has resulted in conflicting observations. Common to both approaches was the need to selectively destroy the acid producing capability of the cells.

Law and Sharpe (1978) had reported that flavor development was not accelerated using lysozyme treated cells. Thompson (et al., 1979) reported that heat shocked _Streptococcus cremoris_ cells,

TABLE 6. Fatty Acid Ratios of Enzyme Modified Cheese

	Butyric Acid Relative Ratio	
Fatty Acid	EMC Cheddar	EMC Swiss
Acetic	0.37	1.11
Butyric	1.00	1.0
Caproic	0.20	0.17
Caprylic	0.12	0.16
Capric	0.31	0.31
Lauric	0.30	0.27
Myristic	0.81	0.65
Palmitic	1.69	1.79
Stearic	0.51	0.70
Oleic	1.24	1.47
Linoleic	0.10	0.08
Linolenic	0.12	0.10

when added to milk in addition to the normal level of culture, produced an intensified and accelerated flavor in both low fat and normal Cheddar. The difference between these results may lie in Thompson's use of intact cells. Enzymes present in the cell wall or interstitial space may be important in flavor development or perhaps a living, metabolically active cell is required in order to generate energy and reducing compounds for flavor development via the organism's normal metabolic pathways.

F. Development of Off Flavors

1. Bitterness. Bitterness is attributed to the proteolytic production of bitter peptides. Proteolytic activity in cheese is due largely to rennet and starter or contaminating microorganisms. A model for proteolytic flavor development is illustrated in Figure 4 (Lawrence et al., 1978).

Bitter cheese is more likely to be produced with rapid acid producing cultures than slower acid producing strains. Rennet presumably hydrolyzes

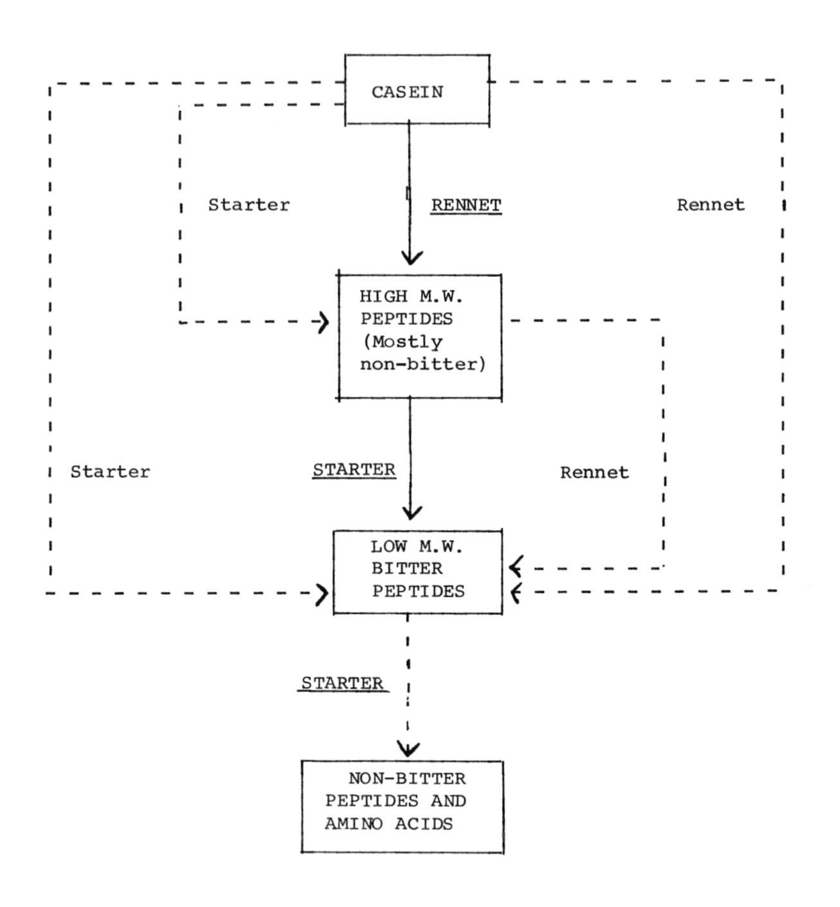

FIGURE 4. Metabolism of casein by starter organisms.

casein to large molecular weight peptides, which are further degraded by the bacteria or rennet to smaller, sometimes bitter peptides. Unless these bitter peptides are further metabolized by the starter organisms, they will accumulate and produce a detectable defect (Lawrence et al., 1978).

2. Effect of Psychrotrophic Organisms. Psychrotrophic organisms are organisms that actively grow at low temperature. While these organisms usually do not survive pasteurization, the lipases and proteases they elaborate do survive. These enzymes are carried into the cheese and, in excess, could produce bitterness and rancidity.

In addition, the breakdown products produced by these enzymes stimulate starter culture acid production which, in excess, produces body and flavor defects (Mikolajckik, 1979).

3. Effect of Spore Formers and Other Spoilage Microorganisms. If present in milk, bacterial spores survive pasteurization and subsequently germinate and produce vegetative cells. These vegetative cells produce cheese with bitter, rancid or spoiled taste defects in addition to a variety of characteristics determined by the particular contaminant. In addition, spoilage microorganisms sometimes produce gas which causes texture, taste and body defects.

4. Transmitted flavors of milk. Some flavor defects in milk that carry over into the cheese include silage and foliage flavors, weed flavors such as those contacted during grazing and barny flavors from improperly ventilated areas. Volatile components pass rapidly from the cow's lungs to the cow's udder and account for many of the above mentioned defects (Shipe et al., 1979). Examples of some of the conditions that result in off flavors in cheese are outlined in Table 7.

Cheese flavor is the result of a complex set of metabolic and non metabolic reactions which, under controlled conditions, produce the desired flavor. Off flavors and defects result when these elements are either out of balance or otherwise interfered with.

TABLE 7. Factors Affecting Flavor in Cheese[a]

Possible causes of bad flavor:
Taint organisms in raw milk
Contamination of pasteurized milk in dairy by unclean equipment
Oxidation of fat in milk before processing
Inadequate heat-treatment of milk
Excessive proportion of mastitis milk
Antibiotics, repressing starter organisms and so allowing taint organisms, especially the Gram-negatives, to grow
Possible causes of lack of flavor:
Absence of aroma organisms in starter
Conditions in cheese nutritionally unfavorable to development of flavor organisms (e.g. absence of growth factors)
Organisms inhibitory to flavor organisms
Substances (antibiotics used therapeutically, antibiotics naturally present in milk inhibitory to flavor organisms)
Over-pasteurization of milk destroying flavor organisms
Physical conditions in cheese not favorable for development of flavor organisms

[a]Davis, J.G. (1976).

REFERENCES

Biede, S.L., and Hammond, E.G. (1979a). J. Dairy Sci. 62, 227.

Biede, S.L., and Hammond, E.G. (1979b). J. Dairy Sci. 62, 238.

Davis, J.G. (1976). "Cheese," Vol. 3, p. 502. American Elsevier Publishing Company, New York.
Davis, J.G. (1965). "Cheese," Vol. 1, p. 44. American Elsevier Publishing Company, New York.
Farnham, M.G. (1950). U.S. Patent 2,531,329.
Harper, W.J. (1957). J. Dairy Sci. 40, 556.
Huang, H.T., and Dooley, J.G. (1976). Biotech. Bioeng. 18, 909.
Kosikowski, F.V., and Mocquot, G. (1958). FAO Agric. Stud. No. 38.
Kosikowski, F.V. (1977). "Cheese and Fermented Milk Foods." F.V. Kosikowski and Associates, New York.
Law, B.A., and Sharpe, M.E. (1975). In "Lactic Acid Bacteria in Beverages and Food." (J.G. Carr, C.U. Cutting and G.C. Whiting, eds.), p. 236. Academic Press, New York.
Law, B.A., Castanon, M.J., and Sharpe, M.E. (1976). J. Dairy Res. 43, 301.
Law, B.A., and Sharpe, M.E. (1977). Dairy Industries International, p. 10.
Law, B.A., and Sharpe, M.E. (1978). J. Dairy Res. 45, 267.
Lawrence, R.C., Heap, H.A., Limsowtin, G., and Jarvis, A.W. (1978). J. Dairy Sci. 61, 1181.
Manning, D.J. (1978a). J. Dairy Res. 45, 479.
Manning, D.J. (1978b). Dairy Industries International, p. 37.
McGugan, W.A. (1975). J. Agric. & Food Chem. 23, 1047.
McGugan, W.A., Emmons, D.B., and Larmond, E. (1979). J. Dairy Sci. 62, 398.
Mikolajcik, E.M. (1979). Cultured Dairy Prod. J., Nov. 1979, p. 6.
Moskowitz, G.J. (1979). In "Microbial Technology," Vol. II, (H.J. Peppler and D. Perlman, eds.), p. 201. Academic Press, New York.
Richardson, G.H., and Nelson, J.H. (1967). J. Dairy Sci. 50, 1061.
Shipe, W.F., et al. (1979). J. Dairy Sci. 61, 855.
Sood, V.K., and Kosikowski, F.V. (1979). J. Dairy Sci. 62, 1865.

Thompson, M.P., Somkuti, G.A., Flanagan, J.F., Bencivengo, M., Brower, D.P., and Steinberg, D.H. (1979). Am. Dairy Sci. Assoc. Abstracts 62, Supp. 1, 68.
Umemoto, Y., Sato, Y., and Kito, J. (1978). Agric. Biol. Chem. 42, 227.
Watts, J.C., and Nelson, J.H. (1963). U.S. Patent 3,072,488.

THE ANALYSIS AND CONTROL OF SOME LESS-DESIRABLE FLAVOR-CONTRIBUTING COMPONENTS OF CORN SYRUP

Donald L. Kiser
Robert L. Hagy
Robert L. Olson

Grain Processing Corporation
Muscatine, Iowa

I. INTRODUCTION

Corn syrup is an important item of commerce. In 1980 over 4 million tons (dry basis) of corn syrup will be sold to food processors and beverage manufacturers (1). The uses are varied. Some areas of use are given in Table I.

Most of these uses require corn syrup having specific qualities. For example, in soft drinks corn syrup should give sweetness and mouthfeel. In ice creams and ice milk products less sweetness is needed but the corn syrup should lend some sweetness as well as smoothness to the dessert. Other uses demand almost no sweetness but benefit from the viscosity increase given by high molecular weight carbohydrates in corn syrup. Corn syrup is useful in some areas because of its partial fermentability. Brewers are aware that the residual carbohydrate remaining after fermentation can influence the body and taste of a beer (2).

TABLE I. Some Food Uses of Corn Syrup

Baking	Frozen foods
Brewing	Ice cream
Candy	Prepared mixes
Canning	Soft drinks
Desserts	Syrups and toppings

ISBN 0-12-169065-2

Corn syrup is prepared by the hydrolysis of corn starch. During hydrolysis, lower molecular weight fragments are produced from long chains of anhydroglucose units. The extent and method of hydrolysis determines sweetness and other properties. Hydrolysates, including high fructose corn syrup, can be produced that have a range of sweetness from very low to that of an equivalent weight (dry basis) of sucrose. Techniques are required that can monitor the extent of hydrolysis and measure specific carbohydrate content of the final product.

The basic steps in the production of corn syrup are shown in Figure 1. In preparation for syrup production, shelled corn is soaked for about 40 hours at 140°F (60°C) in water to which sulfur dioxide has been added. Corn so treated becomes swollen and soft. These soft kernels are next broken open

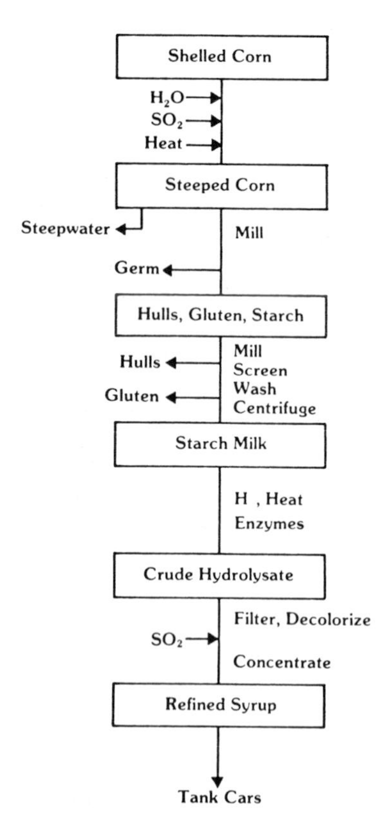

FIGURE 1. Flow chart showing principal stages as corn syrup is produced from shelled corn.

with a special mill and the germ removed by screening and centrifugation, leaving a suspension of starch. Details on a wet milling process have been reviewed by Commerford (3, 4).

Gluten is difficult to remove completely from the starch. Typically, starch milk will contain 0.2-0.3% protein on a dry basis. Residual protein present can contribute to color problems that arise during the concentration step in syrup production (5).

Starch milk can be treated in various ways to produce some unique hydrolysates. Generally, hydrolysis begins with acid and heat treatment. Later, depending on the product desired, specific enzyme treatments can be made. Clean-up and concentration steps are important and if not performed properly can contribute to undesirable flavor. Karkalas showed the importance of carbon and cation exchange treatments in these refining steps (5).

Sulfur dioxide is again added to the process stream at the concentration stage to minimize color formation during heating and evaporation. Most of the added sulfur dioxide is removed during the concentration stage but low amounts, less than 40 ppm, remain.

In the 1950's paper chromatography and thin layer chromatography were the techniques used to characterize the carbohydrate composition of corn syrups. In the 1960's, gas chromatographic techniques were developed that depended on separation of volatile derivatives formed when low molecular weight carbohydrates were trimethylsilylated (6, 7, 8). These silylation techniques were widely used in the wet milling industry but had limited application because of the extreme temperatures required for elution. In all these silylation techniques, derivatives of the higher saccharides were left in the injection port or on the gas chromatography column.

During the past decade, liquid chromatography has developed into a valuable analytical tool for the analysis of carbohydrate solutions. In 1970, Walker and Saunders used ion exchange resin to separate aqueous solutions of low molecular weight carbohydrates (9). Brobst, Scobell and Steele in 1973 developed a technique for separating carbohydrate components in syrups and worts on calcium-form cation exchange resin (10). This procedure is now widely used in the wet milling and brewing industries. The basic procedure has been refined over the years and was recently the subject of a collaborative study (11). The study showed that interlaboratory agreement is good. On samples containing 8 and 50% glucose, the coefficient of variation on the glucose component was 1.4 and 0.5%, respectively.

We will give some examples of how liquid chromatography can be used to advantage by corn syrup manufacturers and their customers to ensure that corn syrups with the desired flavor characteristics are processed.

As mentioned earlier, sulfur dioxide is added to the process stream in two places--at the steep tanks and just before the evaporator. Sulfur dioxide in the steep provides the acidity to help soften the kernel and aids in loosening starch from the protein matrix. Most of this sulfur dioxide of the steeping process is removed in later washing and centrifugation steps in preparation of the starch milk suspension. Sulfur dioxide added before the evaporator helps retard color formation during concentration. Residual sulfur dioxide in high levels can cause a metallic taste in foods and may cause physiological responses (12). Beyond the evaporator, sulfur dioxide in conventional corn syrup is less than 40 ppm. The sulfur dioxide level in syrup intended for canning is less than 10 ppm.

The Corn Refiners Association uses the Monier-Williams distillation-titration procedure for the analysis of sulfur dioxide (13). In this procedure sulfur dioxide is removed from an acidified solution by sweeping the solution with carbon dioxide or nitrogen into a dilute hydrogen peroxide solution. The sulfuric acid formed is titrated with base. The sweeping portion of the assay lasts for an hour, so the procedure, while simple, is not rapid.

An alternate technique for the assay of sulfur dioxide is the method of Dorner (14). In this procedure, sulfur dioxide is released from syrup with sodium hydroxide. After acidification of the sample, sulfur dioxide is measured by adding an excess of iodine and determining the excess with thiosulfate. The titration portion of the method is rapid, but sample preparation is not. Prior to titration, 25 grams of syrup must be dissolved in 25 ml. of water at room temperature. After dissolution, an additional 15-minute period is required for reaction with sodium hydroxide.

Steigman developed a colorimetric method for sulfur dioxide that uses formaldehyde and basic fuchsin (15). West and Gaeke refined the method (16). They used pararosaniline instead of fuchsin and complexed sulfur dioxide with tetrachloromercurate to prevent its loss. They found that the temperature of reactants had little effect on the color developed. They reported that sulfur trioxide, chlorine, ammonia and halogen acids do not interfere. Interference from oxides of nitrogen and from sulfide ions was reported.

Nury, Taylor and Brekke used techniques similar to those of West and Gaeke to assay sulfur dioxide in dried fruits (17). Collaborative studies on dried fruits have shown that the pararosaniline sulfur dioxide procedure is statistically comparable to the Monier-Williams distillation-volumetric method (18, 19).

Because of the success of several authors with pararosaniline for sulfur dioxide assays it seemed that this reagent might provide a convenient assay for sulfur dioxide in corn syrup. A suggested procedure using pararosaniline and some results using that procedure will be given.

Corn syrup can replace at least a portion of sucrose in ice cream with advantage. Ice cream containing corn syrup has a higher melting point and more smoothness than ice cream having sucrose sweetener alone (20, 21). However, undesirable flavors may be detected in ice creams if the syrup level is high (22). These flavors have been attributed to both high- and low-molecular weight components in hydrolysates (23, 24). Work involving the characterization of some of these undesirable flavor contributors will be discussed.

Conventional corn syrup is sold at 70-82% solids. Microorganisms do not grow in carbohydrate solutions that are higher than about 70% solids. However, there are occasions when the growth of yeast and mold needs to be considered. Syrup is shipped in well-insulated tank cars. This syrup is loaded hot, at approximately 100-120°F (38-49°C). Under the right conditions water can condense on the top portion of the tank car and drip back on to the syrup surface. Under certain conditions then, small pools can form on the surface of the warm syrup. When the outside temperature is about 32°F, the rate of temperature decrease of the syrup is about 1°F per day. Thus, if the proper carbohydrate solids develops, the temperature is near ideal for yeast and mold growth. This yeast and mold growth, if occurring, is only on the surface and does not affect taste, but could affect appearance, and in extreme cases could contribute an odor. We will report on yeast and mold growth in carbohydrate solutions under various conditions.

Producers of corn syrup use a variety of analytical tools to assist in tailoring hydrolysates to meet customers' needs and to monitor components that could contribute to an off-flavor in a customer's product.

II. EXPERIMENTAL

A. Carbohydrate Profile

1. Liquid Chromatograph.

a. Column. Stainless steel, 3/8 x 16 inches, jacketed so that column can be heated with water circulated from a bath controlled at 85° ±0.1°C.

b. Column packing. Aminex 50W-X4 ion exchange resin, calcium form, 30-35 microns; Bio-Rad Laboratories. Resin was charged and packed with technique described by Brobst, Scobell, and Steele (10).

c. Pump. Waters Associates Model 6000A, used to supply the eluant, water, at 0.4 ml/minute.

d. Detector. Differential Refractive Index Detector, Model R401, Waters Associates.

e. Integrator. AutoLab System I, Spectra Physics.

f. Sampler. WISP 710A, Waters Associates.

g. Recorder. Input 10 mv, Recordall, Fisher Scientific Company.

2. Sample Preparation. Dissolve syrup in water to give solution having about 20% solids. Inject 10 microliters.

B. Sulfur Dioxide

1. Reagent Preparation.

a. Pararosaniline stock, (PRA). To 0.1 gram pararosaniline hydrochloride in a 100 ml. volumetric flask, add about 50 ml. water. Pipet to this, 20 ml. concentrated HCl. Swirl the solution until all the pararosaniline is dissolved. Dilute to 100 ml. with water and let stand 12 hours or longer before proceeding.

After a few days, dark specks may appear in the PRA stock. In this case, the solution should be filtered through Whatman No. 1 filter paper before further use.

b. Pararosaniline-formaldehyde solution, (PRA-CH_2O). Pipet 5 ml. of PRA stock to about 150 ml. water contained in a 250 ml. volumetric flask. Add by pipet 10 ml. concentrated HCl and 1 ml. formaldehyde. Dilute to the mark with water and mix well. Transfer to a semi-automatic pipet that is set for 5.00 ml.

c. Sodium tetrachloromercurate, (TCM). Dissolve and dilute 54.4 grams of mercuric chloride and 23.4 grams of sodium chloride to 2 liters with water.

d. Sulfur dioxide standard solution, (SO_2). Use certified grade sodium bisulfite. Note SO_2 assay on the Certificate of Analysis on the label of the sodium bisulfite. Divide the percent SO_2 into 100. This gives the weight, in grams, of sodium bisulfite to dilute with water to one liter to make a solution containing 1.00 mg. SO_2 per ml. (For example, if sodium bisulfite assays 66.3% SO_2, 1.508 grams would be dissolved and diluted to one liter.) Pipet 1 ml. of this first solution to a 100 ml. volumetric flask. Dilute to the mark with TCM. Each milliter of this second solution contains 10 micrograms of SO_2.

Standard can be used on the day it is prepared and on the next two days. Discard any standard that is more than three days old.

2. Procedure.

a. Standards. Prepare 25 ml. volumetric flasks containing 0 (blank), 10, 30 and 50 micrograms of SO_2 by adding 0 ml. of SO_2 standard to the first flask, 1 ml. to the second, 3 ml. to the third and 5 ml. to the fourth. Then add 10 ml. TCM to the first flask, 9 to the second, 7 to the third and 5 to the fourth, so that each flask contains 10 ml. of solution. Swirl, then add 5 ml. of PRA-CH_2O solution to each flask from a semi-automatic pipet. Dilute each solution to 25 ml. with water. Mix well then let stand for 30 minutes. Read the absorbancies at 560 mu using the solution that contains no SO_2 as a blank. Plot absorbance versus micrograms of SO_2.

b. Samples. If the expected SO_2 range is 0 to 80 ppm, use a sample weight of 0.4 to 0.5 gram. If the expected SO_2 content is greater than 80 ppm, use a sample weight of 0.2 to 0.3 gram.

Weigh syrup sample to the nearest mg. into a pre-weighed 50 ml. beaker. Add 10 ml. TCM by pipet or semi-automatic pipet. Place a stirring bar in the solution and set the beaker on a magnetic stirrer. Adjust the stirring rate to a low speed. After sample is dissolved, raise the stirring bar from the solution with another magnetized bar, rinse with water and then remove it completely. Pour contents of the beaker into a 25 ml. volumetric flask. Rinse the beaker three times with small amounts of water and add the washings to the flask. Add 5 ml. of PRA-CH_2O solution and proceed as with standards.

The volumes of reagents to add to various flasks are tabulated in Table II.

Table II. Reagent Proportions for Color Development in SO_2 Standards and Corn Syrup Solutions

Reagent	Blank	Standards 10 ug	30 ug	50 ug	Sample[a]
Standard SO_2	0 ml.	1 ml.	3 ml.	5 ml.	0 ml.
TCM	10 ml.	9 ml.	7 ml.	5 ml.	10 ml.
PRA-CH_2O	5 ml.	5 ml.	5 ml.	5 ml.	5 ml.
Water	---dilute all to 25 ml.---				

[a]Sample size is approximately 0.5 g, weighed to nearest mg.

c. Calculations. From a plot of absorbance versus micrograms of SO_2, determine the micrograms of SO_2 present in the sample. Correct to dry basis, if needed.

$$\text{ppm } SO_2 = \frac{\text{micrograms } SO_2 \text{ found}}{\text{grams of sample}}$$

C. Hydroxymethylfurfural and Other Aldehydes

1. Hydroxymethylfurfural, (HMF). The concentration of HMF in selected corn sryups is measured colorimetrically using 2-thiobarbituric acid. The procedures are detailed elsewhere (32, 33).

2. Other Aldehydes. Prepare solution by mixing 1000 g (dry basis) corn syrup with 1500 g water. As detailed elsewhere, vacuum distill volatiles from solution in a flask immersed in a heated water bath (34). Condense higher boiling vapors with a water-jacketed condensor. Collect this condensate in a flask containing 2,4-dinitrophenylhydrazine solution and immersed in an ice bath. Trap components of higher volatility by pulling vapors through two additional vessels, in series, each containing 2,4-dinitrophenylhydrazine solution at 0°C. Add known amount hexanal to syrup solutions to serve as an internal standard. The collected 2,4-dinitrophenylhydrazones are assayed by thin layer chromatography and gas chromatography (34).

D. Mold Growth

Prepare syrup solutions of solids ranging from 7 to 80%. Determine solids using refractive index values. Set tubes or dishes of the syrup solutions, inoculated or air-exposed, into environment of test. Observe at intervals. At termination of test, measure solution turbidity (use % T for low readings) at 600 nm. Correct reading to 1 cm path. Read color on filtered solutions at 470 nm in 1 cm cells. Measure pH and determine carbohydrate profile on filtrate.

III. RESULTS AND DISCUSSION

A. Carbohydrate Profiles

Using liquid chromatography a carbohydrate profile of corn syrup components can be developed in less than 30 minutes. Such a profile graphically shows the molecular weight distribution of saccharides in a hydrolysate. Figure 2 shows a profile of an acid-hydrolyzed 42 DE corn syrup. High molecular weight components elute first, followed by lower molecular weight components. DP in Figure 2 and later figures refers to the degree of polymerization of anhydroglucose units. DP2 is principally maltose but is known to contain low levels of other disaccharides which are reversion products produced during hydrolysis (25, 26, 27).

Customer needs for corn syrup vary. What is proper flavor for one use may be undesirable for another. Each syrup produced needs to be characterized so that every customer receives hydrolysate that is optimum for the intended use.

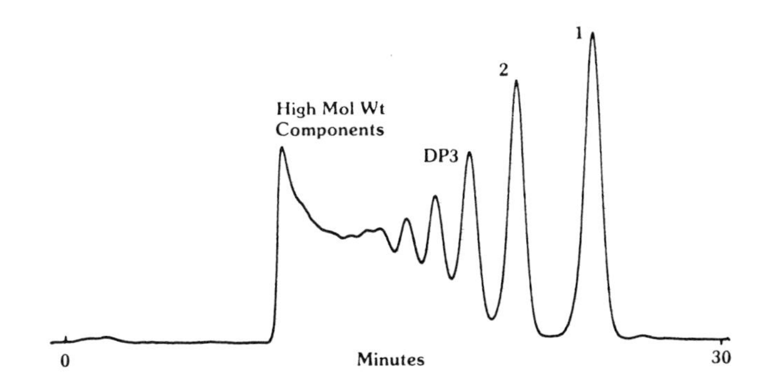

FIGURE 2. Carbohydrate profile of a 42 DE corn syrup; acid hydrolyzed.

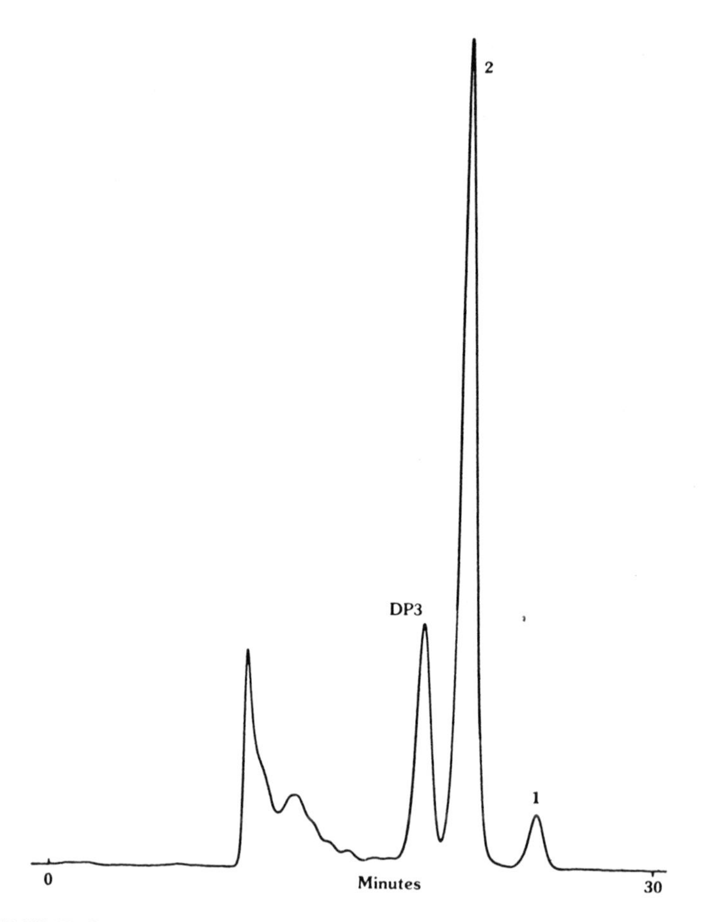

FIGURE 3. Carbohydrate profile of a 43 DE corn syrup; enzyme hydrolyzed with beta-amylase.

1. Hydrolysate Characterization. Hall defines flavor as "the sum of those characteristics of any material taken in the mouth, perceived principally by the senses of taste and smell and also by the general pain and tactile receptors in the mouth, as received and interpreted by the brain" (28). If we use this broad definition, then the extent of the effect of several properties that affect flavor of corn syrup can be estimated from carbohydrate profiles.

a. Sweetness. Sweetness can be estimated from profiles. The relative sweetness values of fructose, glucose, maltose and other low molecular weight carbohydrates have been estimated by many researchers (29). As we look at profiles of hydrolysates developed on calcium-form ion exchange resins, the relative sweetness of components increases with increasing elution time. High molecular weight carbohydrates contribute little to sweetness. Maltose has about six-tenths the relative sweetness of glucose. Fructose has a relative sweetness of approximately 1.7 times an equal weight of glucose. In these elution profiles fructose elutes well beyond glucose. Thus, profiles that show a predominance of low molecular weight components would indicate relatively sweet hydrolysates. As examples, the four syrups profiled in Figures 2 through 5 were set up for sweetness evaluations at 5% solids in water. Each of these samples differed mainly in the amount of glucose present. Seven out of seven tasters named the syrup having the highest proportion of glucose as the sweetest. That syrup is profiled in Figure 5. All seven tasters rated the other 62 DE syrup, Figure 4, as the next sweetest. Six of the seven tasters judged the acid hydrolyzed 42 DE as the least sweet.

b. Non-fermentable content. The non-fermentable portion of corn syrup, when used in brewing, contributes to the body of beer. A corn syrup with a considerable amount of non-fermentable carbohydrate remaining after the brewing will give a beer that is said to be "full-bodied" (2). Some brewers prefer to have a light-bodied beer. These brewers would prefer to have a corn syrup with a relatively low amount of non-fermentable carbohydrate. The yeast fermentable portion of a syrup is approximately the sum of DP1, DP2 and DP3. The actual fermentability achieved by brewers is somewhat less (30).

c. Molecular weight distribution. Molecular weight influences several properties of corn syrup. Knowing the molecular weight distribution from a carbohydrate profile allows us to make evaluations regarding viscosity, osmotic pressure, freezing point depression and the ability of a corn syrup to function as a bodying agent in a prepared food. Materials that are lightly hydrolyzed will have the highest viscosity (20). The hydrolysate of Figure 2 would have higher viscosity than that of Figure 3. Viscosity is important in candy manufacturing.

Those syrups which have a high degree of hydrolysis will show high osmotic pressure. Syrups with high osmotic pressure can serve as preservatives. Freezing point depression is another property related to the molecular weight distribution. This property is of concern to ice cream manufacturers. Ice cream manufacturers and other food processors often depend on

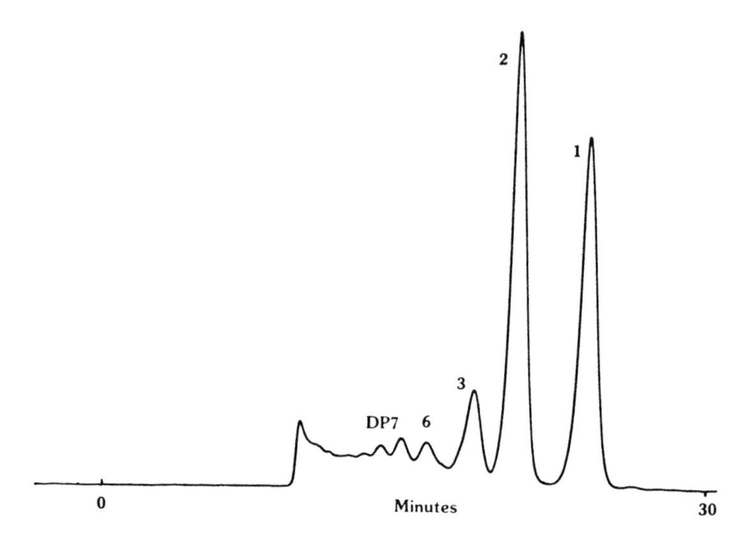

FIGURE 4. Carbohydrate profile of a 62 DE corn syrup; acid-enzyme hydrolyzed.

corn syrup to provide smoothness, mouthfeel and the proper body to their product. Foods that need chewiness and body usually contain syrups with a predominance of high molecular weight components. An example of that type of hydrolysate is given in Figure 2.

d. Browning potential. The flavor and texture of many baked goods depend on the extent of reactions between protein and reducing sugars. For enhancement of this browning reaction, hydrolysates having a large portion of low molecular weight components are chosen. Hydrolysate of Figure 5 would have greater browning potential than hydrolysate of Figure 2.

2. Process Monitoring. Since a carbohydrate profile can be made in about 20-30 minutes, these analyses can be used for in-process monitoring and development work. For example, Table III shows data collected when an enzyme was added to a starch suspension that was previously acid-hydrolyzed to 39 DE. DE titration values indicate the progress of the hydrolysis. However, the carbohydrate profile obtained by liquid chromatography better characterized the hydrolysate at intervals by giving information on the actual proportion of components present, such as glucose content, ratio of DP1:DP2, or total fermentables, DP1 + DP2 + DP3. Carbohydrate profile assays,

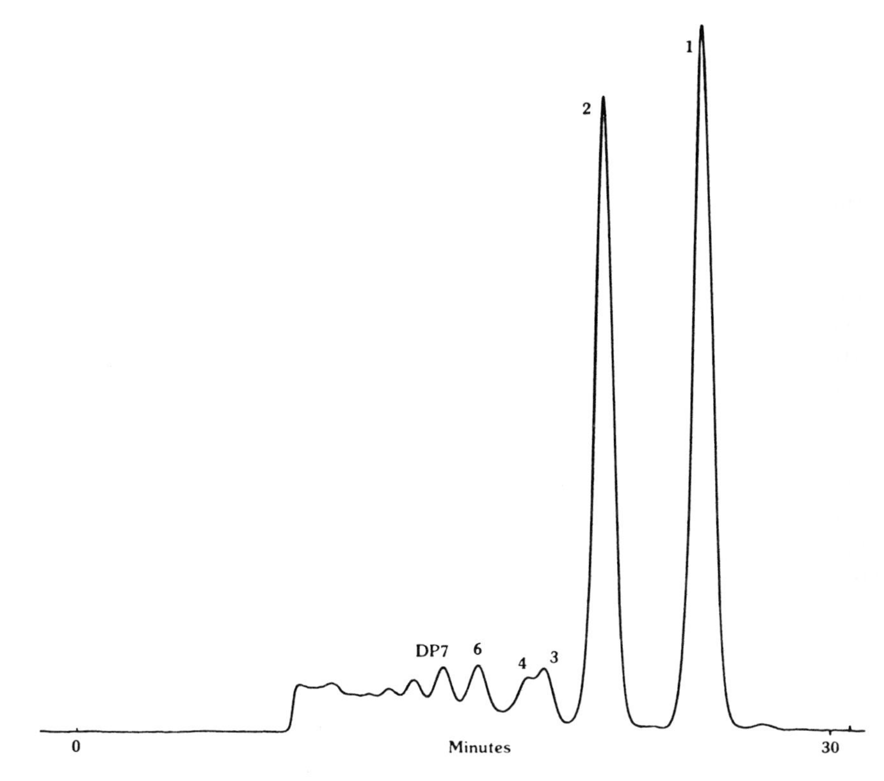

FIGURE 5. Carbohydrate profile of a 62 DE corn syrup.

TABLE III. Carbohydrate Profile Change During Hydrolysis from 39 DE to 65 DE

Hours of conversion	Titration DE	Carbohydrate profile (% d.b.)			
		DP1	DP2	DP3	DP3+
0	38.7	17.4	13.2	10.7	58.7
4	39.9	20.1	13.0	11.7	55.2
8	46.4	26.4	15.3	13.0	45.3
12	51.9	33.9	17.4	11.9	36.8
20	58.5	40.2	21.2	9.2	29.4
26	60.7	45.7	23.1	5.2	26.0
28	64.9	49.0	22.5	5.1	23.4

then, can aid in having closer control of hydrolyses and thereby assure that syrups prepared will have properties that will enhance the flavor of customers product.

B. Sulfur Dioxide

Sulfur dioxide together with formaldehyde and pararosaniline forms a purple-colored solution that gives maximum absorption at 560 nm. This complex develops color rather slowly. As shown in Figure 6, corn syrup solutions containing sulfur dioxide require approximately 20 minutes to reach maximum absorption. There is slight fading of color after 1 hour. Solutions being assayed should be read during the interval 30-60 minutes after mixing. Heating does not significantly affect the rate of color development.

In the case of corn syrup, sulfur dioxide is present in both bound and free states. In the official Corn Refiners Association method, bound sulfur dioxide is released with acid treatment before sulfur dioxide distillation is started. In the precedure outlined here, no sample pretreatment is required. Recoveries of known added quantities of sulfur dioxide are good. To check on recovery, aqueous aliquots of freshly prepared sodium bisulfite solutions were added to aliquots of

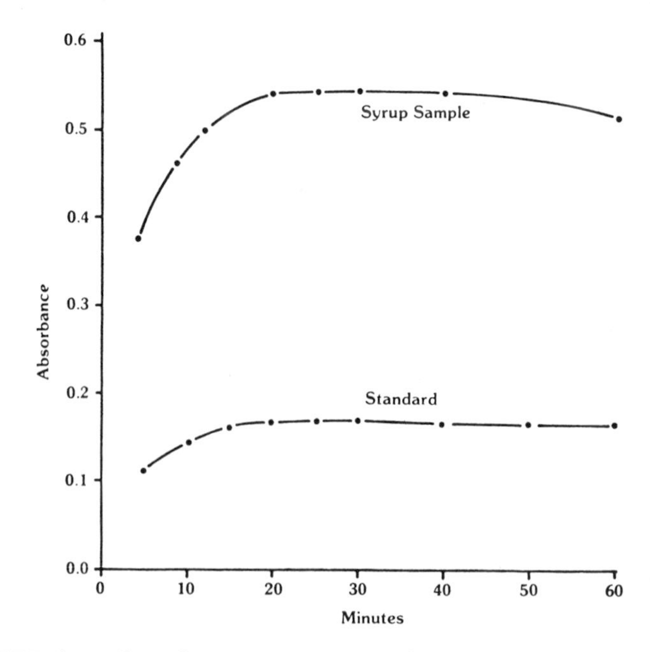

FIGURE 6. Absorbance of SO_2-CH_2O-PRA complex measured after various intervals from time of mixing, at 560 nm.

corn syrup known to assay 19.3 micrograms of SO_2. The sodium bisulfite solution added contained no TCM. This combined solution of corn syrup and sodium bisulfite was allowed to stand, then later assayed for SO_2 level using the PRA technique. Results, comparing the quantities of SO_2 added and recovered, are given in Table IV.

Sulfur dioxide is not stable in aqueous solutions. West and Gaeke pointed out the increased stability of sulfur dioxide in the presence of sodium tetrachloromercurate (16). From the plot given by Nury, et al., it is estimated that the loss of SO_2 in 0.1 M mercurate solution is about 5-12% in 9 days (17). To obtain better estimation of the rate of loss of SO_2 in a standard solution containing 10 micrograms of SO_2 per milliter in 0.1 M mercurate, an aliquot that contained 30 micrograms of SO_2 was assayed at intervals using freshly prepared standards. Results are given in Table V. Loss during the first 10 days averaged about 0.67% of the starting concentration per day. To stay within the experimental error, standards should not be used after they are 3 days old.

To check on precision, one sample was assayed 10 times by 3 analysts at 2 different times (60 assays). The largest source of variance was between analysts. The standard deviation was 0.32 ppm for a 13.5 ppm sample, giving a coefficient of variation of 2.4%.

TABLE IV. Recovery of Known Quantities of SO_2 Added to Corn Syrup

SO_2 Added, ug	Recovered, ug
2.0	2.3
2.0	2.6
4.0	3.8
4.0	5.1
10.0	9.5
20.0	20.7
30.0	30.7

TABLE V. Loss of SO_2 with Time from Stock Solution[a]

Days of storage	Assay of 3 ml aliquot originally containing 30 ug SO_2
0	30.0 ug
2	30.0 ug
4	28.8 ug
6	28.4 ug
8	28.7 ug
10	28.0 ug
12	24.5 ug
20	19.3 ug

[a]Aqueous solution containing 10 ug/ml SO_2 in 0.1M sodium tetrachloromercurate.

C. Hydroxymethylfurfural and Other Aldehydes

1. Hydroxymethylfurfural (HMF). HMF has been reported to be present in corn sweeteners (35). Olson and Winder (36) analyzed 7 sweeteners for HMF content. A colorimetric procedure devised by Keeney and Bassette (32) for milk studies and utilized by Brownley and Lachman (33) for sugar solutions was used. The same sweeteners were also evaluated for flavor characteristics. A 12% solids level was used for taste testing.

The HMF levels found are included in Table VI. Concentrations of HMF varied from about 1 ppm to slightly over 9 ppm. Both syrup solids 4217s and syrup 6433 had a high content of HMF as well as high flavor levels. However, the flavor characteristics of the remaining sweeteners did not necessarily correlate with HMF content. In prepared solutions for taste testing, HMF at 1.2 ppm contributed no flavor in water but did contribute some flavor to a 12% glucose solution. The evaluations indicated that HMF was probably one of several components contributing to typical syrup associated flavors.

2. Other Aldehydes (Ethanal, 2-methylpropanal, 3-methylbutanal). Olson and Winder (36) isolated ethanal as its 2,4-dinitrophenylhydrazone from seven of eight corn sweeteners using the vacuum distillation technique described earlier. Ethanal, 2-methylpropanal, and 3-methylbutanal derivatives were obtained from two of the eight sweeteners. The calculated concentrations of the three aldehydes for syrup solids 4217s are presented in Table VII. The values are expressed on a sweetener solids basis. Concentrations based on a 12%

TABLE VI. Hydroxymethylfurfural Content and Flavor Evaluations of Seven Corn Sweeteners

Corn Sweetener					
Code[a]	Conversion[b]	Refining[c]	Flavor[d]	HMF[e] ppm	SO_2[e] ppm
4117	A	C	Astringent 1, fruity, SO_2	1.93	33
4145	A	C	Astringent 2, acid 1, SO_2	0.97	31
4217	A	C	Astringent 1, acid 1, musty 1, salty	2.32	26
4217s	A	C	Corn-caramel 2½, SO_2	9.09	N.A.[f]
5154	A, E	C	Astringent 1, acid 1, musty 1, SO_2	1.02	20
5325	A	C, D	Astringent 2, acid 1, feedy 1	5.92	3
6433	A, E	C, D	Astringent 1, caramel 1, SO_2	7.65	3

[a]Coding: First digit, when this number is multiplied by 10 it indicates the DE plus or minus 5 DE units; second digit, when this number is multiplied by 10 it indicates the percentage of dextrose (dry weight basis) plus or minus 5%; third digit, when this number is multiplied by 10 it indicates the percentage of maltose (dry weight basis) plus or minus 5%; fourth digit, when this number is multiplied by 10 it indicates the percentage of higher saccharides (dry weight basis) plus or minus 5%; s, indicates a dried syrup solid.

[b]A - acid; E - enzyme.

[c]C - carbon; D - deionized.

[d]Flavor of 12% solutions (solids). Flavor intensity - 1, slight, to 5, pronounced.

[e]Expressed on dry solids basis.

[f]N.A. - not assayed.

TABLE VII. Concentrations of Three Aldehydes Found in Corn Sweetener 4217s, an Acid-Converted, Carbon-Filtered, 42 DE Corn Syrup Solids

Aldehyde	Concentration (ppm)	
	Solids basis[a]	12% Solids basis[b]
Ethanal	0.81	0.10
2-methylpropanal	0.25	0.03
3-methylbutanal	1.71	0.20

[a]Concentration on a dry weight basis for syrup solids 4217s (see Table VI regarding the coding of the syrup solids sample).
[b]Concentration calculated for a 12% solids solution, the solids level at which sweeteners were evaluated for flavor.

syrup solids level were calculated for the purpose of preparing solutions for flavor evaluations. The 12% solids level represented an acceptable level for the tasting of syrup solutions.

Additional work to determine the origin of the three compounds was not done. The three aldehydes may have been produced through some type of browning reaction. The branched chain structure of two of these aldehydes suggested that they may have originated from amino acids. The Strecker degradation of alanine, valine and leucine would give these aldehydes (37). Starch utilized in syrup production contains about 0.3% protein. Thus, one would expect that amino acids would be present for the Strecker degradation.

Water and glucose solutions were prepared to contain the three carbonyls (0.0 to 0.2 ppm) as well as HMF (0.0 to 1.2 ppm) and SO_2 (0.0 to 3.0 ppm) as sodium bisulfite. All compounds contributed to increased flavor in glucose solutions. Intense flavors were noted in mixtures containing ethanal, 2-methylpropanal and 3-methylbutanal. The addition of HMF and sodium bisulfite to the mixtures tended to mask some of the aldehyde flavors. Flavor evaluations of several mixtures were reported by Olson (34). The flavors of some mixtures containing all of these components were similar to those flavors noted when evaluating corn sweetener solutions.

D. Yeast and Mold

Some carbohydrate fermentation by-products, such as diacetyl, ethyl acetate, lactic acid and acetic acid, could contribute to odor or taste in syrup solutions. To check the effect of carbohydrate solids on yeast and mold growth, 62 DE syrup was diluted to various solids and either air-exposed or inoculated with mycelia and spores previously grown on a syrup solution. Results are summarized in Table VIII.

Growth was seen in about four days in most solutions having less than 50% solids. Solutions that were inoculated with a mold that had numerous red subsurface mycelia developed a pink to red rust color. Less intense color was developed in a solution of 7% solids than in solutions of 12-30% solids. Some solutions developed turbidity that apparently was the result of rapid yeast growth. Turbidity measurements showed that the most growth occurred in 22-33% solids solutions. Yeast cells settled eventually giving a sediment at the bottom of the solution. Neither pH nor carbohydrate profile changed significantly in these runs.

Other portions of syrup (82% solids, 62 DE) were placed into wide-mouth jars. Layers of water, 0 to 8 mm depth, were placed on top of the syrup. These were stored loosely-capped, at 30°C, in a large glass vessel having high humidity. Within a few days, no water-syrup interface could be seen. Some observations made at end of nine days are recorded in Table IX. Solids in the top portion of each jar were above 55% by the ninth day. It appears that water on top of syrup mixes into the syrup with little agitation. During the early stage of this mixing, carbohydrate solids are low enough that some osmotolerant organisms can multiply. As the mixing continues, solids reach high enough levels that organisms can survive but not increase.

Little abnormal odor could be detected on moldy solutions at room temperature. A moldy solution heated to 80°C gave a deep (perhaps diacetyl) odor. After longer heating of the solution, at the same temperature, a sharp odor was noted, along with the expected caramel fragrance.

TABLE VIII. Yeast and Mold Growth in Solutions of 62 DE Syrup of Various Solids Levels (30°C 14-23 days)

```
                                         Syrup solids, %
                                 ------------------------------------------
                                 0   10   20   30   40   50   60   70   80

Air Exposed
    Black spores developed         +  + +  +  +       +    o    o    o
    Turbidity developed; yeast        +    +    +   o   o   o    o
    Turbidity highest                           +
Inoculated
    Red solution developed         +  + +  +   +      o    o    o  o
    Red solution most intense           +
    Black spores developed         +  + +  +   +      +    +
    Black spores most numerous                        +
```

+ = growth seen
o = no growth evident

TABLE IX. Mold Growth on Water-layered Corn Syrup[a]

Layer depth, mm			
Syrup	Surface water	Growth 9 Days	Solids in upper portion, 9 days
19.8	0	none	78
15.4	1.4	10% covered white	76
19.8	5.2	100% covered blue green	67
19.2	8.0	100% covered black red subsurface	59

[a]62 DE corn syrup, 82% solids, stored at 30°C.

IV. SUMMARY

A variety of corn syrup products are manufactured. The type of corn syrup that imparts good flavor to one product may contribute to an undesirable flavor in another application. Liquid chromatography is a convenient analytical technique for assuring that, during production, corn syrup with the desired characteristics is tailored for its intended use. Liquid chromatographic molecular weight profiles of hydrolysates can be interpreted to estimate the contribution of sweetness, body and mouthfeel imparted by corn syrup to prepared foods.

Residual sulfur dioxide in corn syrup can be conveniently assayed colorimetrically using formaldehyde and pararosaniline. The coefficient of variation using the procedure described is 2.4%.

Hydroxymethylfurfural, ethanal, 2-methylpropanal, and 3-methylbutanal have been isolated from corn syrup solids using vacuum distillation and traps containing 2,4-dinitrophenylhydrozine (34). These aldehydes may contribute flavor to foods which utilize the corn sweeteners. HMF and sulfur dioxide, in levels up to 3 ppm each, may mask some of the flavor contributed by ethanal, 2-methylpropanal and 3-methylbutanal.

Yeast and mold grow rapidly in dilute solutions of corn syrup, but show no growth in solutions having high solids. Small pools of dilute solutions could be formed on shipped corn syrup if loading conditions permit water vapor to condense at the top of the vessel being loaded and fall on top of loaded syrup. Such pools, with the normal agitation provided by shipment, take on increasingly higher solids level and retard, then stop, any microbiological growth. Moldy dilute corn syrup prepared in our laboratory gave no significant odor at room temperature, but when heated to 80°C, emitted odors not obtained from freshly diluted and heated corn syrup. Corn syrup manufacturers can avoid condensation pools on top of syrup being loaded for shipment by circulating clean, dry air above the syrup (26).

REFERENCES

1. U.S.D.A., Agricultural Marketing Service, Sugar and Sweetener Report 4 (No. 2): 12, 19 (1979).
2. Herwig, W., Wagener, R., Cieslak, M., Chicoye, E., and Helbert, J., 1977 Pittsburgh Conference, Paper No. 113.
3. Commerford, J., in "Products of the Wet Milling Industry", Symposium Proceedings, II-1. Corn Refiners Association, Washington, D.C. (1970).
4. Commerford, J., in "Symposium: Sweeteners" (G. Inglett, ed.), p. 78. Avi Publishing Co., Westport, Conn. (1974).
5. Karkalas, J., Die Stärke 19:338 (1967).
6. Sweeley, C., Bentler, R., Makita, M., and Wells, W., J. Am. Chem. Soc. 85:2497 (1963).
7. Brobst, K., and Lott, C., Cereal Chem. 43:35 (1966).
8. Alexander, R., and Garbutt, J., Anal. Chem. 37:303 (1965).
9. Walker, G., and Saunders, R., Cereal Sci. Today 15:140 (1970).
10. Brobst, K., Scobell, H., and Steele, E., Am. Soc. Brewing Chemists, Proceedings 1973:43.
11. Engel, C.E., and Olinger, P., J. Assoc. Offic. Anal. Chem. 62:527 (1979).
12. Allen, R., and Brook, M., in "Proceedings, Third International Congress on Food and Technology", p. 799. Institute of Food Technologists, Chicago, (1970).
13. Corn Refiners Association, Inc., Standard Analytical Methods of the Member Companies, Method E-66, Washington, D.C. (1978).
14. Dorner, H., Die Stärke 4:84 (1952).
15. Steigmann, A., Anal. Chem. 22:492 (1950).
16. West, P. and Gaeke, G., Anal. Chem. 28:1816 (1956).
17. Nury, F., Taylor, D. and Brekke, J., J. Agr. and Food Chem. 7:351 (1959).
18. Brekke, J., J. Assoc. Offic. Agr. Chemists 46:619 (1963).

19. Nury, F. and Bolin, H., J. Assoc. Offic. Agr. Chemists 48:796 (1965).
20. Newton, J., in "Products of the Wet Milling Industry", Symposium Proceedings, IV-1. Corn Refiners Association, Washington, D.C. (1970).
21. Keeney, P., and Josephson, D., Ice Cream Trade J. 57:28 (1961).
22. Eopechino, A., and Leeder, J., J. Food Sci. 35:398 (1970).
23. Eopechino, A., and Leeder, J., Food Technol. 23:111 (1969).
24. Eskamani, A., and Leeder, J., J. Food Sci. 37:328 (1972).
25. Fetzer, E., Crosby, E., Engel, C., and Krist, L., Ind. Eng. Chem. 45:1075 (1953).
26. Ough, L., Anal. Chem. 34:660 (1962).
27. Brobst, K., in "Products of the Wet Milling Industry", Symposium Proceedings, X-1. Corn Refiners Association, Washington, D.C. (1970).
28. Hall, R., Food Technol. 22:1388 (1968).
29. Pangborn, R., in "Symposium: Sweetners" (G. Inglett, ed.), p. 23. Avi Publishing Co., Westport, Conn. (1974).
30. Hargitt, R., and Buckee, G., J. Inst. Brew. 84:224 (1978).
31. Brownley, C., Jr., and Lachman, L., J. Pharm. Sci. 53:452 (1969).
32. Keeney, M., and Bassette, R., J. Dairy Sci. 42:945 (1959).
33. Brownley, C., and Lachman, L., J. Pharm. Sci., 53:452 (1964).
34. Olson, R., Ph.D. Thesis, Univ. Wisc., Univ. Microfilms, Order No. 70-15, 905, Ann Arbor, Mich., (1970).
35. Kooi, E., in Encycl. Chem. Technol. (R. Kirk and D. Othmer, ed.) p. 919, Vol. 6, 2nd ed., John Wiley and Sons, New York, (1965).
36. Olson, R. and Winder, W., (Abstr.) J. Dairy Sci. 52:883 (1969).
37. Hodge, J., in "Symposium on Foods: The Physiology of Flavors", (H. Schultz, E. Day, and L. Libbey, ed.), p. 465. Avi Publishing Co., Westport, Conn. (1967).

FLAVOR PROBLEMS IN SOY PROTEINS: ORIGIN, NATURE, CONTROL AND BINDING PHENOMENA

John E. Kinsella
Srinivasan Damodaran

Institute of Food Science
Cornell University
Ithaca, New York

I. INTRODUCTION

The important criteria determining the acceptance of new food ingredients are attractive sensory properties (color, aroma, taste, texture), versatile functional properties, adequate nutritional value and cost. Plant proteins, particularly soy protein preparations, meet most of these requirements however their adoption has been limited principally because of the persistent problem of the undesirable off-flavors associated with these proteins. The challenging problem of off-flavors in vegetable proteins (soybean, peanuts, peas, beans, cottonseed) remains to be overcome. Because most information on this problem is available for soybeans and since many of the flavor problem(s) of soy are similar to those of other sources the present review will focus on soy proteins.

The off-flavors associated with soy proteins pose a number of challenges to the food scientist. These include devising processes for prevention or minimization of their formation; removal or masking of unwanted off-flavors, and developing suitable flavors, e.g., meat-like flavors, while allowing for uneven absorption, release, etc., during processing, storage and consumption.

ISBN 0-12-169065-2

II. OFF-FLAVORS IN SOY PROTEINS

Numerous authors have discussed the off-flavors associated with soy protein preparations. Kalbrenner et al. (1) reported data for various commercial samples of soy flours, concentrates and isolates (Table 1). Beany, grassy, bitter, nutty, green, etc., were the most common off-flavors detected by an expert taste panel. Processing may accentuate or cause the development of off-flavors described as toasted, astringent, cardboardy and mealy (1).

TABLE I. Typical Off-Flavor Descriptions of Soy Products

Product	Flavor Score (1-10)	Flavor Description
Raw flours	5.8	Raw beany, fresh beany, bitter, grassy
Soy flours	5.0 - 7.5	Beany, green beany, raw bitter
Concentrates	6.4 - 7.4	Beany, bitter, astringent, stale, toasted, mealy
Isolates	6.8 - 7.7	Beany, bitter, chalky, oxidized, toasted, cerealy

From Kalbrenner et al. (1)

These off-flavor problems have limited the acceptability and hence the greater use of soy proteins in foods. The undesirable flavors associated with soy preparations (Table 1) are perceptible when these proteins are used in a wide variety of foods (breads, meats, beverages, coffee, whiteners, textured products) (1-6). Generally the flavor score of the soy products are not appreciably improved by processing (2) indicating that the causative off-flavors may be bound to the protein.

There is a wide array of off-flavors that belong to different classes of chemicals found in soy protein preparations. These include carbonyls, furans, alcohols, acylhydroperoxides and oxidized lecithin, hydroxy fatty acids and, in some products, phenolic compounds, browning products and amines (2,7-13). Many of these are developed during processing indicating the presence of precursors in the soybean suggesting possible approaches for the control of off-flavor development.

Wolf (5) reviewed the historical aspects of the problem of off-flavors in soy proteins. In the 1960's it was recognized that oxidation of the unsaturated fatty acids in the soybean was the principal source of the off-flavors. Hand (7)

TABLE 2. Compounds Contributing to Off-flavor in Soy Products

Furans	Green beany
Aldehydes	Green beany, grassy, stale, cardboard
Alcohols	Oxidized, grassy/beany
Trihydroxy fatty acids	Bitter
Fatty acid dimers	Bitter
Phenolics	Sweet-nauseating, astringent
Furfurals (-ols)	Cerealy
Browning products	Roasted
Oxidized phosphatidylcholine	Bitter
Volatile amines	Fishy

reported a rapid increase in off-flavors during wet grinding of soybeans and Mustakas et al. (6) showed that peroxide values and rancid odors developed rapidly upon soaking of soybeans.

TABLE 3. Changes in Oxidative Index of Soybean Products During Processing

Sample		TBA Number
Raw soybeans	(8% H_2O)	0.9
Cracked soybeans	(10% H_2O)	2.0
Full fat flakes	(14% H_2O)	10.5
Defatted flakes	(10% H_2O)	11.6
Toasted flakes	(8% H_2O)	7.7

TBA = mg MDA/Kg sample ( ).

In the procedure for preparing defatted soy flakes the beans are initially soaked to 10% moisture. Though it has been assumed that lipoxygenase is not activated by this soaking the data of Mustakas et al. (8) suggest that lipid hydroperoxides could be formed during initial tempering and cracking. Furthermore, the subsequent steam tempering to 12% moisture, with flaking, would not only accelerate lipoxygenase action but the greater surface area may facilitate autoxidation. As soybeans are processed lipid oxidation occurs as indicated by increased thiabarbituric acid (TBA) reacting products (Table 3). Cracked beans when tempered to 14% moisture and formed into thin flakes show increased TBA numbers. Solvent extraction of flakes has little effect on TBA values. Much of the off-flavors remain associated with soy protein during processing.

Comparison of TBA values of soy flakes (Table 4) homogenized in water or acid (to inactivate lipoxygenase) corroborated the data showing that oxidation can occur during grinding of soybeans (9).

TABLE 4. Lipid Oxidation in Processed Soybean Products

Soybean Sample	TBA Number[1] for Sample Homogenized in Water	at pH < 4
Raw soybean	12.0	0.9
Full fat flakes (raw)	84.0	11.1
Defatted flakes (raw)	12.0	9.7
Full fat flakes (toasted)	7.7	3.3

[1]mg MDA/Kg sample (11)

The typical description of the common off-flavors of soy proteins are given in Table 5. Significantly, these are very similar in description and frequency to the types of off-flavors derived from hydroperoxides of linoleic (18:2) and linolenic acids (18:3). These hydroperoxides are facilely formed by the action of soy lipoxygenase on 18:2 and 18:3. Both of these acids occur abundantly in whole soybeans, in soy flour and in sufficient amounts in defatted flour where lipoxygeanse is still active (unless the flour has been adequately heat treated). Thus, lipoxygenase in the bean, which is acti-

TABLE 5. Frequency of Types of Flavors Observed for Hydroperoxides of Linoleic and Linolenic Acid in Water Compared to Those Perceived for Dispersed Soybean Flour (0.25%)

Description of Flavor	Soy Flour	Taste Panel Response (%) 18:2(OOH) (50 ppm)	18:3(OOH) (10 ppm)
Grassy, beany	94	80	90
Bitter	30	16	19
Astringent	34	19	19
Raw vegetable	--	20	16
Cereal-like	20	8	2
Musty, stale	8	35	--
Rancid oil	3	44	5

From ref. (14).

vated upon inbibition of water or disruption of the seed is apparently the major catalyst of lipid oxidation and thus the principal cause of the major off-flavors in soybeans.

The distinctive beany/grassy flavors have been attributed to a few major compounds, however, the range of off-flavors is caused by the interactive effects of several classes of compounds. Wilkins and Lin (13) observed some 80 compounds of off-flavored soy milk. The major components were hexanal (25%), hexanol, hexenal, ethyl-vinyl-ketone and 2-pentyl furan. These collectively impart a grassy, beany flavor to soy milk. The characteristic green, grassy, beany, bitter flavors are attributed mostly to volatile carbonyls (Table 6) derived via lipid oxidation.

TABLE 6. The Major Lipid Oxidation Products Contributing to Off-flavor of Soy Proteins

Compound	Flavor
n-hexanal	Green grassy
cis-hex-3-enal	Green beany
n-2-pentyl furan (cis and trans)	Beany
ethyl vinyl ketone	Green beany
n-alkanols C5, C6, C7	Contribute grassy/beany flavor

The green beany flavor is associated with soybeans during development whereas the bitter component seems to develop upon maturation (15). These flavors are derived from acylhydroperoxides generated from 18:2 and 18:3 by lipoxygenase or via autoxidation (Table 7).

TABLE 7. Off-flavor Compounds Generated from Linoleic and Linolenic Acid by Soybean Lipoxygenase

18:2	n-alkanals (C2-C7); n-prop-1-enal alk-2-enals (C5, C6, C7, C8); alk-2,4-dienals (C9,C10) 2-n-pentylfuran
18:3	n-alkanals (C2, C3); alk-2-enals (C5, C6) hepta-2,4-dienal; nona-2,6-dienal octa-3,5-dien-2-one; nona-2,4,6-trienal

From ref. (16).

III. LIPOXYGENASE AND OFF-FLAVORS

It is generally accepted that lipoxygenase is the major cause of off-flavor development in soybeans and other vegetable protein sources (peanuts, beans, peas) (2,5 ,16,17).

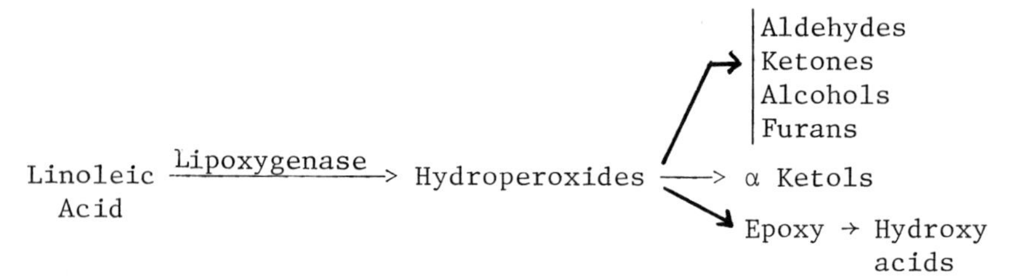

Lipoxygenase (L) catalyzes the oxidation by molecular oxygen of polyunsaturated fatty acids containing the 1,4-pentadiene system to initially yield hydroperoxides. Lipoxygenase contains a non-heme iron which is involved in the free-radical oxidation of fatty acids via the postulated mechanism (18):

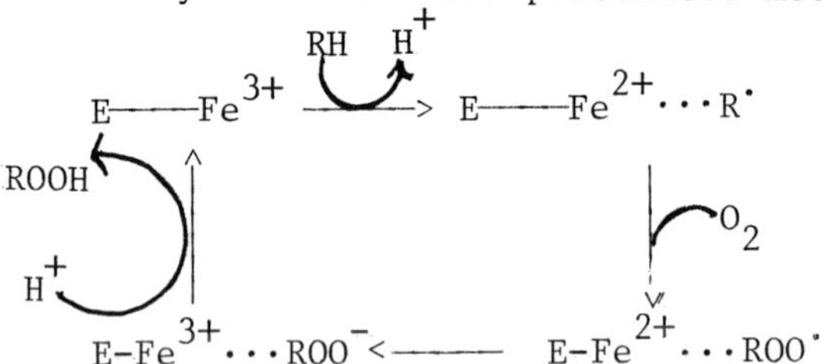

Aerobic oxidation produces hydroperoxides mostly at C13 and C9 in case of linoleic acid (Fig. 1) and at mostly C13 in case of linolenic acid. Lipoxygenase can also function when oxygen is very limiting to yield free radicals (fatty acid, alkoxy), n-pentane, conjugated diene carbonyls, oxo-acids and cross-linked fatty acids. The properties of lipoxygenase from various plants from the point of view of substrate specificity, mechanism of action and immediate products have been reviewed by Veldink et al (18). The enzyme from different plants shows apparent differences in specificity but this may be partly due to variations in conditions. Galliard and Phillips (19) reported that the position of the initial hydroperoxide is influenced by pH of the reaction, i.e. at pH 9.0 the hydroperoxide is formed at C13, whereas at pH 6.6 the hydroperoxide is mostly formed on C9 (19). This influences the nature of the off-flavor carbonyls formed in different plant sources.

Legumes possess high lipoxygenase activity and the enzyme is distributed throughout the cotyledons (20). In the dry bean, oxygen is apparently limiting and enzyme substrate con-

$$\overset{13}{} \quad \overset{11}{} \quad \overset{9}{}$$
$$CH_3-(CH_2)_4-CH=CH-CH_2-CH=CH-(CH_2)_7-COOH$$

$$CH_3-(CH_2)_4-\underset{OOH}{\underset{|}{CH}}-\overset{t}{CH=CH}-\overset{c}{CH=CH}-(CH_2)_7-COOH$$

$$CH_3-(CH_2)_4-\overset{c}{CH=CH}-\overset{t}{CH=CH}-\underset{OOH}{\underset{|}{CH}}-(CH_2)_7-COOH$$

FIGURE 1. Substrate and Initial Products of Lipoxygenase Catalysis in Soybeans

tact is limited by substrate immobility or compartmentalization and thus lipoxygenase is apparently inactive. Upon hydration, oxygen can diffuse into the tissue, enzyme and substrate may gain mobility and oxidation occurs. When the cell structures are disrupted during cracking and milling rampant oxidation occurs.

Soybean possess three isoenzymes of lipoxygenase designated L-1, L-2, L-3 with pH optima at pH 8.3, 6.5 and 6.5, respectively (21,22). Lipoxygenase-1 has been intensively studied (22) and the properties of the different isoenzymes have been summarized by Sessa (16). Soy lipoxygenase can use the free acid or esters of linoleic acid as trilinolein or acylated in phosphatidylcholine (16). There is plenty of substrate for enzymatic or nonenzymatic oxidation in full fat and hexane extracted soy protein preparations. Only trace amounts (ppm) of these products are required to cause off-flavor.

The end-products of lipoxygenase action are mostly aldehydes (Table 6) which arise from the decomposition of alkoxy radicals derived from the acyl hydroperoxides (Fig. 2). These collectively impart the green beany flavor to soy products.

Soy lipoxygenase-2 generates a greater number and quantity of different carbonyl compounds than lipoxygenase-1 (23). These isoenzymes form the 9 and 13 hydroperoxides in equal proportions via an initial peroxy radical. With in vitro preparations these breakdown to yield propanal (>40%), 2-trans-pentenal (11%), 2t-hexenal (∿8%), 2t,6c-nonadienal (2%), 2t,4c-heptadienal (∿20%), 3,5-octadien-2-one (∿8%), 2,4,6-nonatrienal (∿8%) (24).

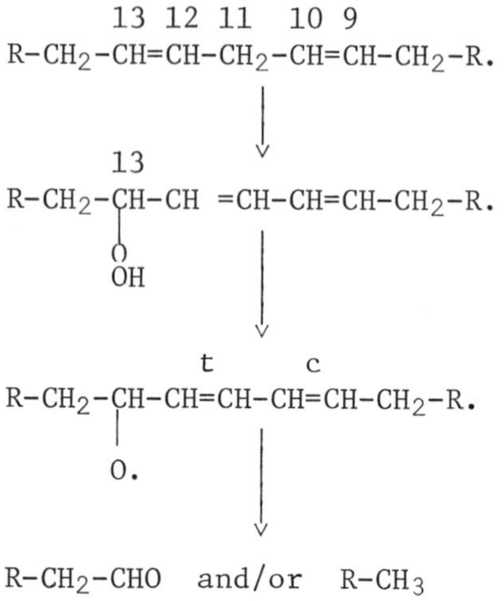

$$
\begin{array}{l}
\quad 13\ 12\ 11\ \ 10\ 9 \\
R\text{-}CH_2\text{-}CH{=}CH\text{-}CH_2\text{-}CH{=}CH\text{-}CH_2\text{-}R. \\
\qquad\qquad\downarrow \\
\quad 13 \\
R\text{-}CH_2\text{-}\underset{\underset{OH}{|}{O}}{CH}\text{-}CH{=}CH\text{-}CH{=}CH\text{-}CH_2\text{-}R. \\
\qquad\qquad\downarrow \\
\qquad\qquad t \qquad c \\
R\text{-}CH_2\text{-}\underset{O.}{\underset{|}{CH}}\text{-}CH{=}CH\text{-}CH{=}CH\text{-}CH_2\text{-}R. \\
\qquad\qquad\downarrow \\
R\text{-}CH_2\text{-}CHO \quad \text{and/or} \quad R\text{-}CH_3
\end{array}
$$

FIGURE 2. Example of Volatile Products Formed by Decomposition of 13-Hydroperoxide of Linoleic Acid

In addition a small amount of singlet oxygen may be formed in the presence of lipoxygenase (25). This may engage in the 1,2 addition to the double bonds and form a dioxetane derivative which upon cleavage can produce 2-trans-hexenal and the 2-trans,6-cis-nonadienal (24).

The hydroperoxides of 18:2 and 18:3 when tested by an expert taste panel give the typical off-flavors associated with soy products (Table 5) especially the characteristic grassy/beany flavor indicating their association with off-flavors. Hydroperoxides of 18:2 decompose to yield alkanals (C4, C5, C6, C7), heptenal and 2n-pentylfuran; octenal; nona-2,4-dienal; deca-2,4-dienal, and pentanol (Table 7) (26,27).

The 2-pentylfuran which can be formed via lipoxygenase mediated or autoxidation (Fig. 3) has a beany flavor (16). The 3-cis-hexenal imparts a green beany flavor. This easily isomerizes to 3-trans-hexenal during thermal processing. This isomer has an oily grassy flavor (11). Ethyl vinyl ketone contributes to the green beany flavor. These compounds are believed to be mainly responsible for beany grassy off-flavor associated with many soy products.

The reported flavor threshold values (ppm in oil) for n-hexanal, 3-cis-hexenal, n-pentyl furan and ethyl vinyl ketone are 0.3, 0.1, 2.0 and ∿5 respectively (11). The values may be an order of magnitude less when tasted in a non-oily medium, e.g. in saliva, when soy products are masticated. Furthermore, synergistic interactions between the numerous volatile and

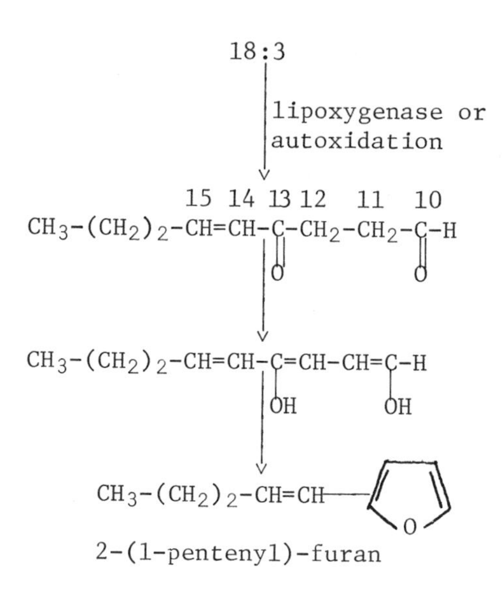

FIGURE 3. Possible Mode of Formation of 2-Pentenyl Furan From the Oxidation of Linolenic Acid

nonvolatile compounds may occur. The threshold value for the grassy beany flavor associated with raw defatted soy flour in aqueous dispersion is around 300 ppm (1) which seems high, but may reflect its binding to the protein.

Acyl hydroperoxides are damaging to enzymes. Therefore plant cells possess hydroperoxide isomerase to metabolize these, converting them mostly to α-ketol derivatives [RCH_2-CH(OH)-C(=O)-CH_2-R'] with some γ-ketols [R-CH(OH)-CH=CH-C(=O)-CH_2-R'] also being formed (18). However in some instances γ-epoxides may be formed. These upon hydrolysis (Fig. 4) yield trihydroxy unsaturated fatty acids (28).

These nonvolatile oxygenated (keto, epoxy, hydroxy) derivatives of fatty acids formed by the action of isomerase may also impart undesirable flavors. Dimers formed from acylhydroperoxides have a bitter persistent taste (29). Unsaturated trihydroxy fatty acids which have flavor threshold values of around 200 ppm (30) have a bitter taste (2). These compounds may contribute to the bitter taste of soy products.

```
   t     c
R-CH-CH=CH-CH=CH-R'
  |
  OOH
            |
            | Isomerase
            v
   t
R-CH-CH=CH-CH-CH-R'
  |         \ /
  OH         O
            |
            | Hydrolysis
            v
R-CH-CH=CH-CH-CH-R'
  |         |  |
  OH        OH OH
```

FIGURE 4. Possible Mode of Formation of Trihydroxy Fatty Acid in Soybean

A. Other Sources of Off-Flavors

In addition to lipoxygenase, autoxidation of unsaturated fatty acids can occur during milling, flaking and processing. Autoxidation would be expected to produce mostly 9 and 13 hydroperoxides of linoleic and probably 6, 9 and 13 hydroperoxides of linolenic acid. These by homolytic scission give rise to several volatile saturated and unsaturated carbonyls (mostly aldehydes, some ketones), and both nonvolatile and volatile alcohols, epoxy, oxo and hydroxy fatty acids. These undoubtedly contribute to the off-flavor of soy products.

Phosphatidylcholine (PC) bound to soy protein (0.08%) develops bitter flavors upon oxidation (lipoxygenase mediated or autoxidation). The bitter threshold level of PC is ∿0.006%. Sessa _et al_. (31) concluded that oxidized PC was a major cause of the bitter flavor of soybean protein products.

Several compounds derived via autoxidation have functional groups (epoxy, hydroxy, carboxy and carbonyl groups) which may interact with functional groups, i.e. NH_2 group, on the proteins resulting in Schiff base formation. It is not known if these complexes cause additional off-flavors but they contribute to browning and loss of protein solubility.

Soy flour contains several phenolic compounds with ferulic, coumaric, syringic and vanillic acids being most prevalent (32). Phenolic acids may contribute bitter astringent flavors to soy flour (33). However Rackis _et al_. (34) stated that phenolic compounds had little significance in flavor of soy products (34). Nevertheless when defatted soy meal or protein preparations are sterilized or heated to high temperature (cooking, retorting) a very unpleasant cooked odor that many consumers find repulsive is generated. This has been attributed to the formation of 4-vinyl phenol and 4-vinyl guaicol by the thermal decarboxylation of p-coumaric and ferulic acids respectively (35). 4-Vinyl phenol has a strong smoky aroma; ethyl guaicol a soy sauce flavor; methyl guaicol a smoky flavor and 4-vinyl guaicol tastes phenolic or bitter. These have flavor threshold values of around 10-90 ppm (32) and contribute an unpleasant flavor to heated soy products.

Finally the sucrose present in soy flour (12% in defatted soy flour) can influence perceived flavor and of course on cooking can give rise to different flavors via the Maillard reaction. This may occur during extrusion and under excessive heating when the free amino acids, sugars and oligosaccharides freely interact and can give rise to many classes of compounds including aldehydes, ketones, furan derivatives, sulfur compounds and pyrazines (12,36). Contaminant flavors may also form in soy proteins during processing particularly in solvent-extracted preparations. Thus, off-flavors like sweet fusel note or catty odors may arise from the formation of mesityl oxide from traces of acetone and hydrogen sulfide derived from sulfur containing amino acids. These of course can be avoided by using purified solvents (37).

IV. CONTROL OF OFF-FLAVORS

A. Lipoxygenase Inactivation

The preponderance of these common off-flavors in soy products are derived from the carbonyls and scission products of the hydroperoxides formed by lipoxygenase. Therefore control of off-flavor formation requires rapid inactivation of lipoxygenase. Because lipoxygenase is heat labile thermal treatments are most commonly employed for its inactivation. Additional methods have also been tested (Table 8).

Christopher et al. (21) reported half-lives of 25 min and 0.7 min for lipoxygenase-1 (pH optimum 9.0) and lipoxygenase-2 (pH optimum 6.8) respectively, at 69 °c. Borhan and Snyder (38) found values of 15 and 0.8 min for these two isoenzymes in crude aqueous extracts of soy flour. Thus the lipoxygenase-1 is significantly more heat stable than lipoxygenase-2 and treatments which inactivate the former isoenzyme should be effective in minimizing off-flavor problems. Maximum stability of the enzyme occurs around pH 6.0 and the enzyme becomes more labile at alkaline or acid pH values. Thus the pH of the solution affects rate of thermal destruction of lipoxygenase. As the

TABLE 8. Some Methods for Minimizing Off-flavor Development in Soy Preparations by Inactivation or Inhibition of Lipoxygenase

Method	Reference	
Grinding in hot water (100°C)	Wilkens *et al.* 1967	(43)
Dry heat or moist heat	Mustakas *et al.* 1969	(6)
Blanching	Nelson 1971	(44)
Grinding at acidic pH	Kon *et al.* 1970	(46)
Grinding with H_2O_2 plus calcium chloride	Paulsen 1963	(47)
Grinding with aqueous ethanol plus heat	Borhan & Snyder 1979	(38)
Grinding with solvent azeotropes	Rackis *et al.* 1975	(59)
Inhibition by acetylenic compounds	Blain & Shearer 1965	(79)
Addition of antioxidants	Yasumoto *et al.* 1970	(52)

pH is increased from 7 to 9 the rate of inactivation increases e.g. ten-fold at 65°C by increasing the pH from 7 to 8 (39). The half-life of lipoxygenase is ∿1 min at pH 4.5 or pH 9 (38). The solubility of the soy protein from such treatments was better following heating at the alkaline pH. Oxygen has no effect on thermal inactivation. The presence of linoleic acid hydroperoxide enhances denaturation (40). The pure enzyme appears to be more labile than the crude enzyme in the bean or meal however this may be due to the poor thermal conductivity of the materials in the bean or meal.

Rapid inactivation of lipoxygenase is effective in minimizing off-flavor development. Because lipoxygenase is heat labile several methods employing heat treatment have been tested experimentally, however few of these approaches have been used commercially.

Full fat soy flour is produced by steaming the beans, drying, dehulling, and grinding. The steam treatment inactivates some of the lipoxygenase and the product may be reasonably bland. However further heating is required to give soy flour devoid of lipoxygenase activity with a stable bland flavor. The extent of heating is controlled to give soy flours with varying amounts of active lipoxygenase (41).

Heating soybeans for 1-2 hours at 70°C is effective in reducing off-flavors because it effectively inactivates the lipoxygenase (42). However heat inactivation causes protein denaturation and generates a cooked or toasted flavor which may not be desirable in certain products. Moist heat treatments i.e. toasting, are commonly employed to improve the flavor of soy flours by elimination of the beany/grassy off-flavors and generating a toasted-like flavor.

Mustakas et al. (6,8) developed a dry heat or steam procedure (Table 9) for preparation of bland soy flour. Lipoxygenase was inactivated by dry heat of the beans, e.g. 100°C, and off-flavors did not develop but the nitrogen solubility index (NSI) was reduced.

TABLE 9. Effect of Heat Treatments of Beans on Flavor of Soy Flour

Treatment	Lipoxygenase Destruction (%)	Peroxide Value	Flavor Score	NSI
Dry heat 82°c,28 min	52	17.0	2.0	77
Dry heat 100°c,62 min	84	2.0	8.0	63
Dry heat 82°C + steaming 5 min	98	0.8	8.0	49
Steam (100°C) 2 min	98	1.1	7.7	39
Wet heat (93°C/ 20 min)	96	4.0	7.3	31

From ref. (6).

A combination of dry heat 66° or 82°C followed by steaming (5 min) was very effective in eliminating rancidity but NSI was significantly reduced. Direct steam heat (100°C for 2-5 min) was effective in inactivating lipoxygenase but also caused extensive denaturation of protein. Wet heat treatments (93°C for 20 min) of beans containing 25% moisture was not fully effective in eliminating off-flavors because the peroxide value remained high.

In the preparation of soy milk grinding the unsoaked beans in water at 100°C for 10 min inactivated lipoxygenase and yielded a good flavored product (43). Nelson et al. (44) developed a blanching procedure for preparing soy flakes with acceptable flavor. Currently available methods for the effective prevention and reduction/elimination of off-flavors from soy proteins are not very compatible with the preparation of highly functional soy proteins because they require conditions that cause denaturation of the proteins. The protein dispersibility index (PDI) of raw unprocessed soybean meal is around 95% whereas toasting (to destroy antitrypsin factors, lipoxygenase and volatilize flavors) reduces the PDI to 15%. Lipoxygenase can be destroyed by a light heat (steam) treatment that reduces PDI to 75-80%. However, more rigorous treatments are necessary for inactivation of trypsin inhibitors and removal of off-flavors (41,45).

Heat treatment of the soybean is desirable for the prevention of off-flavor formation and for the destruction of trypsin inhibitors. However heating, by causing denaturation of the soy proteins, severely reduces the functional properties of these proteins and limits the usefulness of the heated soy proteins for a wide array of applications in food systems. Because heat treatments denature the proteins they must be minimized in preparing functional protein ingredients and less rigorous methods are needed for preparation of bland but functional proteins. Thus other methods for specifically inactivating lipoxygenase need to be explored.

Low pH reduces the activity of lipoxygenase and minimizes formation of off-flavor carbonyls. Thus grinding of soybeans at pH 3.5-4.0 prevents the generation of hydroperoxides, yields bland preparations but these are salty as a result of neutralization and some of the protein is denatured particularly if heat is used (46).

Hydrogen peroxide, at more than a twofold molar excess, irreversibly inhibits lipoxygenase by attacking the non-heme iron, changing it to a high spin state and altering its coordination sphere (18). Since hydrogen peroxide reacts with the catalytic site of the enzyme causing inactivation, H_2O_2 has been used to produce flours with superior flavor characteristics (47). However, its commercial use is not evident. Cysteine may interact with active site of lipoxygenase (18) or with linoleic acid hydroperoxide (48) and thereby reduce off-flavor development. However, it is now surmised that cysteine inactivates lipoxygenase by its ability to generate H_2O_2 (18).

Because the initial hydration of soybeans allows lipoxygenase to become active and generate off-flavors, it is important to rapidly inactivate lipoxygenase in situ if possible. Organic alcohols, including ethanol, reversibly inhibit lipoxygenase and increasing the chain length of alcohol increases the extent of inhibition (49). Eldridge _et al_. (50) reported that soaking soybeans in >2% ethanol (1:4) for 24 hr at 20-25°C inactivated lipoxygenase but the NSI was reduced to 55 and trypsin inhibitor by only 20% of original activity. Flavor panel scores revealed that the best flavor (least amount of grassy, beany flavor) was obtained with 40-50% ethanol treatment where NSI was minimum. Steeping of the beans was slightly more effective in improving flavor than wet milling in the presence of similar concentrations of ethanol.

Borhan and Snyder (38) studied the effects of combined heat and soaking in aqueous ethanol on the lipoxygenase activity in soybeans. Soaking of soybeans in aqueous ethanol solutions (50 g beans/200 ml solution) apparently inactivated lipoxygenase (Table 10). However upon resoaking the beans in water for 24 hr the lipoxygenase activity was restored indicating removal of inhibitory ethanol or reversal of ethanol induced 'denaturation'. Thus resoaking of beans previously exposed to 30% alcohol resulted in regeneration of most of the lipoxygenase activity (38). However, if beans were initially soaked in water and then exposed to 95% alcohol lipoxygenase inactivation was complete. Presumably the hydrated beans facilitated the uptake of the ethanol whereas the dry bean imbibed very little ethanol.

TABLE 10. Lipoxygenase Activity of Soybeans Soaked in Aqueous Ethanol for 24 hr at 4°C

Ethanol Concentration %	Lipoxygenase[1] Activity	Weight Increase (%)
0	56	110
5	47	104
10	33	90
30	3	68
50	8	49
80	45	8

[1]Oxygen uptake μmole $O_2 \cdot min^{-1}$ at pH 9.0 (38).

Increasing the temperature of soaking enhanced the effectiveness of low concentrations of ethanol in denaturing the lipoxygenase. Thus 10, 15 and 25% ethanol eliminated lipoxygenase activity after 24 hr at 55, 45 and 25°C respectively. The NSI of the flours was reduced to ∿60% in these samples (38). The experimental data indicated that soaking of beans in 10-15% ethanol at 45°C for 24 hr was optimum for eliminating lipoxygenase though NSI was 67 and most of the original trypsin inhibitor remained.

Sodium carbonate (0.1M) enhanced the inactivation of the lipoxygenase apparently by increasing the pH to 8 (38). By varying temperature, ethanol concentrations and soaking times, soy flours devoid of lipoxygenase activity with varying NSI were obtained. Increasing ethanol concentrations from 15 to 60% significantly reduced the soaking times required to inactivate the lipoxygenase at 40, 50 and 60°C but the NSI was concurrently reduced (38). Thus, the particular conditions must be decided depending upon the intended end use of the soy preparations.

In addition to the practical scale inactivation of lipoxygenase by thermal treatments, by alcohols, combinations of both or hydrogen peroxide, there are some limited experimental data indicating the possible use of antioxidants (infusion of the bean or addition prior to milling) (51,52) but this approach is currently impractical because of cost.

B. Off-flavor Removal

Because it is practically impossible to prevent the formation of off-flavors in soy protein preparations, practical methods for the removal of off-flavors are necessary. Conventionally prepared raw defatted flakes have bitter astringent flavors. Commercially these are removed by a steam treatment (53,54). Steam treatment reduces solubility of the proteins and thus limits the functional properties of the concentrates and isolates made therefrom. Therefore several researchers have explored the use of non-denaturing solvents for removal of off-flavors. Aqueous ethanol is used to prepare bland soy protein concentrates (55), however, at certain concentrations it may result in excessive denaturation of proteins.

Apolar solvents pentane/hexane which remove >97% of the lipids of soybean (raw full fat soy flour) do not remove all the bitter, beany astringent flavors (11). Further treatment with hydrogen bond breaking solvents e.g. alcohols or hexane ethanol azeotrope (82:18 v/v) is effective in removing additional lipids and most flavor components especially the components causing green-beany off-flavor. Steam deodorization can then be used to make a bland preparation by removing the astringent bitter notes (11).

Extraction with alcohols such as ethanol or propanol can lead to the preparation of bland soy concentrates (56). Eldridge et al. (57) used azeotropic mixtures of hexane and methanol, ethanol or 2-propanol to improve the flavor of flakes with hexane:ethanol being best. However, these caused significant decreases in the extractability of the protein. Others have also used azeotropic mixtures of hexane:ethanol for removal of off-flavors (9,58). The solvent extracted flakes had good flavor attributes (Table 11).

Rackis et al. (59) reported that hexane:alcohol azeotrope extraction of defatted flakes for 3 hr only slightly reduced PDI (4%) while essentially inactivating lipoxygenase and improving flavor. Baker et al. (60) used azeotropes of ethanol, isopropanol and methanol to extract soy flakes/or soy flours. None of the treatments achieved complete elimination of off-flavors and all caused protein denaturation.

Solvent extraction of off-flavors may have limited practical use because of cost; the need to eliminate all residual solvent and significantly because the solvents themselves or the heat treatments required to remove them, may cause excessive protein denaturation. While protein denaturation may not be significant for textured products (61) it is a critical

TABLE 11. Effect of Hexane:Ethanol Azeotrope Extraction and Toasting on Flavor Score of Defatted Soy Flakes

Sample	Flavor Score (1-10)
1. Raw soy flakes	4.0
2. Steam toasting defatted flakes (10 min)	6.6
3. Hexane:ethanol extraction of defatted flakes	7.4
4. Combination of 3 + 2	7.8

From Ref. (58)

limitation when preparing soy proteins for a wide number of applications in the food industry where ease of hydration, protein dispersibility and solubility are important criteria (62, 63).

More research to discover new approaches for the solution of the insidious problem of off-flavors in vegetable proteins is needed. Organic alcohols have much higher (20-250 times) flavor threshold values than the corresponding aldehydes. Because alcohol dehydrogenase has a broad specificity for saturated and unsaturated aldehydes and in the presence of excess NADH the reaction favors alcohol formation (>90%), the use of alcohol dehydrogenase to reduce off-flavors in soy proteins has been suggested (64,65). In aqueous systems e.g. milk, the conversion of aldehydes to alcohol result in a loss in the intensity of perceived flavor due to the increase in FTV (Table 12) and also change in the flavor note since the alcohols are a slightly different flavor than the corresponding aldehydes. This type of interconversion is followed by quantitative and qualitative improvement in sensory qualities (64) though it has not been evaluated for improving soy protein preparations.

Recently Takahashi _et al_. ( ) claimed success in reducing the beany flavor of soy proteins after they had been incubated with aldehyde oxidase from bovine liver. This enzyme, in the presence of dissolved oxygen, converts medium chain aldehydes e.g. hexanal (Km 6μM) to the corresponding acid. However it was not very effective in oxidizing protein bound aldehydes. The practical application of these enzymatic methods is questionable at the present time.

Finally, the use of the plastein reaction for the improvement of the flavor of hydrolyzed soy proteins has been suggested. This procedure has been thoroughly reviewed (67). It appears to have limited practical value at the present time.

TABLE 12. Odor Threshold Concentration of Aldehydes and Corresponding Alcohols in Water

Compounds	Relative Odor Threshold (M)
n-Hexanal:n-hexan-1-ol	2.0×10^{-7}:480×10^{-7}
n-Hex-2t-enal:n-hex-2t-en-1-ol	3.2×10^{-6}: 67×10^{-6}
n-Hex-2,4tt-dienal:n-hex-2,4tt-dien-1-ol	5.0×10^{-6}:240×10^{-6}
n-Hept-2t-enal:n-hept-2t-en-1-ol	3.4×10^{-7}:370×10^{-7}
n-Oct-2t-enal:n-oct-2t-en-1-ol	2.9×10^{-8}:660×10^{-8}

From Ref. (64).

V. FLAVOR PROTEIN INTERACTIONS

Because the problem of off-flavors in soy proteins is due to binding of the off-flavors to the protein additional basic research to elucidate the mechanisms and thermodynamics of flavor-protein interactions is warranted. Knowledge of the nature of these interactions is needed to devise new processes for off-flavor removal that are compatible with retention of the desirable functional properties of the proteins.

In addition to the problem of eliminating off-flavors, a current challenge in the development of new proteinaceous foods is the successful flavoring of the food to simulate desirable products. The ability to mask undesirable flavors and simulate the desired food flavor is significantly influenced by the flavor binding capacity of the protein used. The binding of flavors by proteins, the uneven retention of flavors during processing treatments and storage, the preferential release or uneven retention of some components of a flavor blend during mastication are problems confronting the manufacturer of fabricated foods from protein ingredients.

There are several qualitative observations in the literature concerning the effects of proteins on the volatility of flavor compounds in aqueous systems. Nawar (68) showed that the presence of gelatin decreased the volatility of methyl ketones in aqueous systems. By headspace analyses, Gremli (69) showed that in an aqueous solution of soy protein and aldehydes (10:1 mixture) the volatility of aldehydes decreased compared to in the absence of soy protein. The percent decrease in the volatility increased with the chain length of the aldehyde, indicating that the magnitude of the interaction between the protein and aldehyde depended on the chain length. Franzen and Kinsella (70) studied the influence of different proteins such as α-lactalbumin, bovine serum albumin, leaf protein concentrate, single cell proteins, soy isolates, etc., on the volatility of aldehydes and ketones in model systems. They found that the addition of proteins to model systems containing water and flavor consistently caused a decrease in the headspace concentration. But the decrease in the volatility depended on the source of the protein. This could be due to the difference in the intrinsic structural features of the proteins which may determine the magnitude of the flavor binding or to non-protein contaminants such as lipids.

These studies described the qualitative effects of proteins on the volatility of flavor compounds, but they did not deal with the quantitative aspects of flavor-protein interaction. In order to understand the molecular nature of the binding of carbonyls to soy proteins, Arai et al. (71) studied the interaction of hexanal and n-hexanol with native, denatured and enzymatically hydrolyzed soy protein. The interaction of hexanal and n-hexanol increased with the degree of denaturation of soy protein and these flavors were not removed by vacuum distillation. The increased binding of these carbonyls by denatured, compared to native soy protein, was attributed to the greater exposure of hydrophobic regions in the denatured protein. Hence the resistance of hexanal and n-hexanol to vacuum distillation was ascribed to their strong binding to the denatured protein through hydrophobic bonding. Arai et al. (71) further showed that proteolysis of the denatured protein decreased the amount of hexanal and n-hexanol retained by the protein. These results suggest that the interaction of flavor compounds with proteins is not via mere surface adsorption but through interaction with specific hydrophobic regions in the protein. Using a gel filtration technique, Arai et al. (71) obtained the binding constants 173.4 M^{-1} and 80.3 M^{-1} for hexanal and n-hexanol respectively. These values for binding to partially denatured protein suggest that the binding is relatively weak, and that the resistance to vacuum distillation cannot be interpreted in terms of strong interactions but possibly to entrapment of these compounds in the solid protein matrix. At saturation levels, the amounts of hexanal and n-hexanol bound to partially denatured soy protein was about 0.847 mg/g and 0.889 mg/g protein respectively (71). Assuming that the molecular weight of soy protein is 100,000, the total number of binding sites, calculated from the above values, in partially denatured soy protein is about one. For an oligomeric protein such as soy protein this value seems to be very low and cannot fully explain the inherent flavor problem in soy protein.

In a different approach, using equilibrium dialysis method, Beyeler and Solms (72) studied the interaction of soy protein isolate and bovine serum albumin with many flavor compounds, including aldehydes, ketones, alcohols, etc. For all the compounds studied they obtained a linear relationship between the amount of ligand bound to the protein and the molar concentration of the free ligand. The binding constant K was calculated from the equation: $\bar{\nu} = Kc$; where $\bar{\nu}$ is the number of moles of ligand bound per mole of protein and 'c' is the molar concentration of the free ligand. For example, the binding constant for 2-butanone to a soy protein isolate at 20°C and pH 7.0 was estimated to be 5174 M^{-1}. This value seems to be very high when compared to the binding constant for hexanal, i.e. 173.4 M^{-1} obtained by Arai et al. (71). The validity of using the above equation to calculate the binding constant K is questionable because even for a surface adsorption process the equilibrium binding constant is obtained using the Langmuir equation

$$\frac{1}{x} = \frac{1}{\beta} + \frac{1}{(K\beta)} \cdot \frac{1}{c}$$

where x is the amount adsorbed, c is the concentration of the unbound ligand and β is a constant which represents number of moles of ligand bound to one mole of the acceptor at infinite ligand concentrations. In protein-ligand interactions, which do not follow a mere surface adsorption phenomenon, the amount of ligand bound to the protein molecule cannot be directly proportional to the free ligand concentration. Furthermore, assuming the molecular weight of soy protein to be 50,000, Beyeler and Solms (72) obtained molal ratios of binding, $\bar{\nu}$, up to 1200 for 2-butanone. In a 50,000 molecular weight protein, the number of amino acid residues can be assumed to be 500. This gives a value of more than two moles of 2-butanone bound to each amino acid residue, which is unlikely. The possible reason for this abnormal value may be the error in the experimental determination of the bound ligand. It is inaccurate to estimate the amount of ligand bound to protein from the difference between the total ligand concentration and the free ligand concentration after equilibration, as these authors did unless the ligand does not bind to the membrane. In equilibrium dialysis experiments interaction between the membrane and 2-nonanone can occur (Figure 5). If this fact is not taken into account, then the estimated amount of bound ligand will not reflect the actual amount bound by the protein alone. This could be the reason for the abnormal values obtained by Beyeler and Solms (72).

There are conflicting reports in the literature regarding the nature and magnitude of flavor-protein interaction. There is a need for a systematic approach to understand the mechanism(s) and thermodynamics of protein-flavor interaction. If the objective is to solve the flavor problems in soy flour and soy protein isolates then, one should concentrate on the molecu-

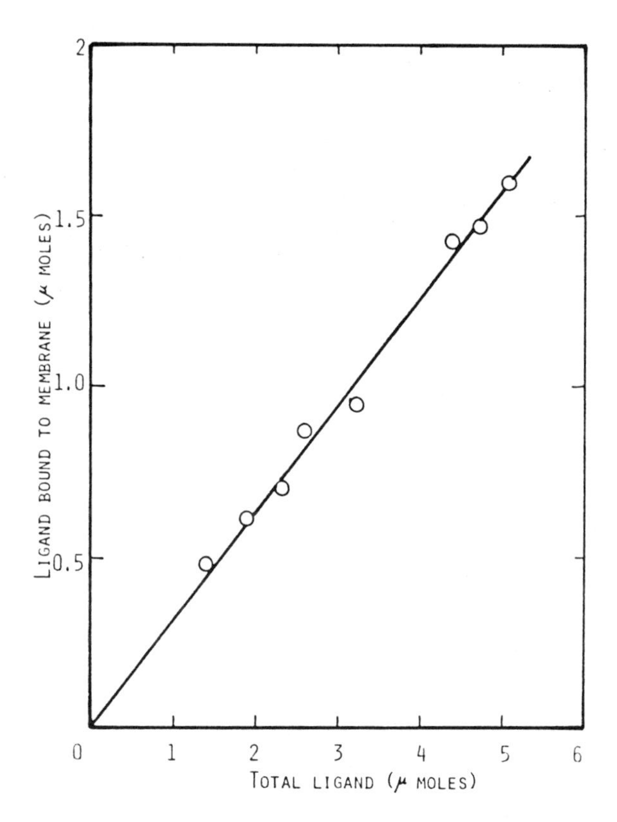

FIGURE 5. Binding of 2-Nonanone to Spectropor-2 Dialysis Membrane

lar nature of the native soy protein (instead of denatured protein) with respect to its interaction with carbonyls, because the carbonyls produced from lipoxygenase activity as the soybean is hydrated, cracked and milled must interact with soy proteins in their native state.

The binding of any organic ligand to a protein is believed to be mostly due to hydrophobic interactions. Most protein molecules are folded in three-dimensional structures which are stabilized by non-covalent forces. In this structure, the hydrophobic side chains are invariably located on the inside of the spherical protein, forming hydrophobic regions, and the more polar groups on the outside. These hydrophobic regions, however, are of importance not only for the stabilization of the protein structure, but also for interaction of the protein with apolar low molecular weight organic molecules, e.g. carbonyls.

The mathematical description of interaction of a ligand with protein can be approached from a general standpoint. Assuming that there are a number of protein molecules 'P' in solution each with 'n' independent, indistinguishable and identical binding sites, then the binding equilibrium between a ligand and the binding sites on the protein can be expressed as

$$P + L \underset{}{\overset{k_1}{\rightleftarrows}} PL_1$$

$$PL_1 + L \overset{k_2}{\rightleftarrows} PL_2$$

$$PL_{i-1} + L \overset{k_i}{\rightleftarrows} PL_i$$

$$PL_{n-1} + L \overset{k_n}{\rightleftarrows} PL_n \quad .$$

According to Scatchard (73) the intrinsic binding constant for the process is given by

$$\frac{\bar{\nu}}{[L]} = nK - \bar{\nu}K$$

where $\bar{\nu}$ is the number of moles of ligand bound per mole of protein, [L] is the molar concentration of the free ligand, 'n' is the total number of binding sites and 'K' is the intrinsic binding constant. According to the above equation a plot of $\bar{\nu}/[L]$ vs. $\bar{\nu}$ will give a straight line with a slope of -K and an intercept 'nK' (Figure 6a). From this the values of 'n' and 'K' can be calculated. The above Scatchard equation can also be expressed in the form of double reciprocal equation (Klotz plot),

$$\frac{1}{\bar{\nu}} = \frac{1}{n} + \frac{1}{nK[L]} \quad .$$

A plot of $1/\bar{\nu}$ vs. $1/[L]$ will give a straight line with a slope of 1/nK and an intercept of 1/n (Figure 6b). Further, the experimental determination of the equilibrium binding constant K as a function of temperature facilitates the determination

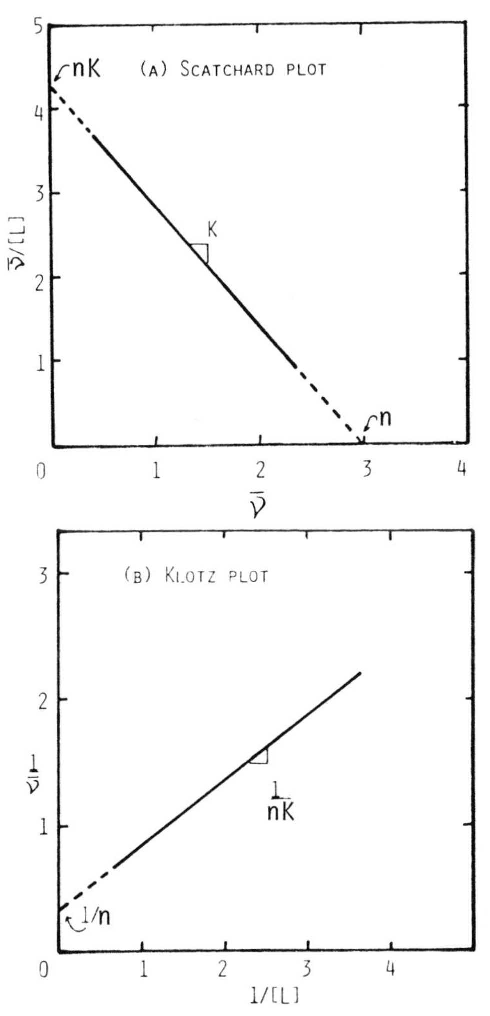

FIGURE 6. Examples of Ideal Scatchard (A) and Klotz Plots (B) of Ligand/Protein Interactions

of the thermosynamic paremeters. Thus

$$\Delta G^\circ = -RT \ \ell n \ K$$

$$\Delta H^\circ = -R \frac{d \ \ell n \ K}{d(1/T)}$$

$$\Delta S^\circ = \frac{\Delta H^\circ - \Delta G^\circ}{T}$$

For a given hydrophobic ligand the value of the binding constant will vary for different proteins. Since the molecular structure of two different proteins can never be the same, the hydrophobicity of the binding sites and hence the magnitude of the hydrophobic interaction with the ligand varies. For example it is unlikely that the binding constant for the binding of 2-butanone to bovine serum albumin and soy protein would be similar as reported (72). Similarly, for a given protein, different ligands have different binding affinities depending on the functional groups and the chain length.

As mentioned earlier, the experimental data required for studying the nature of protein-flavor interaction are $\bar{\nu}$ and [L], i.e. the number of moles of ligand bound per mole of protein at a particular free ligand concentration [L]. The experimental values of $\bar{\nu}$ and [L] can be determined by different methods like equilibrium dialysis, gel filtration, ultrafiltration, etc. But for carbonyl compounds, the methods of choice are equilibrium dialysis (75) and liquid-liquid partition equilibrium (75-77).

We studied the interaction of carbonyls with soy protein using an equilibrium dialysis method (74). Acrylic cells of equal volume, separated by a membrane and clamped together tightly, were used. In a typical experiment 3 ml of 1% soy protein (obtained from the water extraction of low heat soy flour and precipitated at isoelectric pH 4.5) solution was taken on one side of the membrane and 3 ml of the buffer containing a known amount of the ligand on the other side. The cells were shaken at the required temperature for at least 18 hr to attain equilibrium. At the end of the equilibration period, 1 ml of the solution from each side of the membrane was drawn out and placed in vials containing 1 ml isooctane. The ligand from the aqueous phase was extracted into the isooctane phase by shaking. Since the binding is reversible and the aqueous/isooctane partition coefficient of the carbonyls studied is of order of 10^{-3}- 10^{-4}, all the ligand from the aqueous phase could be extracted into the isooctane phase. In fact, a second extraction of the aqueous phase did not contain any ligand. The concentration of the ligand in the isooctane extract was determined by gas chromatography (75). The difference in the concentration of the ligand on either side of the membrane represented the amount of the ligand bound to the protein. Assuming the mean molecular weight of soy protein to be 100,000, the molal ratios of binding $\bar{\nu}$ can be calculated knowing the amount of protein and the amount of ligand bound to the protein. The ligand concentration on the buffer side of the membrane represents the free ligand concentration [L]. Even if some of the ligand is bound to the membrane, it will not affect the values of $\bar{\nu}$ and [L] obtained as above, since the membrane bound ligand does not affect in any way the equilibrium between the free ligand and the protein bound ligand. In fact the $\overline{\nu}$ and [L] values obtained using different membranes in the dialysis cell does follow a single theoretical curve, suggesting that the amount of ligand bound to the membrane does not affect the equilibrium between the free ligand and the protein bound ligand.

Since the purpose of the study was to understand the molecular aspects of the protein flavor binding, rather than the mere estimation of the binding constants for different flavor compounds, we selected 2-nonanone as the representative of carbonyls.

The binding isotherms for the binding of 2-heptanone, 2-octanone and 2-nonanone to soy protein are shown in Figure 7a. The same data is presented in the form of Klotz plot in Figure 7b. The intercept in Figure 7b suggests that there are about 4-5 binding sites for methyl ketones in native soy protein (on the basis of 100,000 molecular weight). The equilibrium binding constants and the free energies of interaction are presented in Table 13. It may be noted that the binding constant increases with the chain length, suggesting that the binding is hydrophobic in nature. For each methylene group increment in the chain the binding constant increases by about threefold, with a corresponding change in the free energy of about -600 cal/CH_2 group. Previously we have shown that in the case of interaction of methyl ketones with bovine serum albumin, the binding constant increased threefold for each CH_2 increment in the chain (75). But, although the magnitude of the change in the binding constant is threefold for each CH_2 group, the absolute value of the binding constant for methyl ketones varies from protein to protein. For example, the intrinsic binding constant for 2-nonanone to BSA is about 1800 M^{-1} (75), whereas with soy protein it is only 930 M^{-1} (Table 13). This is probably due to the differences in the structural properties of these two proteins. The hydrophobicity of the binding sites in the bovine serum albumin may be more than that of soy protein.

TABLE 13. Thermodynamic Constants for the Binding of Carbonyls to Soy Protein at 25°C.

Ligand	Type of Soy Preparation	n	K_{eq} (M^{-1})	ΔG (KCal/mole)
2-Heptanone	Native	4	110	2.781
2-Octanone	Native	4	310	3.395
2-Nonanone	Native	4	930	4.045
"	Part. Denatured	4	1240	4.215
"	Succinylated	2	850	3.992
5-Nonanone	Native	4	541	3.725
Nonanal	Native	4	1094	4.141

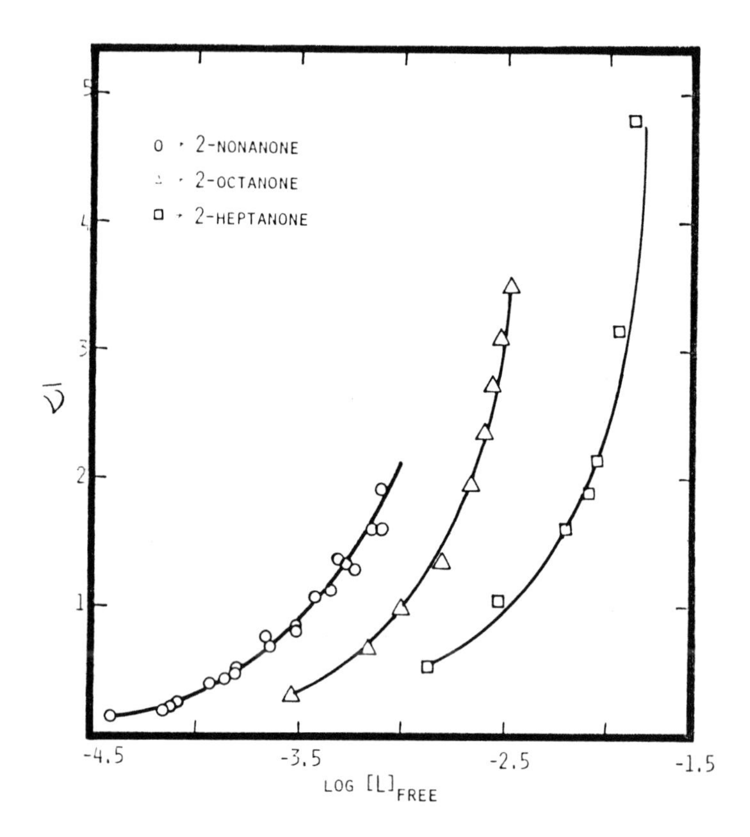

FIGURE 7a. Binding Isotherms of 2-Heptanone (□), 2-Octanone (△) and 2-Nonanone (○) by Soy Protein at pH 8.0, 0.03 M Tris-HCl Buffer and 25°C

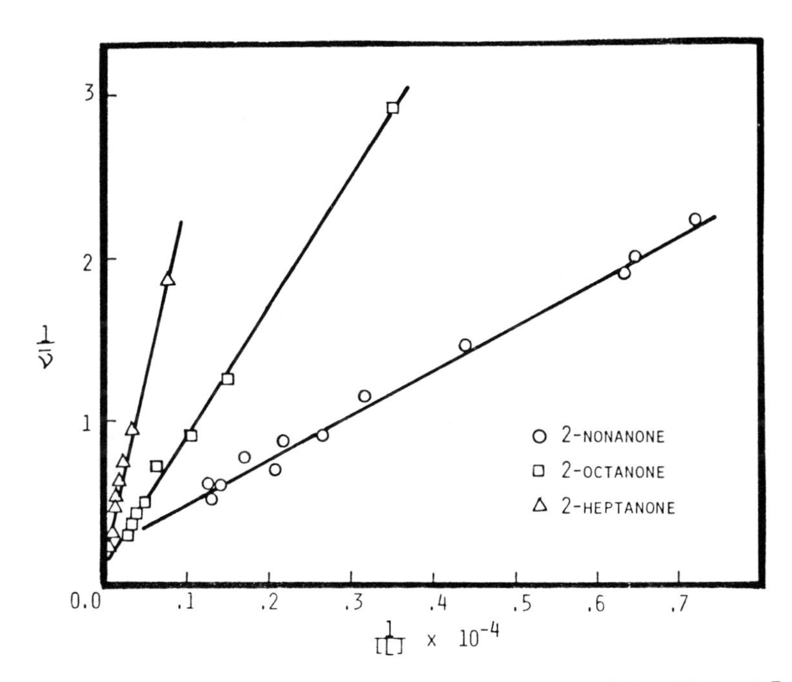

FIGURE 7b. Klotz Plot of the Data Presented in Figure 7a.

The binding affinity of flavor compounds for soy protein is not only a function of the chain length but also depends on the position of the functional groups, such as hydroxyl and keto groups, in the chain. The Klotz plot for the binding of nonanal, 2-nonanone and 5-nonanone is shown in Figure 8a. The relationship between the position of the keto group in the chain and the binding affinity to soy protein is shown in Figure 8b. The binding constant decreases linearly as the keto group is shifted to the center of the chain. This could be due to steric hindrance introduced by the presence of the keto group. In the case of nonanal, the presence of the keto group on the terminal 1-position permits unhindered contact between the entire length of the ligand and the binding site in the protein. With 2-nonanone and 5-nonanone the presence of the keto group in the second and fifth position, respectively, introduces steric hindrance to the hydrophobic interaction between the protein binding site and the ligand. This is reflected in the decreased binding affinity.

The binding affinity of a ligand for a protein is dependent upon the structural state of the binding sites. Any change in the structure of the protein, for example, induced by agents like urea, guanidium-HCl, chaotropic salts or heat, would profoundly affect the binding affinity of the ligand for the protein. The effect of partial denaturation of soy protein on the binding of 2-nonanone is shown in Figure 9. The binding affinity is more in the case of partially denatured soy protein compared to native soy. However, the intercept in Figure 9 suggests that partial denaturation of soy protein does not change the number of binding sites for 2-nonanone. In other words, in both the native and partially denatured soy protein, the number of binding sites for 2-nonanone is about four. This suggests that under the denaturation conditions used (heated at 90°C for 1 hr) there are no major changes in the secondary structure of the protein but certain changes in the tertiary and quaternary structures increase the binding affinity of the existing binding sites without the formation of any new sites. Drastic denaturation may completely unfold the protein and expose new binding sites, as suggested by Arai et al. (71).

The effect of urea on the binding of 2-nonanone to soy protein is shown in Figure 10. The binding affinity decreases as the urea concentration is increased. It is known that in the presence of urea many proteins adopt a highly unfolded conformation in solution. Proteins of multiple subunits are likely to be separated into their constituent polypeptide chains. Such changes in the tertiary and quaternary structure of proteins by urea is believed to be due to its effect on the hydrophobic interactions in the protein, which results in the exposure of the nonpolar groups in the interior of the protein to the solvent. This in fact is reflected in the fluorescence

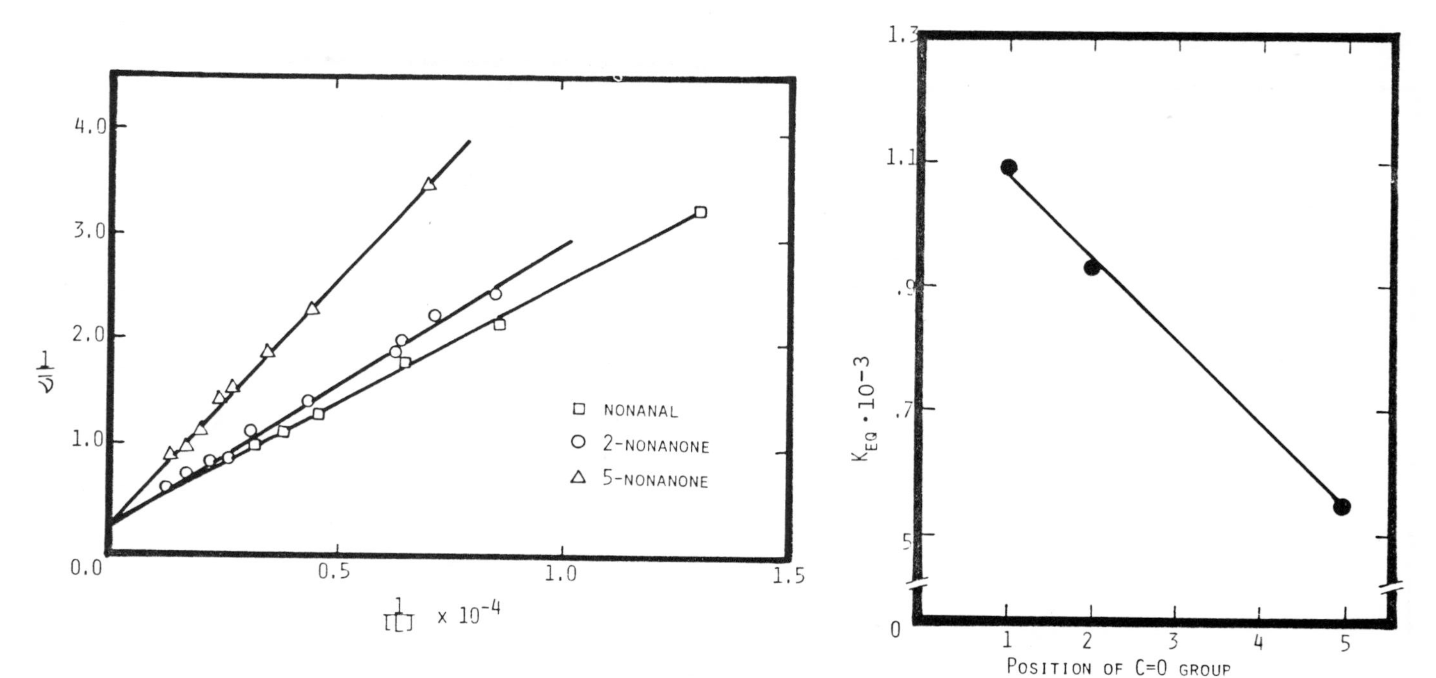

FIGURE 8a. Binding of Nonanal (□), 2-Nonanone (O) and 5-Nonanone (Δ) by Soy Protein at pH 8.0, 0.03M Tris-HCl Buffer and 25°C (Klotz Plot)

FIGURE 8b. Relationship Between the Position of the Keto Group and Binding Constant of a 9 Carbon Ligand by Soy Protein

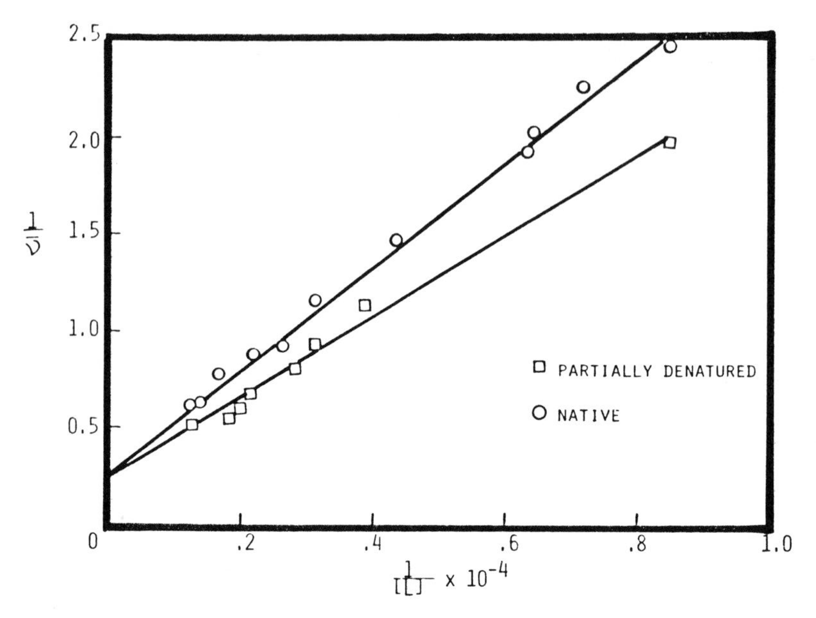

FIGURE 9. Effect of Partial Denaturation of Soy Protein on the Binding of 2-Nonanone.

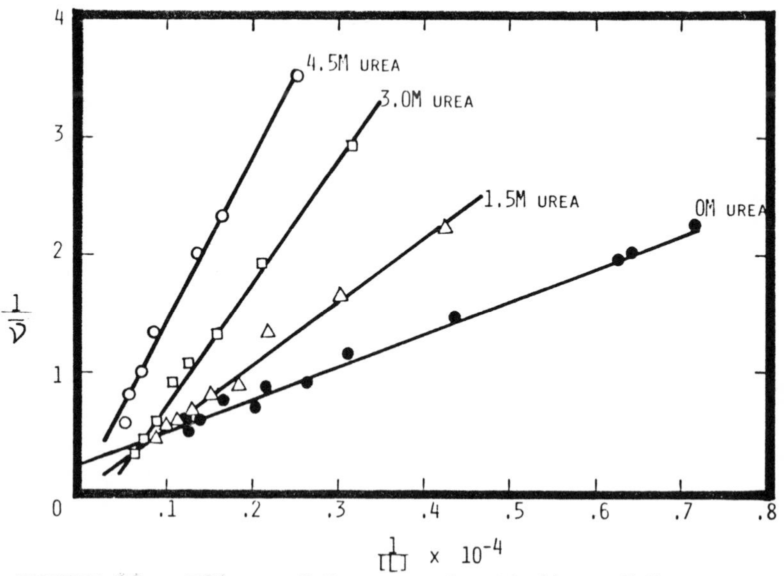

FIGURE 10. Effect of Urea on the Binding of 2-Nonanone by Soy Protein at pH 8.0, 0.03M Tris-HCl Buffer and 25°C

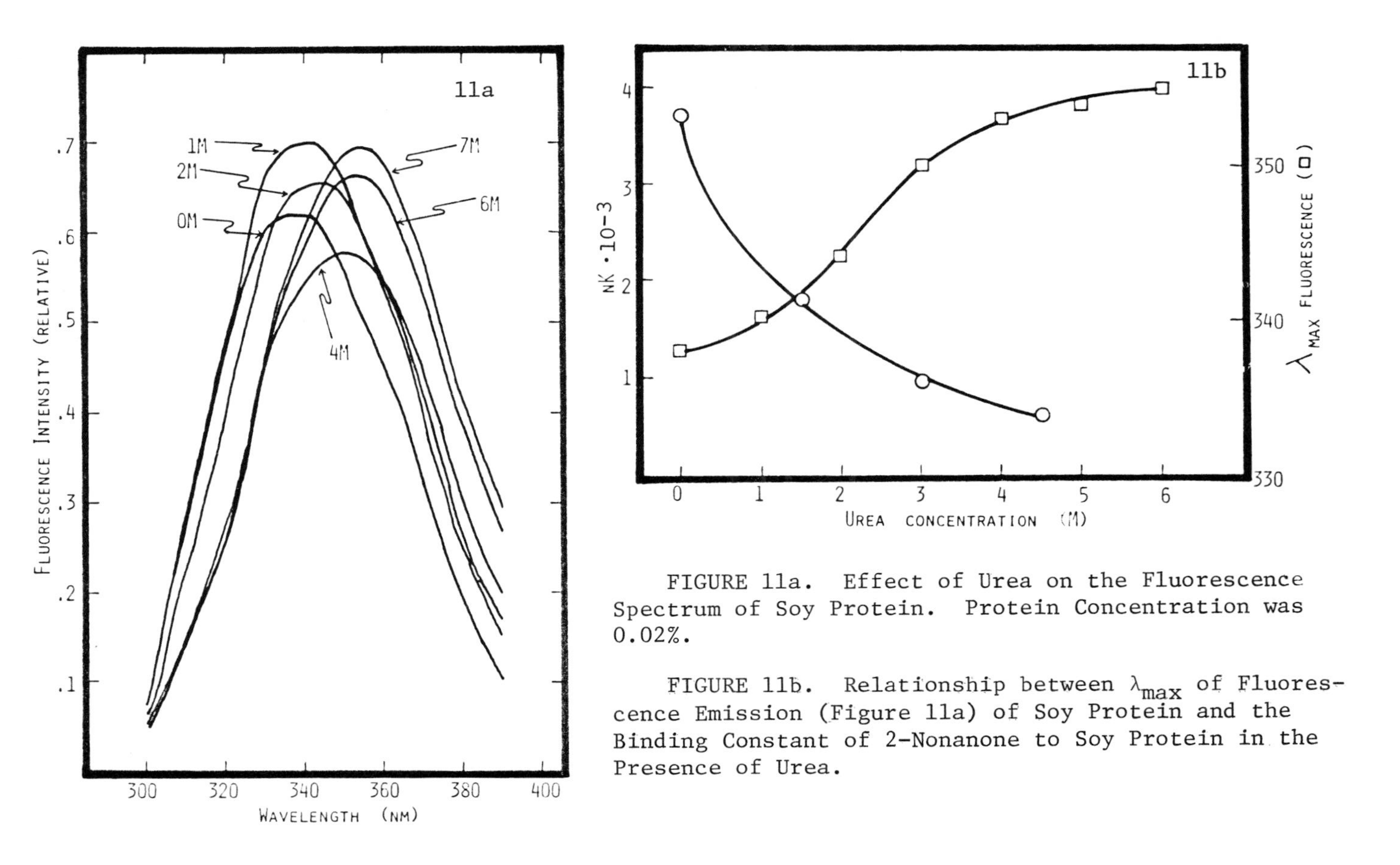

FIGURE 11a. Effect of Urea on the Fluorescence Spectrum of Soy Protein. Protein Concentration was 0.02%.

FIGURE 11b. Relationship between λ_{max} of Fluorescence Emission (Figure 11a) of Soy Protein and the Binding Constant of 2-Nonanone to Soy Protein in the Presence of Urea.

emission spectra of soy protein (Figure 11a). The λ_{max} of fluorescence emission undergoes a red shift as the urea concentration is increased up to 4M. This is due to the unfolding of the protein which exposes the tryptophan residues from the interior of the protein to the aqueous environment. In the event of such structural alterations in the protein by urea, one would expect a decreased interaction between the flavor ligand and the protein. The correlation between the binding affinity of 2-nonanone and the structural changes in soy protein in the presence of urea is shown in Figure 11b. The structural changes in the protein is expressed in the form of shift in the λ_{max} of fluorescence emission. While the λ_{max} of fluorescence emission increases with urea concentration, there is a concomitant decrease in the binding affinity of 2-nonanone to soy protein. The binding affinity decreases by about 50% in the presence of 1.5M urea. The above data suggests that treatment of the soy flour or soy isolate with 2M urea, followed by dialysis to remove urea may help remove the off-flavors from soy protein.

The effect of succinylation of soy protein on the binding of 2-nonanone is shown in Figure 12a. The intercepts in Figure 12a suggest that there are about two binding sites in succinylated soy protein compared to four in the native soy protein (on the basis of M.W. = 100,000). But the intrinsic binding constants for the native and succinylated soy proteins are almost identical, i.e. 930 M^{-1} and 850 M^{-1} respectively. This suggests that the conformational changes in soy protein upon succinylation destroys half of the binding sites originally present. Such conformational changes is clearly reflected in the fluorescence emission spectrum of the succinylated soy protein (Fig. 12b), in which the λ_{max} of fluorescence emission is shifted to a longer wavelength. But, the other half of the binding sites are left intact having the same intrinsic binding constant for 2-nonanone. However, it is obvious that the elimination of the binding sites by succinylation would decrease the binding capacity, i.e. the total amount of ligand bound per mole of protein. In other words, succinylation of soy protein should result in the removal of the off-flavors in soy to a considerable extent. This in fact has been observed in our earlier studies (78).

VI. CONCLUSIONS

The problem of the elimination of off-flavors in soy and other vegetable proteins is a challenging one that has no obvious, all-purpose, practical solution. In keeping with conventional practices steam deodorization and/or solvent extraction are effective in minimizing off-flavors where proteins are used

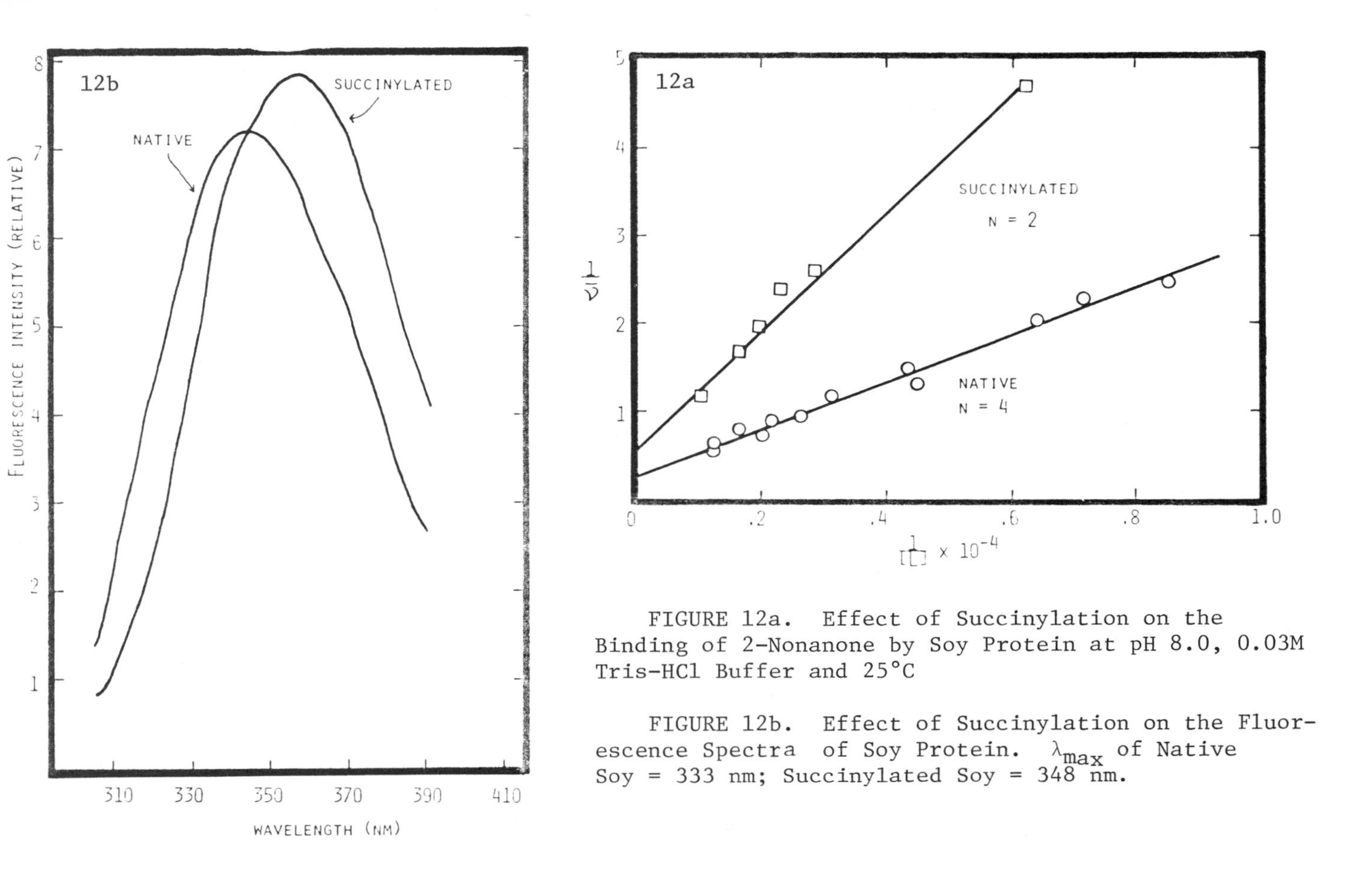

FIGURE 12a. Effect of Succinylation on the Binding of 2-Nonanone by Soy Protein at pH 8.0, 0.03M Tris-HCl Buffer and 25°C

FIGURE 12b. Effect of Succinylation on the Fluorescence Spectra of Soy Protein. λ_{max} of Native Soy = 333 nm; Succinylated Soy = 348 nm.

in applications not requiring soluble proteins e.g. texturized products or where proteins are subsequently modified by chemical or enzymatic processes.

Because lipoxygenase is a major cause of off-flavors, research to develop specific inhibitors of this enzyme that are compatible with food uses is needed. On a more fundamental level research to develop soybean strains that lack lipoxygenase (if such seeds are viable) should be conducted. Currently more information concerning the nature of flavor/protein binding should be obtained to determine if effective mild procedures can be developed to facilitate dissociation of the flavors from the proteins e.g. use of urea, succinylation or chaotropic anions.

In conclusion, the off-flavor problem in vegetable proteins as epitomized in several soy protein preparations, is a major challenge, the solutions to which should be forthcoming when more information becomes available. It behoves appropriate industry concerns and agencies of the federal government in cooperation with the research community (academic, institutional and industrial) to support the basic research needed to solve this critical problem in limiting the full exploitation of this important natural resource.

References

1. Kalbrenner, J. E., Eldridge, A. C., Moser, H. and Wolf, W., Cer. Chem. 48, 595, 1971.
2. Rackis, J., Sessa, D. and Honig, D., J. Am. Oil Chem. Soc. 56, 262, 1979.
3. Johnson, K. W. in Proc. of World Soybean Res. Conf., ed. L. D. Hill, Interstate Pub. IL, 1976.
4. Meyer, E. W. and Williams, L. D. in Proc. of World Soybean Res. Conf., ed. L. D. Hill, Interstate Pub. IL, 1976.
5. Wolf, W. J., J. Ag. Food Chem. 23, 138, 1975.
6. Mustakas, G. C., Albrecht, W. J., McGhee, J. E., Black, L. T., Bookwalter, G. N. and Griffin, E. L., J. Am. Oil Chem. Soc., 46, 623, 1969.
7. Hand, D. B., USDA Bulletin ARS 71-35, May 1967, p. 70.
8. Mustakas, G. C., Albrecht, W. J., Bookwalter, G. N., McGhee, J. E., Kwolek, W. F. and Griffin, E. L. Jr., Food Technol. 24, 1290, 1970.
9. Sessa, D. J., Honig, D. H. and Rackis, J. J., Cer. Chem. 46, 675, 1969.
10. Wolf, W. J. and Cowan, J. C., Soybeans as a Food Source, CRC Press, Inc., Miami, Florida, 1977.
11. Sessa, D. J. and Rackis, J. J., J. Am. Oil Chem. Soc. 54, 468, 1977.

12. Quist, I. H. and Von Sydow, D. C., J. Ag. Food Chem. 22, 1077, 1974.
13. Wilkens, W. F. and Lin, F. M., J. Ag. Food Chem. 18, 333, 1970.
14. Kalbrenner, J. E., Warner, K. and Eldridge, A. C., Cer. Chem. 51, 406, 1974.
15. Rackis, J. J., Honig, D. H., Sessa, D. J. and Moser, H. A., Cer. Chem. 49, 586, 1972.
16. Sessa, D., J. Ag. Food Chem. 27, 234, 1979.
17. St. Angelo, A. J. and Ory, R. L., J. Ag. Food Chem. 23, 141, 1975.
18. Veldink, G. A., Vliegenthart, J.F.G. and Boldingh, J., Prog. Chem. Fats Other Lipids 15, 131, 1977.
19. Galliard, T. and Phillips, D., Biochem. J. 124, 431 1971.
20. Pinsky, A., Grossman, S. and Trop, M., J. Food Sci. 36, 571, 1971.
21. Christopher, J. P., Pistorius, E. and Axelrod, B., Biochim. Biophys. Acta 12, 198, 1970.
22. Axelrod, B., Adv. Chem. Ser. 324, 1974.
23. Fischer, K. H. and Grosch, W., Lebensmitt. Unters-Forsch 165, 137, 1977.
24. Grosch, W. and Laskawy, G., J. Ag. Food Chem. 23, 766, 791, 1975.
25. Finazzi-Agro, A., Veldink, G., Vliengenthart, J. and Boldingh, J., Biochim. Biophys. Acta 36, 462, 1974.
26. Leu, K., Lebensmitt. Wiss. U. Tech. 7, 98, 1974.
27. Heimann, W., Franzen, K. and Rapp, A., Lebensmitt. Unters-Forsch 158, 65, 1975.
28. Heimann, W. and Dressen, P., Helv. Chim. Acta 56, 463, 1973.
29. Evans, C. D., Frankel, E. N., Cooney, P. and Moser, H., J. Am. Oil Chem. Soc. 37, 452, 1960.
30. Bauer, C. and Grosch, W., Lebensmitt. Unters-Forsch 165, 82, 1977.
31. Sessa, D., Warner, K. and Rackis, J. J., J. Ag. Food Chem. 24, 16, 1976.
32. Maga, J. A., Crit. Rev. Food Sci. & Nutr. 10, 323, 1978.
33. Arai, S., Suzuki, H., Fujimaki, M. and Sakurai, Y., Ag. Biol. Chem. 30, 364, 1966.
34. Rackis, J., Sessa, D. and Honig, D., USDA Report ARS 71-35, 1967.
35. Greuell, E. H., Chemisch Weikblad 70, 17, 1964.
36. Maga, J., Lebensmit. und Technol. 10, 100, 1977.
37. Van den Ouweland, G. and Schutte, L. in Flavors of Foods and Beverages, eds. Charalambous and Inglett, Acad. Press, NY, 1978, p. 34.
38. Borhan, M. and Snyder, H. E., J. Food Sci. 44, 586, 1979.

39. Farkas, D. F. and Goldblith, S. A., J. Food Sci. 27, 262, 1962.
40. Svensson, S. G. and Eriksson, C. E., Lebens Wiss Technol. 5, 124, 1972.
41. Rackis, J. in Enzymes in Food and Beverage Processing, p. 251, American Chemical Society, 1979.
42. Johnson, K. W. and Snyder, H. E., J. Food Sci. 43, 349, 1978.
43. Wilkins, W., Mattick, L. and Hand, D., Food Tech. 21, 1630, 1967.
44. Nelson, A. I., Wei, L. S. and Steinberg, M. P., Soybean Dig., 31(3), 32, 1971.
45. Hafner, F. H., Cer. Sci. Today 9, 164, 1964.
46. Kon, S., Wagner, J. R., Guadagni, D. G. and Horvat, R. J., J. Food Sci. 35, 343, 1970.
47. Paulsen, T. W., U. S. Patent 3,100,709, 1963.
48. Gardner, H., Lkeiman, R. and Inglett, G. E., Lipids 12, 655, 1977.
49. Mitsuda, H., Yasumoto, Y. and Yamamoto, A., Arch. Biochem. Biophys. 118, 664, 1967.
50. Eldridge, A. C., Warner, K. and Wolf, W. J., Cer. Chem. 54, 1229, 1977.
51. Siddiqui, A. M. and Tapel, A. C., Arch. Biochem. Biophys. 60, 91, 1956.
52. Yasumoto, K., Yamamoto, A. and Mitsuda, H., Ag. Biol. Chem. 34, 1162, 1970.
53. Wolf, W. J. and Cowan, J. C. in Soybeans as a Food Source, CRC Press, Miami, Florida, 1977.
54. Smith, A. K. and Circle, S. J., in Soybeans: Chemistry and Technology, AVI Pub. Co., Westport, CT, 1978.
55. Meyer, E. W., J. Am. Oil Chem. Soc., 48, 484, 1971.
56. Cowan, J. C., Rackis, J. J. and Wolf, W. J., J. Am. Oil Chem. Soc. 50, 425, 1973.
57. Eldridge, A. C., Kalbrenner, J. E., Moser, H. A., Honig, D. H., Rackis, J. J. and Wolf, W. J., Cer. Chem. 48, 640, 1971.
58. Honig, D. H., Sessa, D. J., Hoffmann, R. L. and Rackis, J. J., Food Technol. 23, 95, 1969.
59. Rackis, J. J., McGhee, J. E. and Honig, D., J. Am. Oil Chem. Soc. 52, 249, 1975.
60. Baker, E. C., Mustakas, G. C. and Warner, K. A., J. Ag. Food Chem. 27, 971, 1979.
61. Kinsella, J. E., Crit. Rev. Food Sci. and Nutr. 10, 147, 1978.
62. Kinsella, J. E., Crit. Rev. Food Sci. and Nutr. 219, 1976.
63. Kinsella, J. E., J. Am. Oil Chem. Soc. 56, 242, 1979.
64. Eriksson, C. E., Lundgren, D. and Valentine, K., Chem. Senses and Flavor 2, 3, 1976.

65. Schmidt, R. H., Farron, L. K., Bateh, R. T. and Arujo, P. E., J. Food Prot. 42, 778, 1979.
66. Takahashi, N., Sasaki, R. and Chiba, H., Ag. Biol. Chem. 43, 2557, 1979.
67. Arai, S., Yamashita, M. and Fujimaki, M., Cer. Foods World 20, 107, 1975.
68. Nawar, W. W., J. Ag. Food Chem. 19, 1057, 1971.
69. Gremli, H. A., J. Am. Oil Chem. Soc. 51, 95A, 1974.
70. Franzen, K. L. and Kinsella, J. E., J. Ag. Food Chem. 22, 675, 1974.
71. Arai, S., Noguchi, M., Koji, M., Kato, H. and Fujimaki, M., Ag. Biol. Chem. 34, 1420, 1970.
72. Beyeler, M. and Solms, J., Lebensmitt. Wiss. U. Technol. 7, 217, 1974.
73. Scatchard, G., Ann. NY Acad. Sci. 51, 660, 1949.
74. Ray, A., Reynolds, J. A., Polet, H. and Steinhardt, J., Biochem. 5, 2606, 1966.
75. Damodaran, S. and Kinsella, J. E., J. Ag. Food Chem. 28, 567, 1980.
76. Ali Mohammadzedeh-K., Feeney, R. E., Samuels, R. B., and Smith, L. M., Biochim. Biophys. Acta 147, 583, 1967.
77. Spector, A. A., Kathryn, J. and Fletcher, J. E., J. Lipid Res. 10, 56, 1969.
78. Franzen, K. and Kinsella, J. E., Ag. and Food Chem. 24, 788, 1976.
79. Blain, J. A. and Shearer, G., J. Sci. Food Ag. 16, 373, 1965.

THE BITTER FLAVOR DUE TO PEPTIDES OR PROTEIN HYDROLYSATES AND ITS CONTROL BY BITTERNESS-MASKING WITH ACIDIC OLIGOPEPTIDES

Soichi Arai

Department of Agricultural Chemistry
University of Tokyo
Bunkyo-ku, Tokyo 113
Japan

I. INTRODUCTION

An old Japanese proverb says, "A good drug tastes bitter". This remind us of quinine, strychinine, etc. as important ingredients of common drugs. Besides alkaloids, a variety of bitter substances occur naturally. These are conventionally classified into terpenoids, flavonoids, glycosides, thiol compounds, amino compounds, minerals and so forth. Some classes of bitter substances are produced through food processing, particularly through heating or roasting. Many foods and beverages are characterized by the bitter flavors that result from such an artificial treatment.

Peptide is added to the above classes of bitter substances. In most cases, bitter peptides occur in protease-treated proteins as well as in fermented foods. The bitter flavor due to peptides is thus recognized as of a flavor produced by food processing.

The history of bitter peptide research dates back to 1950's. A report presented by Murray and Baker (1) seems to be the first that deals with this regard. Shortly after, Raadsveld (2) reported that a bitter flavor of cheese is due primarily to some peptide. A number of related papers followed and, as discussed below, a great deal of information has been presented from independent laboratories in the field of dairy science.

Bitter peptides are formed not only from milk proteins but also from many other proteins by treatment with proteases. However, no examples seem to have been provided as to inten-

ISBN 0-12-169065-2

tional production of such bitter peptides and their positive use in food processing. Rather, a close attention is being paid to how to minimize, remove or mask the bitter flavor that arose unintentionally during enzymatic hydrolysis of proteins.

The present paper reviews the present state of knowledge in this field of research, with some emphasis on our finding that several of acidic oligopeptides have a bitterness-masking activity and could be used as anti-bitter peptides to minimize the bitterness of foods and beverages and even drugs.

It is noted that amino acid residues (other than glycine residue) given in this chapter are all in L-form, unless otherwise specified.

II. BITTER PEPTIDES: OCCURRENCE, ISOLATION AND IDENTIFICATION

Proteases are used in food processing for rarious purposes. Among these is a purpose of deodorizing proteins for food use (3). Our group at the University of Tokyo used pepsin and other proteases to remove the beany flavor from soy protein preparations (4-6). Although the deodorization proceeded during the treatment with pepsin, unintentional formation of a bitter flavor resulted. A number of peptides were isolated as main factors responsible for the bitterness of a peptic hydrolysate of soy globulin and several of them were identified (Table I). It is characteristic that most bitter peptides identified bear leucine residues at the C-terminals. The paper (7) among a series of our papers (7 - 10) dealing with this relevance seems to be the first that discusses amino acid sequences of bitter peptides.

Subsequently, a Kyoto University group presented papers (11, 12) on bitter peptides occurring in a tryptic hydrolysate of casein (Table I). Among these is an extremely bitter peptide having the following sequence:

Gly-Pro-Phe-Pro-Ile-Ile-Val.

This coincides exactly with the partial sequence Gly^{203} - Val^{209} found in bovine β-casein (13). Independent groups have made similar studies on isolation and identification of bitter peptides occurring in protease-treated casein (14-16) and in several types of cheese (17, 18), with the results summarized in Table I.

Other food proteins also are treated with proteases particularly for improvement of their functional properties such as solubility, dispersibility, etc. The solubilization of, for example, fish protein concentrate (FPC) with proteases, however, is often accompanied by the formation of a bitter flavor (19, 20). This is apparently due to oligopeptides (20, 21). Enzymatic hydrolysis of seed proteins offers another example; a series of bitter oligopeptides have been identified in a peptic hydrolysate of zein (22). The data are added in Table I.

Table I. Bitter Peptides Identified in Protein Hydrolysates and Cheese Products

Source	Peptides
Soy globulin hydrolyzed with pepsin (7- 10)	Arg-Leu, Gly-Leu, Leu-Lys, Leu-Phe, Phe-Leu, Arg-Leu-Leu, Tyr-Phe-Leu, Gln-Tyr-Phe-Leu, Ser-Lys-Gly-Leu, Phe-(Ile, Leu_2)-Gln-Gly-Val, Pyr-Gly-Ser-Ala-Ile-Phe-Val-Leu,[a] Phe-(Arg, Asp_2, Gln_2, Gly, Ile, Leu, Lys_2, Pro, Ser, Thr)-Trp-(Ala, Arg, Asp, Gly, Val)-Gln-Tyr-Phe-Leu
Zein hydrolyzed with pepsin (22)	Ala-Ile-Ala, Ala-Ala-Leu, Gly-Ala-Leu, Leu-Gln-Leu, Leu-Glu-Leu, Leu-Val-Leu, Leu-Pro-Phe-Asn-Gln-Leu, Leu-Pro-Phe-Ser-Gln-Leu
Casein hydrolyzed with trypsin (11, 12)	Gly-Pro-Phe-Pro-Ile-Ile-Val, Phe-Ala-Leu-Pro-Gln-Tyr-Leu-Lys, Phe-Phe-Val-Ala-Pro-Phe-Pro-Gln-Val-Phe-Gly-Lys
Casein hydrolyzed with subtilisin (14, 15)	Arg-Gly-Pro-Pro-Phe-Ile-Val, Leu-Val-Pro-Arg-Tyr-Phe-Gly, Val-Tyr-Pro-Phe-Pro-Pro-Gly-Ile-Asn-His
Casein hydrolyzed with papain (16)	Ala-Gln-Thr-Gln-Ser-Leu-Val-Tyr-Pro-Phe-Pro-Gly-Pro-Ile-Pro-Asn-Ser-Leu-Pro-Gln-Asn-Ile-Pro-Pro-Leu-Thr-Gln
Butterkaese (17)	Pro-Phe-Pro-Gly-Pro-Ile-Pro-Asn-Ser
Cheddar cheese (18)	Pro-Phe-Pro-Gly-Ile-Pro, Pro-Phe-Pro-Gly-Pro-Ile-Pro-Asn-Ser, Gln-Asp-Lys-Ile-His-Pro-Phe-Ala-Gln-Thr-Gln-Ser-Leu-Val-Tyr-Pro-Phe-Pro-Gly-Pro-Ile-Pro

[a] "Pyr" refers to pyroglutamyl.

III. CHEMICALLY SYNTHESIZED BITTER PEPTIDES

A Hiroshima University group has been carrying out elaborative experiments of synthesizing a large number of oligopeptides in order to find out a general rule of their structure-bitterness relationship. Table II, adapted from the papers (23 - 27) presented by this group, is a list of typical bitter oligopeptides synthesized. Generally speaking, these are characterized by high contents of hydrophobic amino acid residues. Amino acid sequence also contributes to the bitterness in several cases. This has been confirmed by comparing threshold values of tri-, tetra- and pentapeptides constituted with glycine and leucine residues (Table III). A common rule found here is that any one of these peptides tastes distinctly bitter when it bears a leucine residue at the C-terminal (23). Basic amino acid residues as well may contribute to the bitterness in many cases. To elucidate this point a recent study by the Hiroshima University group is directed toward correlating the structures of Arg-Gly-Pro-Pro-Phe-Ile-Val (14) and its analogues with their bitterness (27). The research group postulates that any of these peptides has both binding and stimulating sites from a molecular aspect. The former is highly hydrophobic as a whole, functioning to bind with a taste receptor. On the other hand, the terminal structure constituted with the arginine residue may contribute to giving a taste stimulus after the binding is completed.

IV. DEGRADATION OR REMOVAL OF BITTER PEPTIDES

The present state of knowledge does not seem enough to define how the amino acid sequence or the location of a particular amino acid residue determines the bitterness. In many cases, however, the importance of a hydrophobic C-terminal structure can be emphasized. Its degradation would therefore result in debittering. Arai et al. (28) tried to apply carboxypeptidase A and *Aspergillus* acid carboxypeptidase to a bitter protein hydrolysate, with the result that its bitter flavor markedly decreases as the enzymatic reaction proceeded to degrade the C-terminal structures of constituent peptides.

Measurement of an overall hydrophobicity of a given oligopeptide can help in discriminating whether or not the peptide tastes bitter. The proposal of the so-called "Q-value" (29) has permitted this discrimination in a simple manner. Independent groups also have put forward the importance of the hydrophobicity, finding a method of removing the bitter flavor due to peptides by means of hydrophobic chromatography (30,31).

Table II. Chemically Synthesized Bitter Peptides (23 - 27)

Dipeptides:

Ala-Phe, Ala-Pro, Ala-Trp, Arg-Arg, Arg-Gly, Arg-Phe, Arg-Pro, Gly-Arg, Gly-Ile, Gly-Leu, Gly-Met, Gly-Phe, Gly-Pro, Gly-Tyr, Gly-Trp, Leu-Gly, Leu-Leu, DL-Leu-DL-Leu, Leu-Phe, Leu-Pro, Leu-Tyr, Leu-Val, Lys-Ala, Lys-Gly, Phe-Gly, DL-Phe-DL-Leu, D-Phe-Leu, Phe-Phe, Phe-Pro, Phe-D-Pro, Pro-Ala, Pro-Arg, Pro-Gly, Pro-Ile, Pro-Leu, Pro-Phe, Pro-D-Phe, Pro-Pro, Trp-Gly, Tyr-Tyr, Val-Ala, Val-Leu, Val-Val

Tripeptides:

Arg-Gly-Pro, Arg-Pro-Phe, Gly-Gly-Leu, Gly-Gly-Phe, Gly-Leu-Gly, Gly-Leu-Leu, Gly-Phe-Gly, Gly-Phe-Phe, Gly-Phe-Pro, Gly-Pro-Gly, Gly-Tyr-Gly, Gly-Tyr-Tyr, Leu-Gly-Gly, Leu-Gly-Leu, Leu-Leu-Gly, Leu-Leu-Leu, Phe-Gly-Gly, Phe-Gly-Phe, Phe-Ile-Val, Phe-Phe-Gly, Phe-Phe-Phe, Pro-Pro-Phe, Tyr-Gly-Gly, Tyr-Gly-Tyr

Tetrapeptides:

Gly-Gly-Gly-Leu, Gly-Gly-Leu-Gly, Gly-Leu-Gly-Gly, Gly-Pro-Pro-Phe, Leu-Gly-Gly-Gly, Leu-Leu-Leu-Leu, Pro-Phe-Ile-Val

Pentapeptides:

Arg-Gly-Pro-Pro-Phe, Gly-Gly-Gly-Gly-Leu, Gly-Gly-Gly-Leu-Gly, Gly-Gly-Leu-Gly-Gly, Gly-Leu-Gly-Gly-Gly, Leu-Gly-Gly-Gly-Gly

Hexapeptides:

Arg-Gly-Gly-Phe-Ile-Val, Arg-Pro-Pro-Phe-Ile-Val, Gly-Pro-Pro-Phe-Ile-Val, Lys-Pro-Pro-Phe-Ile-Val, Orn-Pro-Pro-Phe-Ile-Val, Phe-Pro-Pro-Phe-Ile-Val

Heptapeptides:

Arg-Gly-Pro-Pro-Gly-Gly-Val, Arg-Gly-Pro-Pro-Gly-Ile-Gly, Arg-Gly-Pro-Pro-Phe-Gly-Gly, Arg-Gly-Pro-Pro-Phe-Ile-Val, Arg-Gly-Pro-Pro-Phe-Phe-Phe, Val-Ile-Phe-Pro-Pro-Gly-Arg, Val-Tyr-Pro-Phe-Pro-Pro-Gly

Octapeptide:

Val-Tyr-Pro-Phe-Pro-Pro-Gly-Ile

Decapeptides:

Val-Tyr-Pro-Phe-Gly-Gly-Gly-Ile-Asn-His
Val-Tyr-Pro-Phe-Pro-Pro-Gly-Ile-Asn-His
Val-Tyr-Pro-Phe-Pro-Pro-Gly-Ile-Gly-Gly

Table III. A Relationship between Amino Acid Sequence of Bitter Peptides[a] and Their Threshold Value

Bitter peptide	Threshold value (mM)
Gly-Gly-Leu	10
Gly-Leu-Gly	10
Leu-Gly-Gly	75
Gly-Gly-Gly-Leu	4.5
Gly-Gly-Leu-Gly	25
Gly-Leu-Gly-Gly	19
Leu-Gly-Gly-Gly	13
Gly-Gly-Gly-Gly-Leu	2.2
Gly-Gly-Gly-Leu-Gly	> 13[b]
Gly-Gly-Leu-Gly-Gly	> 13[b]
Gly-Leu-Gly-Gly-Gly	> 13[b]
Leu-Gly-Gly-Gly-Gly	> 13[b]

[a] Dissolved in distilled water.
[b] No exact data were obtained with these pentapeptides because of their limited solubility.

V. ACIDIC OLIGOPEPTIDES: STRUCTURE AND FLAVOR

Although much has been elucidated of hydrophobic peptides with bitter flavor as described above, no information seemed available on the flavor of hydrophilic peptides until about ten years ago. Arai et al. (32, 33) first investigated the flavor of a series of α-L-glutamyl dipeptides in relation to their hydrophobicity. Two independent parameters of hydrophobicity, Δf and ΔW, were obtained. The former was calculated according to the way proposed by Tanford (34), while the latter was experimentally measured according to the theory of paper partition chromatography (35). Assuming an ideal condition, the work (W) required to move 1 mole of a solute from the aqueous phase to the hydrophobic solvent phase under the standard condition is given by

$$W = RT \ln A_2/A_1[(1/R_f) - 1]$$

where A_1 and A_2 are the cross-section areas of the aqueous and the solvent phases, respectively. For two different solutes showing Rf_1 and Rf_2, the difference, ΔW, is expresses as

$$\Delta W = RT \ln (1 - Rf_1)Rf_2/Rf_1(1 - Rf_2)$$

since the constants, R, T, A_1 and A_2 are cancelled. Applying this method of measurement to glutamyl dipeptides, Arai et al. (33) obtained a series of ΔW values. Table IV shows the ΔW values of twelve samples as well as their Δf values. It was found that these samples could be classified into three groups according to flavor (Table IV). Interestingly, the four dipeptides belonging to the least hydrophobic group do not taste bitter but have a flavor resembling that of monosodium glutamate (MSG), although their threshold values, 150 - 300 mg %, are much higher than the threshold value of MSG (30 mg %).

Fujimaki et al. (36) found that a similar MSG-like flavor occurred in FPC treated with Pronase under a controlled condition. Noguchi et al. (37) fractionated a low-molecular-weight fraction from this FPC hydrolysate and disclosed that an acidic oligopeptide fraction, though carrying no free glutamic acid, had an MSG-like flavor. A number of peptides were isolated from this fraction, several of which were determined for amino acid sequence by means of mass spectrometry (Table V). It was also confirmed that acidic oligopeptides rich in hydrophilic amino acid residues tended to have an MSG-like flavor.

Table IV. Flavors of α-L-Glutamyl Dipeptides in Relation to Hydrophobicity (33)

Dipeptide	Δf[a] (cal/mol)	ΔW[b] (cal/mol)	Flavor
Glu-Asp	< 0	-199	MSG-like
Glu-Glu	< 0	0	MSG-like
Glu-Ser	40	-47	MSG-like
Glu-Thr	440	43	MSG-like
Glu-Gly	0	0	Flat
Glu-Ala	730	299	Flat
Glu-Pro	2600	360	Flat
Glu-Val	1690	480	Flat
Glu-Tyr	2870	422	Bitter
Glu-Phe	2650	1073	Bitter
Glu-Leu	2420	1073	Bitter
Glu-Ile	2970	1170	Bitter

[a] Relative to the hydrophobicity of Glu-Gly. cf. (34).

[b] Relative to the hydrophobicity of Glu-Gly. cf. (35).

Table V. Acidic Oligopeptides Occurring in Fish Protein Concentrate Hydrolyzed with Pronase and Their Sensory Properties (37)

Acidic oligopeptide[a]	Yield[b]	Flavor	Threshold[c]
Ala-Glu	17	Flat	
Asp-Ala	28	Flat	
Asp-Gly	36	Flat	
Asp-Leu	9	Bitter	
Glu-Asp	47	MSG-like	200
Glu-Glu	78	MSG-like	150
Glu-Gly	21	Flat	
Glu-Ser	20	MSG-like	200
Ile-Asp	13	Bitter	
Ile-Glu	14	Bitter	
Ser-Asp	12	Flat	
Thr-Glu	23	MSG-like	300
Val-Asp	9	Flat	
Val-Glu	10	Flat	
Asp-Glu-Ser	5	MSG-like	300
Glu-Asp-Glu	4	MSG-like	300
Glu-Asp-Val	4	Bitter	
Glu-Gly-Ala	5	Flat	
Glu-Gly-Ser	7	MSG-like	200
Glu-Gln-Glu	6	MSG-like	300
(Glu, Ile)-Asp	5	Flat	
Ile-Glu-Glu	17	Bitter	
Ser-Glu-Glu	6	MSG-like	200
(Asp, Gly, Ser)-Glu	2	?	
(Asp, Glu, Gly)-Asp	2	?	
(Asp, Glu, Ser)-Asp	3	?	
(Glu, Ile, Leu)-Glu	4	?	
(Asp, Glu, Gly, Ser)-Asp	2	?	
(Asp, Glu, Gly, Ser)-Glu	2	?	
(Asp, Glu, Ser, Thr)-Glu	1	?	
(Asp, Glu_2, Gly, Ser)-Thr	1	?	

[a] Determined primarily by mass spectrometry after acetylation and permethylation.

[b] mg from 100 g protein.

[c] mg/100 ml water (pH 6).

VI. BITTERNESS-MASKING WITH ACIDIC OLIGOPEPTIDES

Noguchi et al. (38) carried out an experiment to obtain an acidic oligopeptide fraction in a much higher yield. For this purpose, a glutamic acid-enriched plastein (39) was used as material and hydrolyzed with Pronase. The resulting hydrolysate comprised acidic, basic and neutral oligopeptides in a weight ratio of 8:3:10, respectively. An aqueous solution of the acidic oligopeptide fraction had no bitter flavor at the concentration of 8 mg/ml, whereas that of the neutral oligopeptide fraction gave a strongly bitter flavor at the concentration of 10 mg/ml. When both solutions were mixed with each other in the same proportion, the resulting mixture no longer showed any bitterness (FIGURE 1). The acidic oligopeptide fraction used in this experiment was found to consist primarily of di-, tri-, tetra- and pentapeptides as listed in Table VI. This result indicates that some of these peptides have an ability to mask the bitterness. Arai et al. (40) confirmed that the addition of each of them to hydrolysates of soy globulin and casein was more or less effective in masking their bitter flavors. According to the effectiveness, the peptides could be classified into four groups (Table VI). The four peptides belonging to the very effective group have a potent activity for masking the bitter flavor of both protein hydrolysates. The peptides in the ineffective group were not well evaluated for the bitterness-masking activity because each has a bitter flavor of its own.

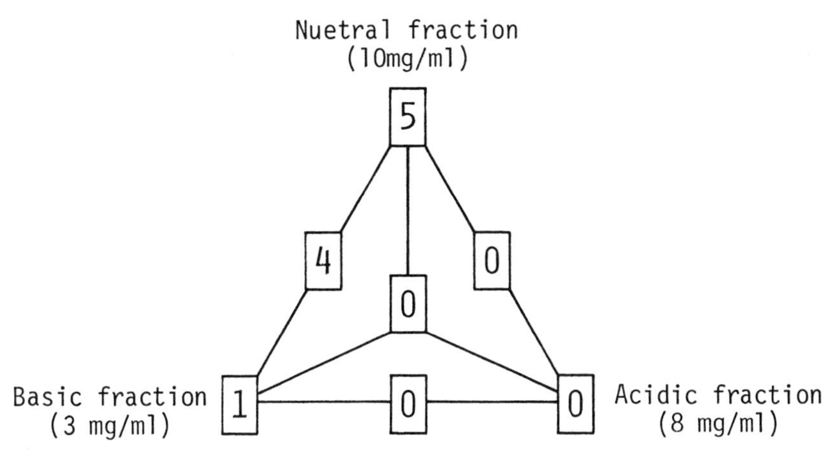

FIGURE 1. Bitter flavor scores of acidic, basic and neutral oligopeptide fractions and of their mixtures (38). Scores: 5, very bitter; 4, bitter; 3, middle; 2, weakly bitter; 1, slightly bitter; and 0, flat.

Table VI. Effects of Acidic Oligopeptides on Masking the Bitter Flavors of Protein Hydrolysates (40)

	Bitterness after addition of each peptide[a]	
Acidic oligopeptide	Peptic hydrolysate of soy globulin[b]	Tryptic hydrolysate of casein[c]
Very effective group		
Glu-Asp	±	+
Glu-Glu	±	+
Glu-Gln-Glu	+	++
Glu-Glu-Glu	+	+
Ser-Glu-Glu	+	++
Effective group[d]		
Glu-Ser	++	++
Glu-Thr	++	++
Glu-Gly-Glu	++	++
Glu-Gly-Ser	++	+++
Glu-Glu-Glu-Glu	++	++
Glu-Glu-Glu-Glu-Glu	++	++
Weakly effective group		
Glu-Gly	+++	+++
Glu-Ala-Gly	+++	++++
Glu-Ala-Ala	+++	++++
Ineffective group		
Glu-Arg	++++	+++++
Glu-Ile	++++	+++++
Glu-Leu	++++	+++++
Glu-Lys	++++	+++++
Glu-Phe	++++	+++++
Glu-Trp	++++	+++++
Glu-Tyr	++++	+++++

[a] To a protein hydrolysate solution (1 %) was added an equal volume of each peptide solution (5 mM) and the resulting mixture evaluated for bitterness. Scales: +++++, very bitter; ++++, bitter; +++, middle; ++, weakly bitter; and +, slightly bitter.

[b] For preparation see the reference (8).

[c] For preparation see the reference (11).

[d] The effect of free L-glutamic acid was comparable with that of a peptide in this group.

VII. FURTHER APPLICATIONS

Noguchi et al. (38) selected Glu-Glu as one of the most potent anti-bitter peptides and tried applying it to various bitter substances. When an aqueous solution of Glu-Glu was added to aqueous solutions of bitter substances at respective concentrations, their bitterness was apparently lessened or minimized (Table VII). The data indicate that the bitterness-masking by this peptide is a general phenomenon, applicable not only to protein hydrolysates but also to a wide range of bitter foods and others.

An attempt was made to use Glu-Glu as an additive to foods and beverages. The result is shown in Table VIII, indicating that in any case the bitter flavor can be masked to a greater or lesser extent depending on the type of sample. A similar bitterness-masking effect was observed when Glu-Glu was applied to a number of bitter ingredients of drugs (Table IX).

Table VII. The Effect of α-L-Glutamyl-L-glutamic Acid (Glu-Glu) on Masking the Bitterness of Several Selected Substances (38)

Bitter substance	Concentration (%)	Glu-Glu	Bitterness[a]
Glycyl-L-leucine	0.2	With[b]	-
		Without[c]	++++
L-Isoleucine	1.0	With	-
		Without	++++
Magnesium chloride	2.0	With	+
		Without	++++
Chlorogenic acid	0.06	With	+
		Without	++++
Caffeine	0.06	With	±
		Without	++++
Phenylthiocarbamide	0.006	With	±
		Without	+++
Brucine	3×10^{-6}	With	+
		Without	+++

[a] See Table VI.
[b] To each of the bitter substance solutions was added an equal volume of 0.1 % Glu-Glu in distilled water.
[c] Instead of the Glu-Glu solution, an equal volume of distilled water was added.

VIII. CONCLUSIONS

In most cases, enzymatic hydrolysis of proteins gives rise to a bitter flavor. Responsible factors are known to be primarily a variety of highly hydrophobic oligopeptides produced during the hydrolysis. Sometimes, basic amino acid residues of peptides also contribute to their bitterness.

There is the case that a protein hydrolysate, while containing bitter peptides, apparently shows no bitterness when adequate amounts of hydrophilic acidic oligopeptides are contained together. This may be a result of their functioning to mask the bitterness. Glu-Glu is one of the most potent anti-bitter peptides, which can be applied to masking the bitterness of foods, beverages and even drugs as well as that of protein hydrolysates.

Studies from the standpoint of taste physiology are needed to explain the mechanism of such bitterness-masking by acidic oligopeptides.

Table VIII. The Effect of α-L-Glutamyl-L-glutamic Acid (Glu-Glu) on Masking the Bitter Flavor of Selected Food Samples (41)

Sample	Glu-Glu	Bitterness[a]
Summer-orange juice (fresh)	With[b]	-
	Without[c]	++
Vegetable juice[d] (canned)	With	-
	Without	+
Extract from green tea[e]	With	+
	Without	++++
Instant coffee[d] (1 % solution)	With	+++
	Without	+++++
Cocoa[d] (2 % suspension)	With	+
	Without	+++
Caramel[d] (1 % solution)	With	+
	Without	++

[a] See Table VI.
[b] To each of the samples was added an equal volume of 0.1 % Glu-Glu in distilled water.
[c] Instead of the Glu-Glu solution, an equal volume of distilled water was added.
[d] A commercial product.
[e] A commercial brand.

Table IX. The Effect of α-L-Glutamyl-L-glutamic Acid (Glu-Glu) on Masking the Bitterness of Selected Drug Ingredients (41)

Sample	Concentration (%)	Glu-Glu	Bitterness[a]
Thyradin	1.0	With[b]	±
		Without	++
Methiocil	1.0	With	±
		Without	++
Gemonil	1.0	With	±
		Without	++
Kanamycin	1.0	With	-
		Without	+
Aminobenzyl penicillin	1.0	With	-
		Without	+
Chloramphenicol palmitate	1.0	With	+
		Without	++
Aderoxal	0.5	With	+
		Without	++
Aleviatin	0.5	With	+
		Without	++
Lincosin	0.5	With	+
		Without	+++
Acetylspiramycin	0.3	With	-
		Without	++
Erythromycin	0.3	With	-
		Without	+
Gilurytmal	0.01	With	±
		Without	++

[a] See Table VI.
[b] Crystalline Glu-Glu (2.7 mg or 0.1 m mole) was added to each sample solution (1 ml).

REFERENCES

1. Murray, T. K., and Baker, B. E., *J. Sci. Food Agric. 3,* 470 (1952).
2. Raadsveld, C. W., *Proceedings of 13th International Dairy Congress 2,* 676 (1953).
3. Fujimaki, M., Arai, S., and Yamashita, M., *in* "Food Proteins: Improvement through Chemical and Enzymatic Modification" (R. E. Feeney and J. R. Whitaker, eds.), p. 156. American Chemical Society, Washington, D. C., (1977).
4. Fujimaki, M., Kato, H., Arai, S., and Tamaki, E., *Food Technol. 22,* 889 (1968).
5. Noguchi, M., Arai, S., Kato, H., and Fujimaki, M., *J. Food Sci. 35,* 211 (1970).
6. Arai, S., Noguchi, M., Yamashita, M., Kato, H., and Fujimaki, M., *Agric. Biol. Chem. 34,* 1569 (1970).
7. Fujimaki, M., Yamashita, M., Okazawa, Y., and Arai, S., *Agric. Biol. Chem. 32,* 794 (1968).
8. Fujimaki, M., Yamashita, M., Okazawa, Y., and Arai, S., *J. Food Sci. 35,* 215 (1970).
9. Yamashita, M., Arai, S., and Fujimaki, M., *Agric. Biol. Chem. 33,* 321 (1969).
10. Arai, S., Yamashita, M., Kato, H., and Fujimaki, M., *Agric. Biol. Chem. 34,* 729 (1970).
11. Matoba, T., Nagayasu, C., Hayashi, R., and Hata, T., *Agric. Biol. Chem. 33,* 1662 (1969).
12. Matoba, T., Hayashi, R., and Hata, T., *Agric. Biol. Chem. 34,* 1235 (1970).
13. Ribadeau-Dumas, B., Bringon, G., Grosclaude, F., and Mercier, J.-C., *Eur. J. Biochem. 25,* 505 (1972).
14. Ichikawa, K., Yamamoto, T., and Fukumoto, J., *J. Agric. Chem. Soc. Japan 33,* 1044 (1959).
15. Minamiura, N., Matsumura, Y., Fukumoto, J., and Yamamoto, T., *Agric. Biol. Chem. 36,* 588 (1972).
16. Clegg, K. M., Lim, C. L., and Manson, W., *J. Dairy Res. 41,* 283 (1974).
17. Huber, L., and Klostermeyer, H., *Milchwissenschaft 29,* 449 (1974).
18. Hamilton, J. S., and Hill, R. D., *Agric. Biol. Chem. 38,* 375 (1974).
19. Fujimaki, M., Yamashita, M., Arai, S., and Kato, H., *Agric. Biol. Chem. 34,* 1325 (1970).
20. Hevia, P., Whitaker, J. R., and Olcott, H. S., *J. Agric. Food Chem. 24,* 383 (1976).
21. Hevia, P., and Olcott, H. S., *J. Agric. Food Chem. 25,* 772 (1977).
22. Wieser, H., and Belitz, H. D., *Z. Lebensm.-Unters. Forsch. 159,* 329 (1975).

23. Okai, H., Ishibashi, N., and Oka, S., *in* "Proceedings of the 8th Symposium on Peptide Chemistry" (T. Kaneko, ed.), p. 66. Protein Research Foundation, Osaka, (1971).
24. Kanehisa, H., Fukui, H., Okai, H., and Oka, S., *in* "Prodeedings of the 12th Symposium on Peptide Chemistry" (H. Yajima, ed.), p. 81. Protein Research Foundation, Osaka, (1975).
25. Kanehisa, H., Okai, H., and Oka, S., *in* "Proceedings of the 13th Symposium on Peptide Chemistry" (S. Yamada, ed.), p. 72. Protein Research Foundation, Osaka, (1976).
26. Okai, H., *in* "Peptide Chemistry" (T. Nakajima, ed.), p. 139. Protein Research Foundation, Osaka, (1977).
27. Kouge, K., Kanehisa, H., Okai, H., and Oka, S., *in* "Peptide Chemistry" (N. Izumiya, ed.), p. 105. Protein Research Foundation, Osaka, (1978).
28. Arai, S., Noguchi, M., Kurosawa, S., Kato, H., and Fujimaki, M., *J. Food Sci. 35,* 392 (1970).
29. Ney, K. H., *Z. Lebensm.-Unters. Forsch. 147,* 64 (1971).
30. Roland, J. F., Mattis, D. L., Kiang, S., and Alm, W. L., *J. Food Sci. 43,* 1491 (1978).
31. Lalasidis, G., and Sjöberg, L.-B., *J. Agric. Food Chem. 26,* 742 (1978).
32. Arai, S., Yamashita, M., and Fujimaki, M., *Agric. Biol. Chem. 36,* 1253 (1972).
33. Arai, S., Yamashita, M., Noguchi, M., and Fujimaki, M., *Agric. Biol. Chem. 37,* 151 (1973).
34. Tanford, C., *J. Amer. Chem. Soc. 84,* 4240 (1962).
35. Pardee, A. B., *J. Biol. Chem. 190,* 757 (1951).
36. Fujimaki, M., Arai, S., Yamashita, M., Kato, H., and Noguchi, M., *Agric. Biol. Chem. 37,* 2891 (1973).
37. Noguchi, M., Arai, S., Yamashita, M., Kato, H., and Fujimaki, M., *J. Agric. Food Chem. 23,* 49 (1975).
38. Noguchi, M., Yamashita, M., Arai, S., and Fujimaki, M., *J. Food Sci. 40,* 367 (1975).
39. Yamashita, M., Arai, S., Kokubo, S., Aso, K., and Fujimaki, M., *J. Agric. Food Chem. 23,* 27 (1975).
40. Arai, S., Noguchi, M., Yamashita, M., and Fujimaki, M., *in* "Peptide Chemistry" (T. Nakajima, ed.), p. 143. Protein Research Foundation, Osaka, (1977).
41. Unpublished.

TASTES AND ODORS IN PUBLIC WATER SUPPLIES

Harry W. Tracy
Steven D. Leonard

Water Quality Division
San Francisco Water Department
Millbrae, California

I. INTRODUCTION

A. *All the Water That's Fit to Drink*

The delivery of potable water free from off tastes and odors is a key to good consumer relations for a water utility. Unfortunately, despite the best efforts of water quality engineers and scientists, tastes and odors do develop. These tastes and odors elicit very definite responses from consumers, who describe them as "musty," "earthy," "vile," "grassy," "medicinal," and "astringent." It is very difficult to maintain the confidence of millions of water users when water is not palatable and is, at least in their minds, not fit to drink.

Water utilities were developed to provide water to operate the systems of urban environments and to provide the inhabitants with drinking water. Water is an important beverage in peoples' diets and is a major ingredient in food products. If water consumers do not consume the water provided them because it either tastes or smells disagreeable, then water utilities have failed in a significant portion of their mission.

As a public utility, the San Francisco Water Department is sensitive to this mission and has strived over the last forty-plus years to provide water that is palatable, as well as potable. This goal is not always easily accomplished, however. The logistics of operating a large utility serving hundreds of

ISBN 0-12-169065-2

millions of gallons per day through hundreds of miles of pipe are not always conducive to insuring good-tasting water at each customer's tap. Factors beyond the control of system operators, and unknown variables in the environment, can cause problems faster than they can be identified. Detection methods, despite the increased level of technology in analytical instrumentation, still emphasize human sensory perception. Thus, more often than not, it is the consumer who gives the water a bad rating.

There are no hard and fast scientific analyses or techniques that can fully account for the taste or smell of water. The San Francisco Water Department depends primarily on experience when confronted with a taste or odor problem. In this paper, some of that experience -- particularly as it relates to the identification and treatment of off tastes and odors -- will be related.

B. San Francisco's System

To better understand the size of the problems and their potential complexity, it is necessary to have a frame of reference vis-a-vis the San Francisco Water Department's system.

The operating plant of the San Francisco Water Department was formed by the merger of two water systems (Fig. 1). In 1937, the City and County of San Francisco purchased the Spring Valley Water Company. At the same time, San Francisco was completing the Hetch Hetchy Water and Power System, which was to deliver Sierra Nevada water from Hetch Hetchy reservoir to San Francisco by gravity. Today, this combined system serves over one million suburban consumers in Alameda, Santa Clara, and San Mateo counties, as well as more than 700,000 consumers in San Francisco.

Most of the system's water (74%) originates at Hetch Hetchy reservoir in Yosemite National Park. Six local impoundments, the heart of the old Spring Valley system, provide the rest of the water. Combined, the two systems can ultimately produce in excess of 450 million gallons per day (MGD), with an average production of 230 MGD for the calendar year 1979.

Water quality varies sharply between the Hetch Hetchy supply and the local reservoirs. Hetch Hetchy is of excellent quality: it is soft (hardness 4-5 mg/l as $CaCO_3$) and meets the 1975 Environmental Protection Agency (EPA) Primary Drinking Water Standards with only chlorination and pH

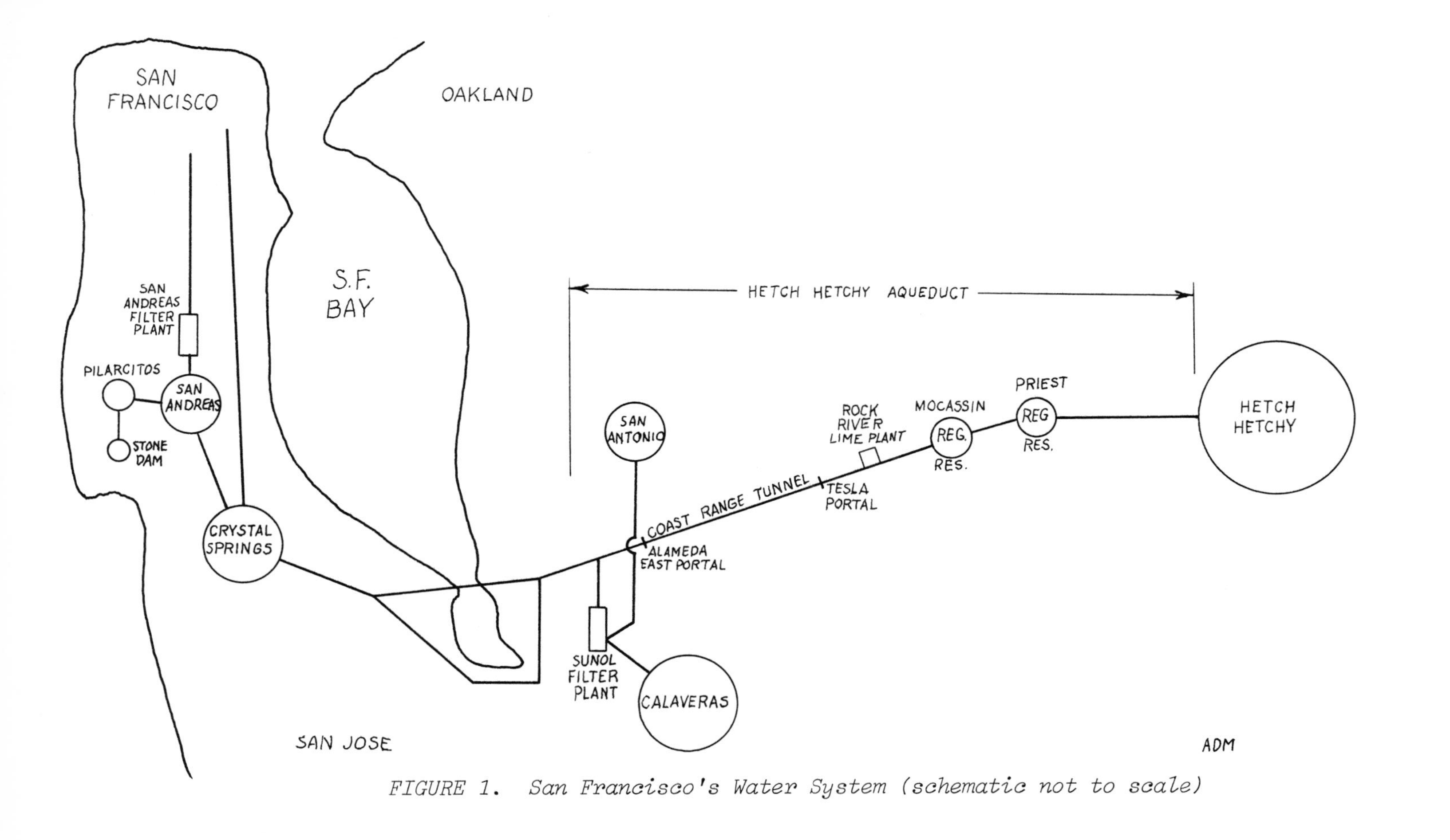

FIGURE 1. San Francisco's Water System (schematic not to scale)

adjustment (using quicklime, CaO). The local supplies vary chemically (hardness 45-100 mg/1) and generally have lower water quality characteristics.

The most obvious physical differences between the local sources and Sierra sources are the sediment loads and levels of biological activity. The runoff in local watersheds generates large amounts of suspended organic materials and sediment, which in turn cause high turbidities in the impounded waters. In addition, seasonal growths of algae are common locally. Both of these factors have dictated the building of two filtration plants to treat these waters. The two plants (160 MGD and 80 MGD capacities for San Antonio-Calaveras and San Andreas, respectively) provide full treatment (flocculation, sedimentation, and filtration), including the capability of using powdered activated carbon.

Following treatment, the water leaves the treatment plants and is distributed to smaller utilities, which deliver the water to their customers through their own distribution systems. Some of these utilities perform additional treatment, such as rechlorination or fluoridation. San Francisco does not sample these systems, nor does it provide treatment or operational support beyond the meter but does assist utilities with their system water quality problems.

The San Francisco Water Department distributes water within San Francisco and has full control of water distribution operations, water quality sampling, and treatment within San Francisco.

II. IDENTIFYING TASTES AND ODORS

A. *Complaints*

Through the years, the Department's primary source of information regarding tastes and odors has come in the form of consumer complaints. (See Table I, which lists the most common complaints received by utilities in the United States and Canada.) The causes, where identified, are varied and have implicated everything from microorganisms in reservoirs to poor operation of mains. Most of the obvious problems have seldom been repeated, but others are difficult to control when they sporadically occur.

TABLE I. Causes of Consumer Complaints (United States and Canada) (1)

Complaint	Ground Waters	Ground and Surface Waters	Surface Waters
	Complaints %	Complaints %	Complaints %
Chlorine	0.1-100	1-100	0.1- 99
Monochloramine	2		4- 90
Dichloramine		2- 10	4- 50
Nitrogen trichloride		5- 10	5
Organic chloramines	5	5- 20	1- 90
Chlorophenolic bodies	5	2- 40	2- 90
Algae		1- 80	1-100
Actinomycetes		1- 70	1- 80
Iron bacteria and corrosion products	1-100	1- 90	1-100
Sulfur bacteria and reaction products	2- 10	5- 60	1- 13
Hydrogen sulfide		2- 10	
Stale water at dead ends			5- 30
Corrosion products		20	5 & 11
Undetermined causes			5
Unjustified complaints			30

B. Testing for Tastes and Odors

Ideally, a utility would like all tastes and odors to be caused by an identifiable source in the raw water or in the treatment process. In these instances, the operator has a chance to identify the problem and take remedial action. Too often, this is not the case.

The Department's experience is that tastes, especially those caused by metallic ions, develop as corrosion products in the distribution network and are not typically found at the treatment stage. Since taste is seldom the problem in treatment, the emphasis in the plant is on detecting odors. Thus, San Francisco relies heavily on the sensory perception of its operators to identify problems at the treatment stage.

The operators continually test the raw and treated water in the plant for odors caused by problems in raw water. Odor problems can develop in all seasons. Raw water is monitored on a weekly basis by the Department's laboratory staff and on a daily basis by the filtration plant operators for physical characteristics, including odor. The operators use the Threshold Odor Number (TON) procedure outlined in Standard Methods of Water and Wastewater Analysis (15th edition) (2); this procedure indicates the number of dilutions of an odiferous sample with odor-free water necessary just to extinguish the odor. This procedure and the similar TON and OII (odor intensity index) procedures outlined in ASTM Standards (Part 31) (3) are applicable primarily to gross problems.

In practice, logistical problems hamper this type of testing. The Standard Methods procedure suggests trial panels of five to ten judges. Such panels are simply impractical in water treatment plants. Instead, the normal procedure is to have the water examined by one operator, who elicits additional opinions in suspect samples. If an obvious odor problem is detected in the plant, the operators can then take corrective measures. One drawback to this procedure is that personnel who work in a laboratory or near a water treatment plant frequently become desensitized to off odors because of their exposure to treatment or laboratory chemicals.

Often, odor problems develop not in the plant, but in the transmissions mains or in the distribution reservoirs after the treatment process. In these cases, the Department relies on field samples and consumer complaints to identify problems. The best judge of an odor problem is usually the consumer who has been habituated to the water in his or her local area. In investigating these complaints, the investigator attempts to ascertain sufficient information to begin to determine the source of the problem. Since most consumers receive a mixture of waters from various reservoirs, the source is seldom identified easily; in fact, the distribution system is so complex that it is not unusual for next-door neighbors to be served from different sources.

Even with the recent advances in analytical instrumentation, San Francisco still depends on human senses for analysis. The gas chromatographic techniques that have been used successfully to isolate and identify odor-causing chemicals still have limited applicability. The logistics of an operating system dictate that the testing be continuous and rapid enough to identify odiferous water before it is delivered.

Some waterworks do operate chemical detection systems in conjunction with their treatment processes. Since 1967, Kansas City has been running daily gas chromatographs on its raw water in addition to routine taste and odor measurements (4). The data are analyzed by a computer to provide information on previous episodes and successful treatment strategies. This approach seems to have more potential uses on river sources contaminated by sewage, industrial pollutants, and agricultural wastes than on natural waters; for naturally-occurring odors, the human nose can best detect the low levels of chemicals or groups of chemicals involved.

Work in recent years has led to isolation and identification of two odor-producing chemicals produced by microorganisms (5). These microorganisms -- geosmin and 2-methylisoborneal (Fig. 2) -- are responsible primarily for woody and earthy odors.

III. CAUSES OF OFF TASTES AND ODORS

A. *Impounded Water*

The majority of San Francisco's watersheds are owned by the City and County of San Francisco. The remainder are privately-owned grazing lands and national park lands. There is very little chance of man-made or industrial pollutants entering these supplies. For these reasons, the Department has little experience with raw water pollutants. Other utilities, however, have experienced problems from industrial products which have extremely low detection levels. Table II shows the taste and odor threshold levels for several common industrial pollutants.

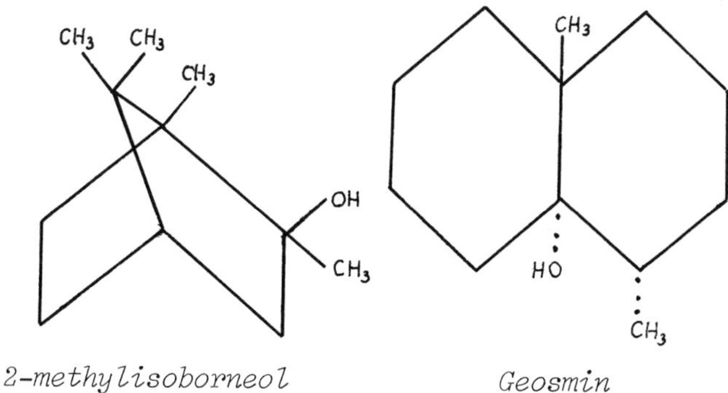

FIGURE 2. Natural taste and odor compounds(5)

TABLE II. Concentration of Some Chemicals Causing Taste and Odor Problems (6)

Chemical	*Concentration Detectable (ppb)*[a]
Formaldehyde	*50,000*
Picolines	*500-1,000*
Phenolics	*250-4,000*
Xylenes	*300-1,000*
Refinery hydrycarbons	*25- 50*
Petrochemical wastes	*15- 100*
Phenylether	*13*
Chlorinated phenolics	*1- 100*

[a]*Concentrations were determined by taking the median of 4-12 observations.*

1. Nuisance algae. The vast majority of taste and odor episodes in San Francisco have been directly or indirectly attributable to the activity of microorganisms in raw water impoundments. Of the two most troublesome groups of organisms, algae and actinomycetes, algae cause the most dramatic problems.

Algal odors can be classed into two basic groups: those caused by green and blue-green algae, generally described as "aromatic," and those caused by diatoms and pigmented flagellates, typically characterized as "grassy" or "fishy" (7).

In San Francisco's system, blue-green algae cause the most problems. When environmental conditions are right, *Anabaena* and *Aphanizomenon* can produce surface-covering blooms in a matter of several days. As the population reaches peak density, definite odors (described as "grassy," "geranium-like," or "vile") are noticeable in the raw water. It has been demonstrated that the geosmin isolated from pure cultures of the blue-green alga *Oscillatoria tenius* is the same chemical produced by actinomycetes, which is known to cause an "earthy" taste in water (8).

As the bloom declines, other tastes and odors develop, the result of the algae's characteristic retention of most odor products in their cells until they die and disintegrate (7). Work done by several authors suggests that these later tastes can be caused by the utilization of the blue-green algal metabolites by other microorganisms, such as the actinomycete *Streptomyces* (8).

Blue-green algae can cause secondary problems, especially during the summer months, when reservoirs are thermally stratified. If the bloom dies off suddenly, decomposition products of such a die are often worse than the initial bloom. Several by-products of bacterial decomposition of blue-green algae have been identified, including mehtyl-, isobutyl- and n-butyl-mercaptan, as well as dimethyl sulfide, which has a definite fishy odor (9).

New information suggests that blue-green algae may have the capability to grow in unexpected areas. *Oscillatoria* has been found growing in darkened tunnels within a southern California utility; the presence of this alga may well explain several mysterious taste and odor problems (10).

2. *Actinomycetes.* Actinomycetes are widely distributed in nature and account for a significant portion of the microbial population of lake and river muds. They are decomposing organisms and do not rely on photosynthesis for energy. In some instances actinomycetes alone appear to cause taste and odor problems. Since they do not require sunlight to grow, they can produce tastes on bottom muds as well as up in the water column and are, therefore, often hard to isolate. Counting methods have not always been reliable in correlating actinomycetes numbers with odor levels.

3. *Crustacea.* Other organisms have demonstrated extreme nuisance potential. Planktonic crustaceans (copepods and *Daphnia spp*) often grow in local supplies, causing particular problems in water sources that are served without filtration, since the chlorination of large numbers of these organisms causes extremely fishy tastes.

B. *Raw Water Transmission*

Over the years, the Department has experienced two taste and odor problems caused by pipeline-encrusting organisms. The first case occurred in the Coast Range Tunnel, a twenty-four mile tunnel section of the Hetch Hetchy aqueduct. Downstream consumers complained of tastes and odors variously described as "rotten egg-like," "metallic," and "putrid." An investigation shows that iron bacteria *Crenothrix* and *Gallionella* were growing on the wall of the tunnel. These organisms were utilizing the dissolved iron and manganese from ground water that was infiltrating the tunnel. As the biomass grew large and hydrodynamically unstable, it sloughed off into the system, where it rotted and produced a most disagreeable problem.

A virtually identical problem was caused by an encrusting fresh water sponge which was discovered living on the walls of a raw water intake screen and pipeline on the San Andreas Reservoir. The sponge was thriving in the inlet and pipeline because the currents enhanced its food supply

Both problems were corrected by the construction of chlorination stations. The intake screens are kept sponge-free by periodic back flushing with super-chlorinated water.

IV. MANAGEMENT OF RAW WATER PROBLEMS

A. Reservoir Control

Isolated nuisances, such as those that encrust in pipes, are easily corrected using fixed water treatment facilities. Control of planktonic or bottom-dwelling algae in large impoundments requires a more flexible approach. The number of variables and logistical problems have caused the Department's scientists and engineers to utilize different biological control strategies for each situation.

Experience has shown that the keys to taste and odor control in the raw water are control of algae production and management of the watershed. A 1976 survey by the American Water Works Association showed that most utilities still rely primarily on copper-based algicides, such as copper sulfate penthydrate, or manufactured copper chelates (Table III).

San Francisco has also relied primarily on $CuSO_4$ for algae control. The reservoirs are treated by spraying the surface with a solution of copper sulfate or copper sulfate mixed in a 1:1 ratio with citric acid. The latter is added to the treatment in lakes with alkalinity exceeding 40 mg/l to decrease the precipitation rate of copper ions and thus improve treatment effectiveness.

Intensive limnological data colleced on impounding reservoirs in the 1970's has provided better information on algae

TABLE III. Reservoir Treatment(1)

Treatment	No. of Utilities	Range (mg/l)	Median of Range (mg/l)
$CuSO_4$	33	0.05-5	0.4
Cl_2	8	0.1-6	0.9
C	3	1-20	--
Aeration	4	--	--

population patterns and has provided more reliable models on which to make treatment decisions. Using a "limited treatment" philosophy, the Department has reduced its usage of copper sulfate in the seven local impoundments from a 1963-72 average of 153,000 lbs/year to 30,000 lbs/year for the period since 1974 without any major episodes of taste or odor problems from those sources. The "limited treatment" approach adopted by the Department stresses alternate methods of controlling algae or operational schemes that avoid trouble spots.

B. *Aeration*

In recent years, a better understanding of watershed and runoff management, as well as the installation of a compressed air destratification system in several reservoirs, has helped greatly in reducing the Department's use of algicides for control of nuisance organisms. In 1973, water quality engineers designed and installed a compressed air destratification system in Pilarcitos Reservoir. In this system, compressed air is released from a perforated pipe at the bottom of the reservoir. The convection current created by the rising bubbles helps prevent the thermal stratification of the reservoir. With such a system, the lake circulates aerated water year-round, so anaerobic conditions do not form at the bottom. This prevents the production of hydrogen sulfide and other obnoxious decomposition products and prevents the efficient recycling of nutrients back into the water column (11). The reduced level of nutrients such as iron and phosphates helps retard algae production.

This installation has been a success. Blue-green algae blooms, which previously occurred two and three times a year, are now rare, usually occurring only when the compressor breaks down.

C. *Preimpoundment Treatments*

One of the most effective algae control projects carried out in this Department may be the fortuitous benefit of a runoff management program initiated in the early 1950's at Crystal Springs reservoir. In an effort to control runoff turbidity, the Department installed an alum treatment system and a series of preimpoundment ponds. A retrospective review of the reservoir's biology has strongly suggested that this treatment scheme has led to the elimination of hetrocystic blue-green algae from Crystal Springs. Records dating back to

1913 showed *Anabaena* or *Aphanizomenon* blooms at least two of every three years until 1952. Since 1952, there has not been one recorded blue-green algae bloom.

Current research suggests alum treatment and preimpounding of runoff water as a method of nutrient (specifically phosphate) control (12). The Department's work supports the theory that this preimpoundment treatment has reduced the taste and odor problems through elimination of blue-green algae growths.

D. *Biological Control*

Biological control of taste and odor problems appears to be a possibility in the near future. Work using bacteria and viral parasites of blue-green algae may eventually provide an effective control technique to supplement chemical algicide treatment (13). It has been demonstrated in the laboratory that a gram positive bacteria, *Bacillus cereus*, is responsible for the natural destruction of geosmin and may have application in a taste destruction process within the treatment plant (14).

V. CONTROL OF TASTES AND ODORS IN THE TREATMENT PROCESS

A. *Raw Water Operations*

In those fortunate instances when the cause of taste and odot problems can be clearly identified in the impounded water, the operator has several treatment strategies at his disposal. San Francisco's local reservoirs have several draw off levels in the reservoir. Using limnological data, the purification engineers select water from a strata that is not affected by the current microbe bloom. This is especially useful in blue-green algae blooms, since they are buoyant and tend to occupy the top ten to twenty feet of the water column. In most instances, the operator can draw off the unaffected bottom water until the bloom has subsided and the problem no longer exists.

San Francisco's experience has been that algicide treatment is sometimes not indicated for a large, established bloom. In these situations, a natural die off is often preferable to a massive kill since, in the latter situation, a massive die off of algae leads directly to increased populations of decomposing bacteria in the hypnolimnion; this, in turn, leads to dissolved oxygen depletion and the production of H_2S and other putrefaction products. A natural die off produces problems which are more easily treated. The treatment of choice for these volatile and rapidly-oxidized products produced through bacterial action is spray aeration before chemical treatment. Aeration, in addition to reducing volatile odor compounds, helps to oxidize soluble iron and manganese compounds that interfere with the treatment process and can be responsible for taste problems in the distribution system.

B. *Chlorine as Cure and Cause of Problem*

San Francisco uses gas chlorine as the primary disinfectant for all waters. When properly applied, chlorine tends to reduce or eliminate taste and odor problems in water. However, chlorination or insufficient chlorination can lead to some of the most noticeable taste and odor problems. Typically, raw water chlorinated beyond its "break point" has a lower taste and odor potential (15).

The "break point" occurs when each additional unit of applied chlorine produces an additional unit of free chlorine residual. If water high in ammonia or other nitrogen products is not chlorinated to the "break point," a taste and odor problem can develop due to the production of chloramine compounds (mono- and dichloramine and nitrogen trichloride) in the chlorine-ammonia reaction. The sensory threshold limits of chlorinated ammonia products have been classified as follows:

Monochloramine (NH_2Cl)	5.0 ppm
Dichloramine ($NHCl_2$)	0.8 ppm
Nitrogen trichloride (NCl_3)	0.02 ppm (16)

Paradoxically, the cause of the vast majority of customer complaints of chlorine taste and odor results from insufficient chlorine, not excess chlorine. Over the years, the Department has learned that most people cannot detect chlorine when the free chlorine residual is less than 0.8-1.0mg/l. In fact, many people cannot taste chlorine under 1.5-1.7 mg/l. The usual complaint of chlorine occurs in waters with 0-0.4 mg/l free chlorine residual. The Department's normal corrective action when faced with such a problem is to increase the local chlorine levels to 0.6-0.8 mg/l, which destroys the dichloramines and nitrogen trichlorides, thus alleviating the problem.

San Francisco is fortunate that all of its water comes from watersheds that are protected from industrial and sewage pollutants. Industrial pollutants, notably phenolic compounds, are extremely potent taste and odor problems when chlorinated (Table II).

An unusual taste and odor problem has been traced to wooden reservoir supports treated with creosote (17). Apparently, the timber's proximity to the chlorinated water is enough to cause a very potent problem even though it is not actually in contact with the water. To prevent this situation, San Francisco no longer uses coating materials that contain phenolic compounds.

C. Other Treatment Chemicals

Many utilities rely on oxidant chemicals such as potassium permanganate, chlorine dioxide, and ozone in their treatment processes to control taste and odor problems. These treatments are less frequently used than chlorine, but have the advantage that they destroy taste and odor products without producing chlorinated taste and odor products. Table IV shows the number of utilities in the United States and Canada using specific types of in-plant treatment.

TABLE IV. Plant Treatment for Taste and Odor Control (United States and Canada) (1)

Plant Treatment	Number of Utilities	Range of Doses (mg/l)	Median Dose (mg/l)
Chlorination	72	0.5-19	3
Superchlorination	4	0.5-36	13
Sodium chlorite for chlorine dioxide	8	0.12-2	0.4
Ammonia for chloramine	29	0.13-3	0.4
Activated carbon			
Powdered	55	0.25-60	5
Granular	6		
Potassium permanganate	13	0.1-15	0.3
Copper sulfate	4	0.02-4	
Aeration	1		
Polyphosphate	12	0.4-4	1.3
pH adjustment with lime	19	1-61	12.5
$Ca(OH)_2$	1	15.7-32.3	
Na_2CO_3	4	2-21	6
NaOH	3	1-25	

D. *Filtration*

The coagulation, sedimentation, and filtration processes are extremely important in controlling an in-plant taste and odor problem. It is essential that algae cell fragments and sediment particles that have absorbed taste and odor chemicals be removed from the water to prevent further development of the problem. The Department's experience has shown that in-plant odor episodes can develop in normally nonproblematic water. In one instance, the lack of proper flocculator action caused dead algae cells to concentrate in surface foam in the plant, producing an odor during decomposition. The problem was corrected when the flocculator speed was increased to prevent the cells from collecting at the surface.

E. *Carbon Adsorption*

If all the usual processes are ineffective, powdered carbon can be applied. The carbon is applied as a slurry to the water during treatment and then removed in the filters. Although powdered carbon is usually effective at controlling tastes, difficulty in handling and problems with carbon sludge disposal make it a method of last resort.

The use of granulated activated carbon (GAC) may become a routine odor control technique in water treatment in the future. Regulations proposed by the EPA prescribe GAC treatment for utilities that have significant exposure to organic chemical pollutants (18). If these regulations are implemented, many utilities will have odor reduction as a by-product of organic chemical removal.

VI. TASTES AND ODORS ASSOCIATED WITH DISTRIBUTION AND STORAGE

Tastes and odors commonly develop after treatment during transmission, storage, and distribution. In these instances, the operator's information comes primarily from consumer feedback. There are numerous known and many more unknown causes of sensory degradation of water after treatment. In some cases, the problems are inherent in the treated water, but take time to develop. Others relate directly to the physical design and/or operation of the distribution system.

A. "Iodoform" Taste and Odor

The most perplexing and widespread taste and odor problem that San Francisco experiences is an annual short spell of "iodoform" or "medicinal" water from the Hetch Hetchy aqueduct. In late winter or early spring, the Department invariably receives complaints that the water is "medicinal," "iodine-like" or "chlorine-tasting." These complaints typically start in suburban systems and gradually move north toward San Francisco's distribution system.

The mechanism of the iodoform development is not known, but it apparently has a rather specific development time. Suburban consumers on the southern extreme of the transmission network do not experience this problem, which is evident from a point ten to twenty miles further along the pipeline all the way to San Francisco. Lab tests have established that the taste potential was present at Alameda East Portal, where the water entered the transmission system.

The typical taste complaint comes from consumers who drink water after it has been refrigerated or left overnight in a glass. An investigation of the problem has suggested that the occurrences of the taste were associated with waters at pH levels above 9.0. During this period, only the Hetch Hetchy source was in service, and the chlorine residual was relatively constant (0.6-0.7mg/l). The addition of calcium oxide (quicklime) to water at Rock River Lime Plant for pH and corrosion control was inconsistent during this period due to variable lime quality and delivery problems associated with the vendor. All justified consumer complaints were examined, and it was found that the samples all had pH values in excess of 9.3 and above.

This iodoform problem is difficult, because it is short-lived and because it moves through the system in slugs and is therefore not conducive to study or control. It can, however, have serious consequences: in one instance, several thousand pounds of ground beef processed by a suburban food processor had to be discarded because it was processed with "iodine-like" water.

The iodoform problem has been described by other utilities, but there is no agreement as to its causes. The consensus is that it occurs most commonly in periods of winter runoff or spring snow melt. Others have correlated the problem with water tempature (1) and lime sludges (19).

B. Problems in Distribution Reservoirs

Other less exotic taste and odor problems develop in the course of storage and distribution. The storage of finished water in uncovered reservoirs is a major source of post-treatment taste and odors due to microbial growth. Bacteriological decomposition of sediment can produce tastes and odors in covered distribution reservoirs because of poor circulation patterns in reservoirs.

The poor circulation patterns are caused by design flaws in the reservoir structure or by operations that cause the reservoir to "float on the line." Some of San Francisco's reservoirs have common inlet and outlet pipes. In these instances, only water near the pipe is actively circulated; in the extremities of the reservoir, there is little or no circulation, and it is not uncommon to find significant sedimentation in these areas.

The poor circulation also affect chlorine residuals in extremes. Without adequate chlorine levels, sediments become biologically active, supporting mixed populations of bacteria, protozoans, and micro-invertebrates (such as nematodes and crustaceans). Decomposing bacteria thrive in this ecological niche and can produce significant taste and odor products. These problems can be easily corrected by building inlets and outlets at opposite ends of the reservoir and by increasing the reservoir's operational range of volume change.

C. *Problems in Distribution Water Mains*

The design, operation, and maintenance of a utility can either produce or prevent taste problems. Good design criteria dictate that low circulation or dead-end mains be minimized in a water system. In areas of low water circulation, populations of sulfur, iron, and other bacteria flourish because of a lack of chlorine levels to suppress their growth. Bacteria can utilize dissolved organic materials, local sediment, or inorganic materials in the pipe to support growth. As by-products of growth, they produce sulfides, sulfates, and metal corrosion products, all of which can produce off tastes and odors. These materials either build up to the point where a nearby consumer can taste or smell their by-products or they create situations in which a sudden surge of water use releases these materials into a wide area of the distribution system. The problems are minimized by good design (circulating systems with few dead ends) and regular maintenance through a main flushing program.

VII. LOCALIZED PROBLEMS

A. *Cross Connections*

Cross connections, or the mixing of potable and nonpotable water sources, creates the most potentially dangerous conditions. These are usually created either when low main pressure causes back siphonage from a consumer's premises to the main or because the consumer's plant pumps force water back through the service into the main. In addition to producing a variety of taste and odor problems, these cross connections may cause serious health hazards through introduction of toxic chemicals or bacteria-laden water. Backflow conditions are prevented through maintenance of proper main pressures and through the use of backflow prevention devices.

The most common example of a taste and odor problem caused by a cross connection occurs when a consumer relies on a hose nozzle, rather than the hose bib, to turn off a garden hose. Tastes which are leached out of the plastic hose (generally described as "phenolic" or "plastic-tasting") are literally "pumped" into the house pipes by the expansion and contraction of the hose as it warms and cools during the day. The solution is to turn off the hose at the hose bib.

B. *Household Plumbing*

When a water inspector is called on to check a taste or odor complaint, he cannot rule out problems with the consumer's plumbing system as a potential source of the trouble. The Department's experience demonstrates that many complaints are the result of bad plumbing. The most frequent complaint is an off taste created by metallic ions which are released by chemical or electrolytic corrosion. These ions are readily noticeable to many consumers. (See Table V, which lists the experimental taste thresholds of common metals in distilled and spring water.)

Metallic tastes are most frequently generated by the use of both iron-galvanized and copper pipe without proper insulating joints. The electrolytic reaction can produce levels of copper above the taste threshold. In addition, iron pipes can cause taste problems when soft water is allowed to stand overnight. Typically, early morning samples of this water taste terrible, but the problem generally corrects itself, at least temporarily, when the lines are flushed through normal use.

TABLE V. Taste Thresholds for Metals in Water (20)

	Threshold Frequency in Distilled Water [a]		Threshold Frequency in Spring Water	
	5%	50%	5%	50%
Zinc sulfate	4.3[b]	18	6.8	27
Zinc nitrate	5.2	22		
Zinc chloride	6.3	25	8.6	33
Cupric chloride	2.6	6.6	5.0	13
Ferrous sulfate	0.04	3.4	0.12	1.8
Hydrous ferric oxide	0.7	8.8		
Manganous sulfate	3.6	45		

[a]Threshold values are for the metal ion, not the salt.
[b]Ion concentration (ppm)

C. Problems with Plastic Service Pipe

San Francisco has experienced several problems with plastic services that have been installed since the late 1960s. The Department received numerous complaints from consumers whose metal service pipes were replaced with plastic pipe. Samples of the polyethylene pipe were forwarded to the National Sanitary Foundation for analysis, which confirmed that tastes and odors were caused by the pipe; apparently, an error in the formulation of the material allowed a plasticizer to leach out and create off tastes and odors.

In recent years, the Department has also experienced several instances of paint thinner and gasoline smell in household water. The Department's investigators have invariably identified the polyethylene service pipe as the source of the offending odors. Investigations by local utilities into plastic pipe have conclusively demonstrated that gasoline can penetrate polyethylene and polybutylene pipe containing water at 40 psi pressure (19). The Department's current practice is not to install plastic pipe in areas that might be subject to spills of hydrocarbon solvents.

D. In-House Purification Devices

More and more frequently, industrial consumer and homeowners are installing in-line or faucet-mounted water purification systems. The typical system includes a particle filter followed by a granular activated carbon cartridge. These devices work well for the most part, removing taste and odor-causing chemicals in the water. However, they can also be the cause of problems if not properly maintained. Since these systems remove chlorine as well as organic chemicals from the water, the carbon beds become a perfect breeding ground for bacteria. In the absence of a chlorine residual, bacteria grow well on the organic materials adsorbed on the filter. As the filter loses its ability to adsorb organic materials (through clogging of the active sites on the carbon), the bacteria produce by-products that cannot be handled by the filter and are thus passed along to the user. If these devices are used, therefore, they must be monitored regularly to insure that they are, in fact, capable of providing taste and odor reduction.

VIII. CONCLUSION

The San Francisco Water Department, like many other public water utilities, functions to fulfill the water needs of a modern metropolitan society. Part of its task is thus to produce the millions of gallons of water needed daily for industry and the running of domestic water systems. In addition, however, the Department must be mindful that it is also producing a beverage. In this latter connection, it is not enough that the utility deliver an ample quantity of water that is safe to drink; it must deliver water that people believe is drinkable.

The utility's ability to meet this goal is influenced to some extent by the fact that the drinkability of water is a matter of personal preference. This fact was emphasized dramatically in a recent taste testing of Bay Area water by a panel of "experts" (21). This panel rated various samples of San Francisco water from near the top to the very bottom of all samples tested. Although all the samples came from the same source, they were served by different distribution systems. None of the samples had off tastes; the differences in ranking merely reflected the reality that the panel simply liked some water samples better than others.

Personal preference does not end the matter, however. Producing millions of gallons of a beverage daily is a formidable task, given the number of variables involved in the production, purification, and distribution of drinking water. The inexact nature of testing and control of the raw ingredients -- in this case, the raw water from rain and snow-melt runoff -- makes the water operator's job a difficult one. The unpredictability of microorganism activities and the difficulty of monitoring sometimes allow potential problems to develop into actual ones. These are the time where the experience of the treatment personnel makes the difference between success and failure of the aesthetic goals set by the utility.

The whole philosophy of a utility, not just that of its water treatment personnel, contibutes to the delivery of palatable water. A well designed, maintained, and operated distribution system can prevent good quality water from deteriorating after it leaves the treatment plant. To maintain this type of in-house cooperation requires constant communication among the many segments of the utility. If the system operating personnel are well advised on causes of tastes and odors, they can take the necessary actions to prevent or correct the problems. This, in turn, will ultimately lead to greater consumer confidence through greater consumer acceptance of the product.

ACKNOWLEDGMENTS

The authors would like to acknowledge the contributions made to the production of this manuscript by Patricia Doyle, Annette Loosli, Robert Hewitt, Allan McDonald, and Joan Jenkin of the Water Quality Division of the San Francisco Water Department.

REFERENCES

1. American Water Works Association Research Foundation, *Handbook of Taste and Odor Control Experiences in the U.S. and Canada* p. A-11 (1976).
2. American Public Health Association (APHA), American Water Works Association and Water Pollution Control Federation, *Standard Methods for the Examination of Water and Wastewater*, 14th Edition, APHA, Washington, D.C. (1976).
3. American Society for Testing and Materials, *1978 Annual Book of ASTM Standards (part 31) Water*. Philadelphia, Pa. (1978).
4. Popalisky, J.R. in *Handbook of Taste and Odor Control Experiences in the U.S. and Canada*, p. SVIII-2. American Water Works Association, Denver, Colorado (1976).
5. Rosen, A.A., Mashni, C.I., and Safferman, R.S., *Soc.Water Treatment and Examination Proc.* 19,106 (1970).
6. Baker, R.A., *J. Amer. Water Works Assoc.* 55,913 (1963).

7. Maloney, M.E. *J. Amer. Water Works Assoc.* 55,481 (1963).
8. Medsker, L.C., Jenkins, D. and Thomas, J.F. *Enviro. Sci. and Technol.* 2,461 (1968).
9. Jenkins, D. et al, *Enviro. Sci. and Technol.* 1,731 (1967).
10. Cohen, R. Metropolitan Water District of Southern California (personal communication).
11. Symons, J.M., Irwin, W.H. and Robeck, G.G. in *Water Quality Behavior in Reservoirs.* p. 299. U.S. Public Health Service (1969).
12. Lee, G.F., Jones, R.A., and Rast, W. in *Proceedings of Conference on Phosphorus Management Strategies for the Great Lakes*, Cornell University (1979).
13. Safferman, R.S. and Morris, M-E, *J. Amer. Water Works Assoc.* 56,1217 (1964).
14. Silvey, J.K., Henley, D.E., Hoehn, B. and Nunez, W.C. in *Proceedings AWWA Technology Conference*, Atlanta, Georgia (1975).
15. Rebhum, M., Fox, M.A. and Sless, J.B. *J. Amer. Water Works Assoc.* 63,219 (1971).
16. White, G. *Handbook of Chlorination*, Van Nostrand-Reinhold Co., New York, New York (1972).
17. Conners, J. in *Handbook of Taste and Odor Control Experiences in the U.S. and Canada*, p. XVIII-2, American Water Works Association, Denver, Colorado (1976).
18. Environmental Protection Agency in *Federal Register*, Vol. 41 No. 136 (1976).
19. Carns, K., East Bay Municipal Utilities District, Oakland, CA. (personal communication).
20. Cohen, J.M., Kamphake, L.V., Harris, E.K., and Woodward, R.L. *J. Amer. Water Works Assoc.* 52,660 (1960).
21. Reubenstein, S. in *San Francisco Chronicle* Vol. 116, No. 122 (1980).

EVALUATION AND CONTROL OF UNDESIRABLE FLAVORS IN PROCESSED CITRUS JUICES

Steven Nagy
Russell Rouseff

Florida Department of Citrus
University of Florida, AREC
Lake Alfred, Florida

I. INTRODUCTION

Citrus juice for the consumer market is more acceptable the closer its flavor matches that of freshly juiced fruit; yet the difficulty of preserving that fresh flavor is well known. Commercially processed citrus juice is susceptible to flavor change when stored at warm temperatures for prolonged storage periods. Unless the juice is kept near freezing a disagreeable odor and off-flavor soon develop. Over four decades investigators have attempted to delineate the components responsible for this detrimental flavor change and to develop chemical tests to monitor processing and storage abuses.

Changes occurring in citrus juice after processing and upon subsequent storage may be divided into two types: (a) loss of original flavor and (b) development of flavors foreign to fresh juice (1). Several theories have been proposed to account for flavor change in commercial citrus juices, namely, (a) off-flavors were caused by oxidation of juice components from the incorporation of oxygen during processing (2), (b) changes in lipid constituents cause off-flavor development (3-6), (c) flavor changes were due to changes in peel oil constituents (7, 8), and (d) a combination of degradation products were responsible for the malodorous properties of aged juice (9).

In addition to problems associated with the off-flavor compounds that develop during storage, there are a few naturally occurring compounds that possess undesirable flavor properties. These compounds are responsible for bitterness in citrus juices. While some bitterness is acceptable in grapefruit and lemon juices, any degree of bitterness in orange juice is objectionable. Bitterness in citrus can be primarily attributed to either limonin or the flavanone glycoside, naringin or a combination of the two. Early research in the bitterness area centered around the isolation of these two compounds. Once citrus bitterness had been

ISBN 0-12-169065-2

identified with specific compounds the problem was to: (a) develop methods to accurately measure the bitter forms of the compounds, (b) establish taste thresholds and relative bitterness for these compounds, and (c) determine the effects of variety and fruit maturity on juice bitterness. Even though bitter compounds are found as natural components of citrus juice, efforts to reduce the levels of these compounds in juice have been attempted either by: (a) altering the manner in which juice is extracted (some juicing techniques enhance the extractability of bitter components from various fruit parts) and/or (b) remove or structurally modify the bitter compounds after juicing but before packaging.

II. EFFECTS OF ATMOSPHERIC OXYGEN

The amount of oxygen incorporated in the juice and in the container headspace during extracting, finishing and canning is related to the development of off-flavors (7, 10, 11). However, while deaeration of citrus juice during processing has been shown to be of importance in preventing deterioration initially, it does not protect against the ultimate development of off-flavors during long storage at normal warehouse temperatures (1, 12). Current industrial practice has been to keep air in juice as low as possible, and this has been accomplished by use of deoiling-deaeration equipment and by the injection of live steam into the headspace of the can during closure (14). Although vacuum deaeration and live steam injection substantially reduce the oxygen content of the product, there is still some oxygen dissolved in the juice and entrapped within the headspace atmosphere of the container. Kefford and co-workers (15) described canned citrus juice as one in which there is competition for oxygen among a number of reactions, including corrosion reactions, ascorbic acid oxidation, and oxidations contributing to flavor and color change. Since the oxygen content of canned juices disappears rapidly (16, 17), the consensus of many workers (1, 5, 12) is that oxidation does not influence flavor change to any substantial degree. In fact, Kefford and co-workers (15) questioned the value of deaeration and indicated that it appeared to cause little difference in storage stability or flavor change of a citrus juice product.

III. LIPID CONSTITUENTS

Early studies by Nolte and Von Loesecke (18) on fresh and stored canned Florida Valencia orange juice implicated the lipid fraction as one of the causative factors in off-flavor development. These workers showed that temperature-aged juice differed from fresh juice by increased acidity, increased saponification and peroxide values, and contained carbonyl compounds. Curl (3) compared the flavors before and after storage of whole tangerine juice and tangerine juice from which the suspended matter, which includes the lipids, had been removed. Curl concluded that the suspended matter was responsible for much of the characteristic flavor but also contained precursors of the off-flavors that developed on storage. Curl and Veldhuis (4) and Swift (5) confirmed these studies by showing that the suspended matter of citrus juice was the principal contributor to off-flavor in aged orange juice. In a more recent investigation, Nagy and Nordy (6) studied the effects of storage temperatures and length of storage on different lipid classes of glass-packed, single-strength orange juice (SSOJ). These workers found that during storage the phospholipid group degraded to free fatty acids and other products. Since no lyso phospholipid derivatives were found, it appeared that phospholipid breakdown was complete and that both fatty acids associated with the phospholipid molecule were split off by random nonenzymatic degradation (19). Figure 1 shows that over a 16-month storage period at 29.4^{o}C phosphatidyl choline (PhC) decreased 91%, phosphatidyl ethanolamine (PhE) 88%, phosphatidyl inositol (PhI) 88% and phosphatidyl serine (PhS) 46%. The production of large quantities of unsaturated fatty acids from phospholipid breakdown predisposes commercial orange juice to off-odor and off-flavor development. The primary degradation products of lipids are not in themselves off-flavored but are thought to constitute an unstable system which ultimately leads to the production of substances which impart a rancid note to orange juice flavor. Many fatty acid oxidation products, for example, alk-2,4-dienals, 2-octenal and n-hexanal from linoleic acid possess malflavorous properties. Askar et al. (20) showed that storage of orange juice resulted in an increase in n-hexanal, n-hexenal, n-octanal and n-decanal; all of which were not necessarily detrimental to flavor. Although some lipid oxidation products contribute to flavor change, they apparently are not the major off-flavor components in citrus juice. It is apparent from the works of Swift (5), Nagy and Nordby (6), Askar et al. (20), Nagy and Dinsmore (21) and Tatum et al. (9) that the major off-flavor components of high temperature stored juice are non-lipid in nature. With increase in storage time, lipid oxidation products are formed and additively contribute to the overall malflavorous property of aged citrus juice (22).

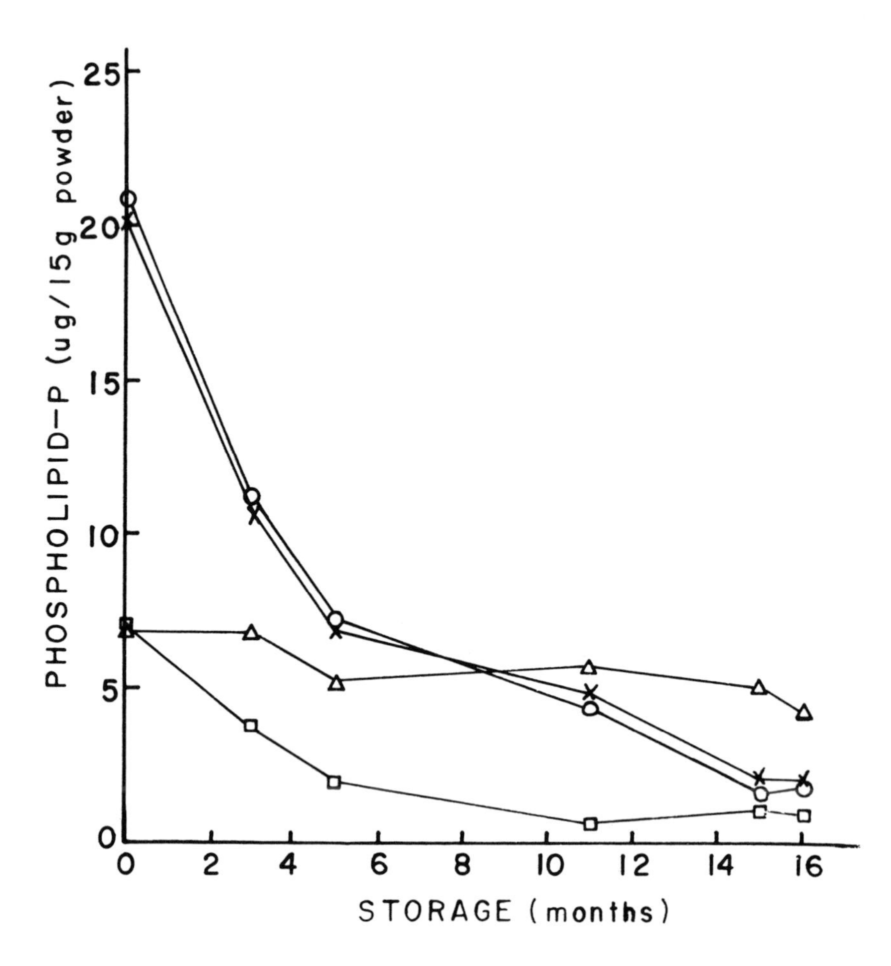

Figure 1. Effect of storage at 85°F on the concentrations of PhC (O), PhE (X), PhI (D) and PhS (▲) over a 16-month period; phospholipid concentrations expressed as phospholipid phosphorus. (From J. Agric. Food Chem. 18, 593 (1970). Copyright by American Chemical Society.)

IV. PEEL OIL CONSTITUENTS

A major group of investigations (1, 7, 8) has been based on the premise that changes in peel oil constituents are responsible for flavor change in citrus juice during storage. Blair and co-workers (23) proposed that flavor deterioration of canned orange juice is complex, and they empirically differentiated between a "terebinthine" (resembling terpentine) flavor factor and other factors designated collectively as "staleness". The terebinthine factor was dominant in juices stored above 21°C whereas juices stored below 21°C exhibited a variable taste quality which was characterized as stale. Blair et al. (23) suggested that the acid of the juice promoted a series of hydration-dehydration reactions with the terpenes of the peel oil. d-Limonene, the principal component of peel oil, is thought to undergo a series of hydration-dehydration reactions to produce polyhydroxy compounds, alpha-terpineol, 1,4-cineole and other products. Since 1, 4-cineole is produced when acidity and storage temperatures are high, it is thought to be responsible for the typical, pungent off-flavor of storage-abused citrus juices (23).

In contrast to the claim of Blair et al. (8, 23) that changes in peel oil constituents were predominately responsible for off-flavors, other investigators have tended to refute that proposition. Curl (3) and Curl and Veldhuis (4) concluded from their experiments on orange and tangerine juices that peel oil did not contribute to off-flavor development and, in fact, masked some of the off-flavors that did develop. Experiments conducted by the Continental Can Company (24) showed that the flavor of canned single-strength orange juice stored at room temperature continued to deteriorate as rapidly with 0.0016% peel oil as with 0.0125% peel oil. Their results indicated that some factor or combination of factors, other than peel oil, was involved in off-flavor development.

V. DEGRADATION COMPOUNDS AFFECTING FLAVOR CHANGE

Considerable interest in the identification and organoleptic evaluation of the degradation compounds that form during high-temperature storage of juices has been shown by the citrus industry because of the need to continually monitor the quality and flavor stability of commercial products. Numerous studies (9, 20, 21, 23, 25-28) have been conducted to identify changes in the volatile flavor constituents of canned and bottled single-strength citrus juices during storage. An extensive study by Tatum et al. (9) on the identification of degradation compounds in temperature-abused, canned SSOJ clarified the roles of three major off-flavor components. These investigators found alpha-terpineol, 4-vinyl quaiacol and 2,5-dimethyl-4-hydroxy-3(2H)-furanone to be the major off-flavor compounds in stored SSOJ (Table 1). When these three off-flavor compounds were added separately to a control juice, flavor changes were detected at the concentration levels shown in Table 2. Alpha-terpineol had previously been reported (23, 25, 27, 28) as a

TABLE 1
Degradation products in canned SSOJ after 12 wk at 35°C.

Compound	GLC retention time (min)[a]
Furfural	12.5
alpha-Terpineol	21.0
3-Hydroxy-2-pyrone[b]	41.0
2-Hydroxyacetyl furan[b]	41.0
2,5-Dimethyl-4-hydroxy-3(2H)-furanone[b]	43.0
Unidentified	43.0
cis-1,8-p-Menthanediol	47.5
trans-1,8-p-Menthanediol	52.0
4-Vinyl guaiacol	54.5
Benzoic acid	72.0
5-Hydroxymethyl furfural	75.0

[a] Retention times determined on 9-ft x 1/2-in. 20M column (20%).

[b] Resolved on 9-ft x 1/4-in. UCW-98 column (20%).

(From J. Food Sci. 40, 707 (1975). Copyright by Institute of Food Technologists.)

TABLE 2

Concentration of degradation products causing detectable flavor change when added to SSOJ[a].

Compound	Conc (ppm)	Significance of difference
alpha-Terpineol	2.5	$p < 0.001$
	2.0	$p < 0.05$
4-Vinyl guaiacol	0.075	$p < 0.001$
	0.050	$p < 0.01$
2,5-Dimethyl-4-hydroxy-3(2H)-furanone	0.10	$p < 0.01$
	0.05	$p < 0.05$

[a] Controls stored at -18°C.

(From J. Food Sci. 40, 707 (1975). Copyright by Institute of Food Technologists.

degradation product of aged orange juice. This compound was characterized as imparting a stale, musty or piney aroma to fresh juice (9). Cis-1,8-p-menthanediol (Table 1) forms from alpha-terpineol by an acid-catalyzed hydration reaction (8) and might actually be a useful degradation compound because it lowers the content of the objectionable alpha-terpineol. 4-Vinyl guaiacol was the most potent off-flavor compound. When added to fresh juice, this compound imparted an "old fruit" or "rotten" flavor to juice. 2,5-Dimethyl-4-hydroxy-3(2H)-furanone was responsible for the pineapple-like odor of aged orange juice. Its flavor was not objectionable but suppressed or masked the full orange-like aroma of SSOJ when found at concentrations exceeding 0.05 ppm.

Nootkatone was found in control and storage-abused juices by Tatum et al. (9). This compound was of interest because of its grapefruit-like aroma; however, Tatum and co-workers (9) concluded that the grapefruit-like aroma of aged juice was not primarily due to the nootkatone level.

Two volatile components of concern to processors of frozen concentrated citrus juice are diacetyl and acetylmethylcarbinol (29). These end products of bacterial growth (Lactobacillus and Leuconostoc) impart to citrus concentrate an off-flavor characteristic of buttermilk. When there is a microbial buildup of these lactic acid organisms in a processing plant, the diacetyl concentration in a juice concentrate also increases. If the diacetyl concentration of a product is allowed to increase until it is 0.8 ppm or higher, the generation rate of lactic organisms has reached an exceedingly high rate (29). When this occurs, it is of utmost importance to shut down the citrus processing line and thoroughly clean the production area involved (29).

VI. COMPOUNDS USEFUL IN MONITORING STORAGE ABUSE

A considerable range of chemical, chromatographic and spectroscopic tests (25, 27, 29-31) have been developed to indicate storage abuse of citrus products. Nevertheless, there has remained a need for an analysis which would serve as a useful index of overall deterioration, in particular by showing some correlation with other evidence of deterioration, such as off-flavor, and by having the sensitivity to detect its onset at a very early stage. Of the various known compounds formed when juice deteriorates - aldehydes, ketones, peroxides, oxygenated terpenes, furanones, phenolic derivatives and furfural - the last of these has the most satisfactory features for this type of analysis. Furfural is an excellent compound to monitor because: (a) the furfural content of freshly processed juice is virtually zero whereas large amounts accumulate in storage-abused juice (25, 27, 28), (b) furfural can be distilled readily from a juice and enriched at least 6-fold relative to its original level, (c) furfural readily lends itself to colorimetric assay (32), and (d) correlations between off-flavor and furfural are very good even though the compound itself does not contribute to flavor change.

The relationship of furfural content to storage temperature is shown in Figure 2 for canned SSOJ. After 16 weeks the furfural levels per liter of juice were 533 ug ($30^{o}C$), 131 ug ($21^{o}C$), 66 ug ($16^{o}C$), 32 ug ($10^{o}C$) and 19 ug ($5^{o}C$). It was evident that furfural levels correlated in a predictable manner to storage temperatures. The relationship of furfural content to storage temperature for glass-packed orange juice is shown in Figure 3. The furfural content vs. storage time profiles for the five temperatures were similar to those for canned juice (Figure 2). Glass-packed orange juice appeared to accumulate slightly more furfural at higher temperatures ($16^{o}C$) than canned orange juice (33).

Table 3 shows the relationship between the increase in furfural content with time and the organoleptic evaluation of canned and glass-packed orange juice (33). With canned and glass-packed juices stored at $10^{o}C$, no significant taste differences were observed within the 16-week storage period.

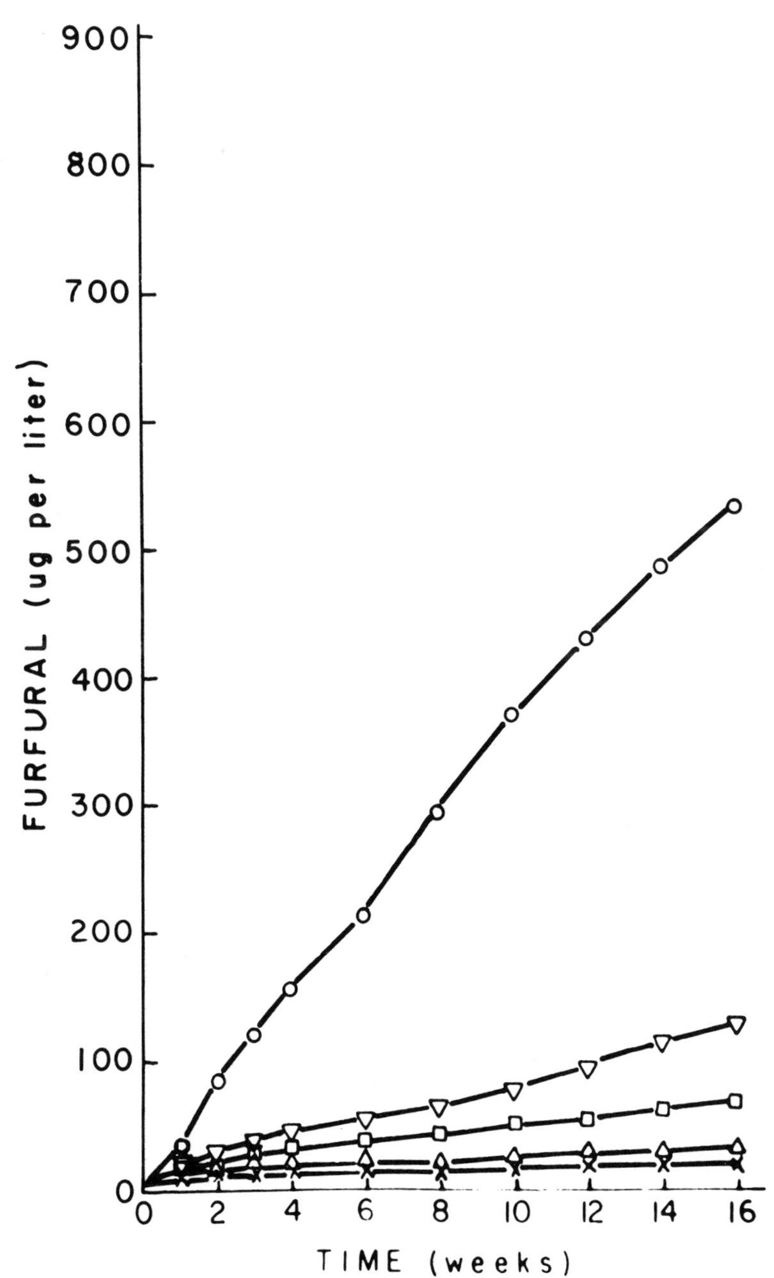

Figure 2. Increase in furfural content in canned orange juice over a 16-week period at 5° (X), 10° (▲), 16° (□), 21° (▽), and 30° (O). (From J. Agric. Food Chem. 21, 272 (1973). Copyright by the American Chemical Society.)

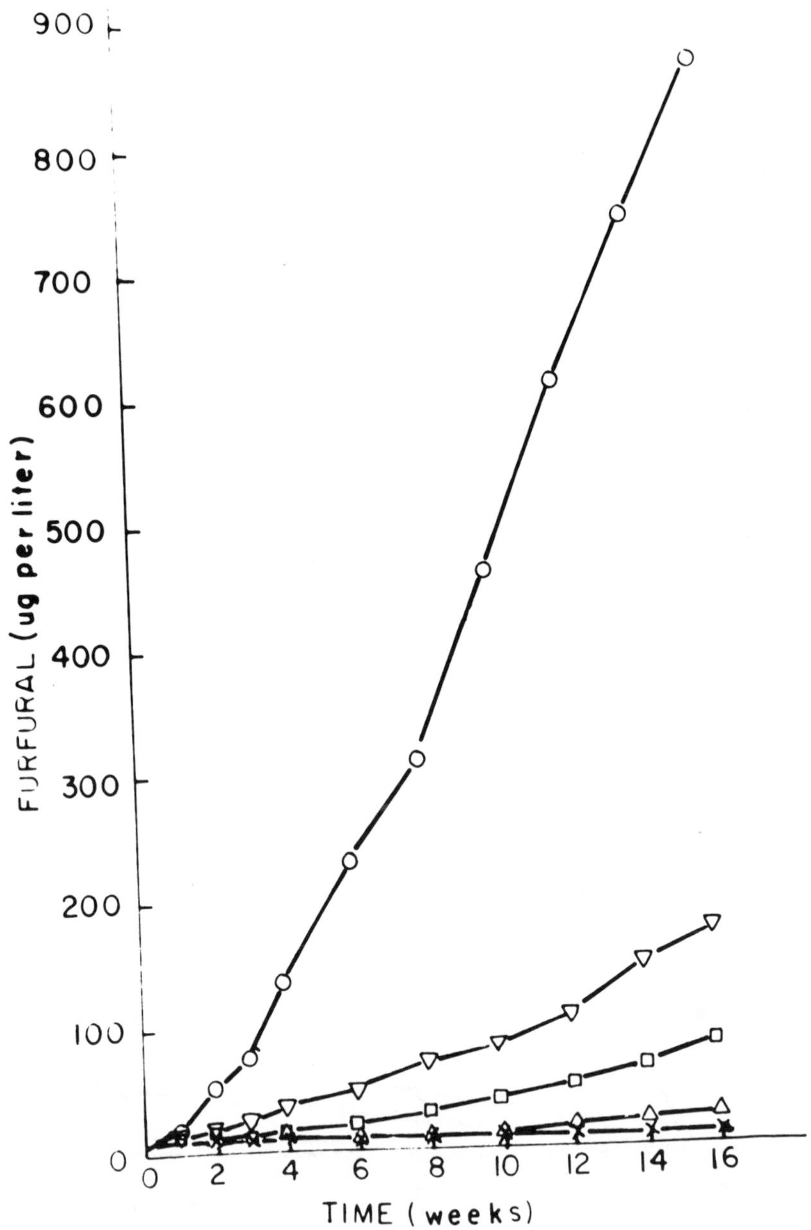

Figure 3. Increase in furfural content in glass-packed orange juice over a 16-week period at 5° (X), 10° (Δ), 16° (□), 21° (▽), and 30° (O). (From J. Agric. Food Chem. 21, 272 (1973). Copyright by the American Chemical Society.)

For the 16°C juices, highly significant taste differences were observed at the 10th (55 ug/l furfural) and 12th (52 ug/l) week for canned and glass-packed juices, respectively. Juices stored at 21°C showed flavor changes at the 8th week (62 ug/l can; 64 ug/l glass) whereas those stored at 30°C developed off-flavors within a 2-week period. The data in Table 3 indicated that furfural values within the approximate region 50-70 ug/l correlated with a difference in flavor in comparison to controls at the significance of difference level of $p < 0.001$. Subsequent work by Maraulja et al. (34) showed that correlation coefficients between flavor scores (10-point hedonic rating) and the furfural content of canned Hamlin and Valencia orange juices were highly significant. These workers reported that the rate of flavor deterioration of canned juices was dependent on both storage temperature and storage time; however, temperature was more significant than time.

A study of canned and glass-packed grapefruit juices by Nagy and co-workers (35) revealed a linear relationship between storage time and furfural content that was similar to orange juice (21, 33). In grapefruit juice, a minimum level of furfural could be correlated with the onset of off-flavor; however, that level was much higher when contrasted to the orange juice level. When the level of furfural exceeded a value of about 175 ug/l in canned grapefruit juice, a taste panel observed a highly significant difference in flavor whereas in glass-packed juice a significant flavor difference was observed when furfural exceeded about 150 ug/l. Canned grapefruit juice retains its typical flavor for longer storage periods than canned orange juices (34, 35). Apparently, grapefruit juice possesses flavor characteristics which tend to mask off-flavors. This masking would function by increasing the furfural-flavor change threshold values for grapefruit juice (34).

Subsequent work by Dougherty et al (36) on canned grapefruit juice showed strong correlations between furfural content, flavor score and storage temperature (Figure 4). Samples stored at 100°F for 6 weeks showed a rapid increase in furfural with a concomitant decrease in flavor quality (flavor score of about 4). Similar trends were observed for samples stored at 90°F and 80°F; however, the rate of furfural buildup and flavor deterioration were slower. Samples stored at 70°F decreased in flavor and increased in furfural only slightly - even after 6 months.

TABLE 3
Relationship of furfural content to extent of flavor changes in canned and glass-packed orange juice stored at 10, 16, 21 and 30°C.

Storage time weeks	10°		16°		21°		30°	
	FF[a] ppb	Significance of difference	FF ppb	Significance of difference	FF ppb	Significance of difference	FF ppb	Significance of difference
			Canned orange juice					
1	12[b]	N.S.[c]	15	N.S.	18	N.S.	31	N.S.
2	17	N.S.	27	N.S.	32	N.S.	85	$p < 0.001$
3	21	N.S.	32	N.S.	39	N.S.	120	$p < 0.001$
4	24	N.S.	35	N.S.	47	$p < 0.01$	156	$p < 0.001$
6	24	N.S.	38	N.S.	54	$p < 0.01$	215	$p < 0.001$
8	25	N.S.	42	$p < 0.05$	62	$p < 0.001$	291	$p < 0.001$
10	27	N.S.	51	$p < 0.001$	77	$p < 0.001$	370	$p < 0.001$
12	29	N.S.	55	$p < 0.001$	94	$p < 0.001$	432	$p < 0.001$
14	31	N.S.	61	$p < 0.001$	115	$p < 0.001$	480	$p < 0.001$
16	32	N.S.	66	$p < 0.001$	131	$p < 0.001$	533	$p < 0.001$

TABLE 3 continued

Storage	10°		16°		21°		30°	
time weeks	FF ppb	Significance of difference	FF ppb	Significance of difference	FF ppb	Significance of difference	FF ppb	Significance of difference
			Glass-packed orange juice					
1	9	N.S.	10	N.S.	14	N.S.	18	N.S.
2	9	N.S.	12	N.S.	24	N.S.	55	$p < 0.001$
3	11	N.S.	14	N.S.	27	$p < 0.05$	74	$p < 0.001$
4	11	N.S.	19	N.S.	37	$p < 0.01$	134	$p < 0.001$
6	12	N.S.	23	N.S.	48	$p < 0.01$	239	$p < 0.001$
8	13	N.S.	30	N.S.	64	$p < 0.001$	312	$p < 0.001$
10	13	N.S.	39	$p < 0.01$	81	$p < 0.001$	461	$p < 0.001$
12	15	N.S.	52	$p < 0.001$	105	$p < 0.001$	610	$p < 0.001$
14	19	N.S.	63	$p < 0.001$	145	$p < 0.001$	739	$p < 0.001$
16	25	N.S.	80	$p < 0.001$	173	$p < 0.001$	859	$p < 0.001$

[a] FF - Furfural.

[b] Furfural value represents the mean of three determinations.

[c] Not significant at $p < 0.05$.

Reprinted with permission from J. Agric. Food Chem. 21, 272 (1973). Copyright by the American Chemical Society.

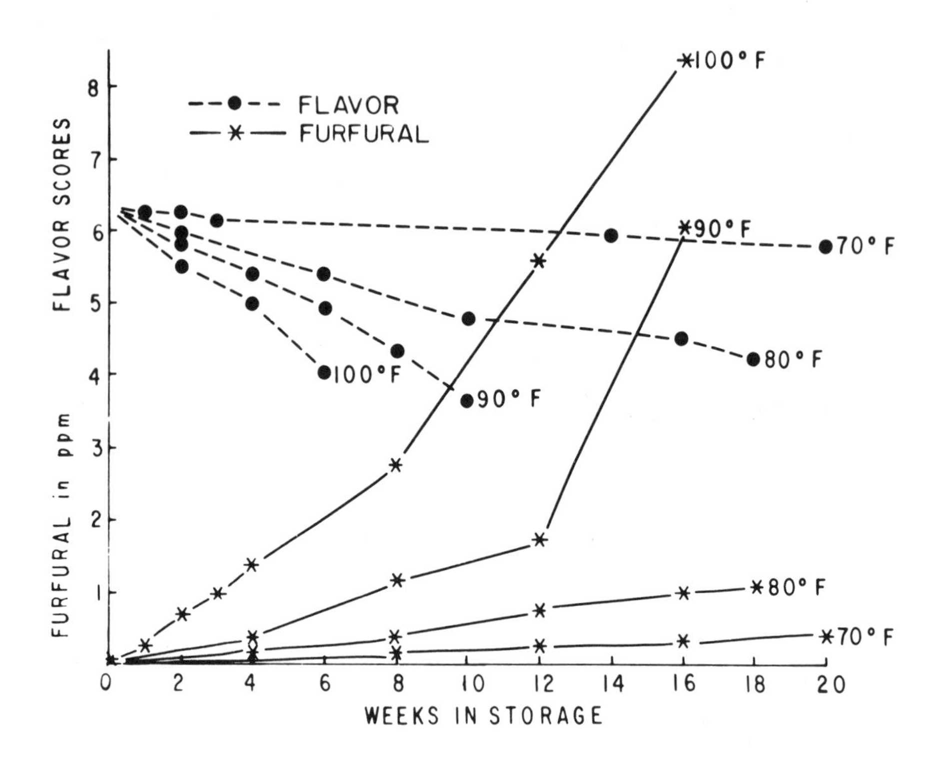

Figure 4. Effect of storage, time and temperature on the flavor score and furfural content of canned grapefruit juice. (From Proc. Fla. State Hortic. Soc. 90, 165 (1977). Copyright by Fla. State Hortic. Soc.)

VII. COMPOUNDS RESPONSIBLE FOR JUICE BITTERNESS

Bitterness has always been a natural but generally undesirable flavor property of citrus. Since several primitive varieties of citrus are often characterized by what might be considered excessive bitterness, it is probable that bitterness has been selectively bred out or reduced in most modern varieties.

Surprisingly, there is little structural similarity between the two chemical groups responsible for bitterness in citrus. The flavanone glycosides have the normal C_6-C_3-C_6 flavanoid structure plus a glucose and/or rhamnose sugar(s) attached at the 7-hydroxy group of the aglycone. It is extremely interesting to note that the flavanone aglycones are non-bitter, and only with the addition of a sugar does bitterness appear (37). However, not all flavanone glycosides are bitter; for bitterness is also dependent upon the linkage of the two sugars. If the sugars are linked via the 1 and 2 positions, the resulting flavanone glycoside is bitter. The glucose-rhamnose disaccharide is called neohesperidose (2-rhamnopyranosyl-1-glucopyranose) (38). If the sugars are linked in the 1 and 6 positions, the resulting isomeric flavanone glycoside is tasteless. The disaccharide in this case is called rutinose (6-rhamnopyranosyl-1-glucopyranose) (39). There are three major bitter flavanone glycosides in citrus: naringin, neohesperidin and poncirin. Each has a corresponding non-bitter rutinoside.

Limonoids, on the other hand, have planar steroid-like structures. While some limonoids are bitter, others are tasteless. They have no attached sugars, therefore their bitterness cannot be derived from sugar linkages. No specific portion of the limonin molecule has been associated with bitterness, although it is known that bitterness can be removed by opening the A or D lactone rings. Limonin is the major limonoid found in citrus (40), and is the compound primarily responsible for limonoid bitterness. Nomilin is also bitter but is found in much lower concentrations in juice of mature fruit (41).

The onset of bitterness is different for limonoids and flavanone neohesperidosides. Any immediate bitterness found in freshly extracted juice is due to flavanone neohesperidosides, whereas the bitterness that develops after the juice is allowed to stand and/or is pasteurized is due to limonin. This delayed bitterness is due to the conversion of the tasteless limonin precursor, identified as limonoic acid A-ring lactone (42), to the bitter limonin. The latent bitterness of some varieties, which are highly prized on the fresh fruit market, may preclude their use in processed juice.

VIII. METHODS OF ANALYSIS

The study and control of citrus bitterness has been hampered by the lack of good quantitative methods to determine the amounts of specific compounds responsible for bitterness. Numerous colorimetric tests for the presence of citrus flavanones have been developed (43-46). These involve the addition of a reducing agent, such as magnesium or sodium borohydride, to juice followed by the addition of HCl to form a characteristic color. However, the results are not very specific and represent a measure of total flavanone glycosides, both bitter and non-bitter, in addition to the flavanone aglycones.

Probably the most widely used colorimetric test is the one developed by W. B. Davis in 1947 (47). This test is based on the reaction of dilute alkali with flavanones to form the corresponding chalcones which give a characteristic yellow color at 427 nm. It has been shown (48) that the Davis test should be specific for only those flavanones with a protected 7-hydroxyl and a free 4'-hydroxyl, i.e., naringin, narirutin, prunin, and has been used as a general measure of naringin. Thus, other flavanone glycosides and aglycones, such as hesperidin, hesperitin, poncirin, etc. should not be detected. In practice, compounds which do not meet the above structural criteria, such as hesperidin and other non-flavanone juice components, do in fact form a weak yellow color. Investigators, such as Hendrickson and Kesterson (49), have used the Davis test to determine hesperidin in oranges, and Hagen et al. (50) found that the Davis test gave higher values in grapefruit juice when compared to other methods. None of the colorimetric methods could distinguish between the bitter and non-bitter flavanone glycosides, thus the values obtained are only an approximate measure of bitterness.

With the advent of thin-layer chromatography (TLC), several methods were developed to separate the bitter and non-bitter flavanone glycosides (51-54). Even though some methods required lengthy sample preparations good semi-quantitative results were obtained. Specific high pressure liquid chromatographyic (HPLC) methods have been developed for naringin and narirutin (55), hesperidin (56), and several other flavanone glycosides (57). These methods offer the advantages of greater analysis speed and ease of quantitation. With recent improvements (58), it is now possible to separate naringin, narirutin, neohesperidin and hesperidin in less than 15 min.

The determination of limonin in citrus juices has been a difficult problem because of the variety of fruit analyzed, the chemical complexity of the juices, low concentration levels and the lack of specific means of detection. Limonoids do not possess strong chromaphores and lack a characteristic UV spectrum. Therefore, they are usually separated and concentrated before analysis. A few spectrophotometric techniques were developed (59, 60) which require an extremely lengthy sample preparation prior to the actual analysis. Many TLC methods have been developed (42, 61-65) with varying clean-up procedures. As expected, those procedures with the greatest sensitivity, least susceptibility to interferences, and the ability to resolve several limonoids take the greatest amount of sample preparation.

Several HPLC procedures have been developed (66, 68) which reduce analysis time to less than 10 min and can be automated. However, they like most of the TLC methods, still employ chloroform extraction and concentration steps prior to analysis. The method of Rouseff and Fisher (68) offers several advantages in that it is rapid, highly sensitive and capable of resolving other limonoids in addition to limonin.

Thus, with recent improvements in methodology it is now possible to evaluate the relative contribution of individual citrus bitter components because the concentration of each can now be rapidly and accurately measured.

IX. TASTE THRESHOLDS AND RELATIVE BITTERNESS

Bitterness is considered an undesirable flavor in all citrus juices except grapefruit juice where a little bitterness is actually desirable. However, the level at which bitterness is first detected, as well as the level considered excessive, varies considerably. Fellers (69) evaluated a large population of grapefruit juice users and found that the level of preference decreased as the level of added naringin increased. Some of this data is shown in Table 4. Juices

TABLE 4

Evaluation of naringin bitterness as a factor in preference of Florida grapefruit juice.

	Naringin conc. (ppm)				
	300	700	1100	1500	1900
Mean rating	3.7	3.6	3.4	3.4	3.3
Bitterness					
Too bitter	17[a]	31	33	46	51
Just right	59	55	50	44	37
Not enough bitterness	24	14	17	10	12
Don't know	--	--	--	--	--
(Number of respondents)	(201)	(198)	(198)	(197)	(194)

[a] Percentage of tasters.

Source: Fellers, P. J., Florida Department of Citrus, Unpublished data (1975).

were rated on a six-point scale from poor = 1 to excellent = 6. With nothing else changed the overall preference for the juice decreased regularly as naringin concentrations increased. When just bitterness was evaluated the proportion of the group that thought the juice was "too bitter" increased regularly with increasing naringin concentration. However, the surprising part of this data is that even at 1900 ppm naringin, 37% of the group thought the juice just right and 24% thought the original 300 ppm juice not bitter enough. This suggests that the tolerances and preferences for naringin bitterness varies widely.

Guadagni et al. (70) found in a panel of 27 judges trained in detecting bitterness that individual taste thresholds for limonin bitterness in orange juice varied widely. The most sensitive individual had a taste threshold of 0.5 ppm in the juice whereas the least sensitive individual had a threshold of 32 ppm. It can be seen from Table 5 that as the

TABLE 5

Thresholds of individuals for limonin bitterness in orange juice.[a]

Limonin threshold (ppm)	Cumulative % of panel
0.5	8
1.0	17
2.0	30
3.0	49
4.0	62
5.0	70
6.0	75
10.0	91
32.0	99.5

[a] pH 3.8, B/A = 14.8

Source: Guadagni, D. G., Maier, V. P., and Turnbaugh, J. G., J. Sci. Food Agric. 24:1277 (1973).

limonin concentration is increased a greater portion of the panel could detect its bitterness. In order for a simple majority of the panel to detect the presence of bitterness in the juice, limonin concentration had to be between 3 and 4 ppm.

In a study of 30 possible analytical indicators in 108 orange juice samples, Carter et al. (71) found that of all the indicators, limonin concentration gave the greatest simple correlation coefficient (r = -0.666) to flavor score. This is even more interesting in that the average limonin concentration was less than 1.0 ppm which is considerably below the average taste threshold for this compound in orange juice. Maximum limonin concentration of any single juice was 4 ppm which just exceeds the average taste threshold (See Table 5). Based on Guadagni's work (70), Maier (72) speculated that a sizable percentage of consumers may have low bitterness thresholds and may avoid or restrict their consumption of citrus juices even though the limonin content of these juices is below the average taste threshold of the population in general. The work of Carter et al. (71) appears to confirm this speculation.

Other factors in addition to limonin or naringin alone have been found to alter the preception of bitterness. Guadagni et al. (73) were the first to note the additive nature of limonin and naringin. Solutions in which individual limonin and naringin concentrations were below their respective taste thresholds were judged bitter (at the 99% confidence level). In addition, Guadagni et al. (70, 74) found the presence of sugars or other sweetners would increase bitterness thresholds. Juice pH was also found to alter taste thresholds and was suggested as a means of altering the preception of juice bitterness.

Citrus juices contain a number of related compounds which may increase or reduce the preception of bitterness. Citrus juices contain a number of flavanone glycosides in addition to naringin which are also bitter. However, as illustrated in Table 6, the occurrence and relative bitterness of these compounds varies considerably. The sweet oranges, lemons, tangerines, etc. are characterized by almost a total lack of bitter flavanone neohesperidosides. All their flavanone glycosides are in the non-bitter rutinoside form. Any bitterness in these varieties is due to limonin. Grapefruit contain a mixture of the bitter neohesperidosides and non-bitter rutinosides. Naringin is the predominant neohesperidoside followed by the equally bitter poncirin, and the much less bitter prunin (naringinen 7-β-glucoside). Thus, naringin is the predominant but not the sole source of flavanone glycoside bitterness.

TABLE 6

Occurrence and relative bitterness of citrus flavanone neohesperidosides.

	Relative bitterness[a]	Relative concentrations[b] Sweet orange	Lemon	Grapefruit
Neoeriocitrin	$<$ 2	--	--	--
Neohesperidin	2	--	--	tr
Prunin	6	--	--	1
Naringin	20	--	--	10
Poncirin	20	--	--	2
Quinine dihydrochloride	100	--	--	--

Source: [a] Horowitz, R. M. in "Biochemistry of Phenolic Compounds" (Harborne, J. B., ed). Academic Press, New York, 1964, p. 545.

[b] Albach, R. F., and Redman, G. H., Phytochemistry 8: 127 (1969).

While other limonoids have been found in the seeds of many citrus varieties (40), limonin was the only limonoid thought to exist in juice. Rouseff (41) found the presence of nomilin in both orange and grapefruit juices, however, the concentrations of this bitter limonoid was only 1/10 to 1/20th of that found for limonin. Furthermore, the relative bitterness of this limonoid has not been established so it is difficult to determine what the relative contribution, if any, nomilin adds to total limonoid bitterness.

Thus, the wide range of sensitivity to both limonin and naringin bitterness, as well as the presence of other compounds which may alter the preception of bitterness, has made it extremely difficult to set objective standards for juice bitterness based on the concentrations of specific bitter compounds.

X. EFFECTS OF MATURITY AND VARIETY

Any immediate bitterness found in fresh citrus is due to the bitter flavanone neohesperidosides. It has been proposed (37) that all citrus species can be placed into one of two groups depending on whether all their flavanone glycosides were in the neohesperidoside or rutinoside forms. Thus, any variety that contained both forms must be considered a hybrid rather than a true species. The validity of this hypothesis was demonstrated by Albach and Redman (75).

Grapefruit, which has both glycosidic forms is suspected of being a hybrid of the pumelo (all neohesperidosides) and the sweet orange (all rutinosides). Thus, flavanone glycoside bitterness might be reduced or eliminated by crossing bitter hybrids, such as grapefruit, with varieties whose flavanone glycosides are all in the non-bitter rutinoside form.

Limonin and limonoids appear to be ubiquitous in citrus (40). However, in most varieties the limonin level is below the taste threshold in mature fruit. As seen in Figure 5,

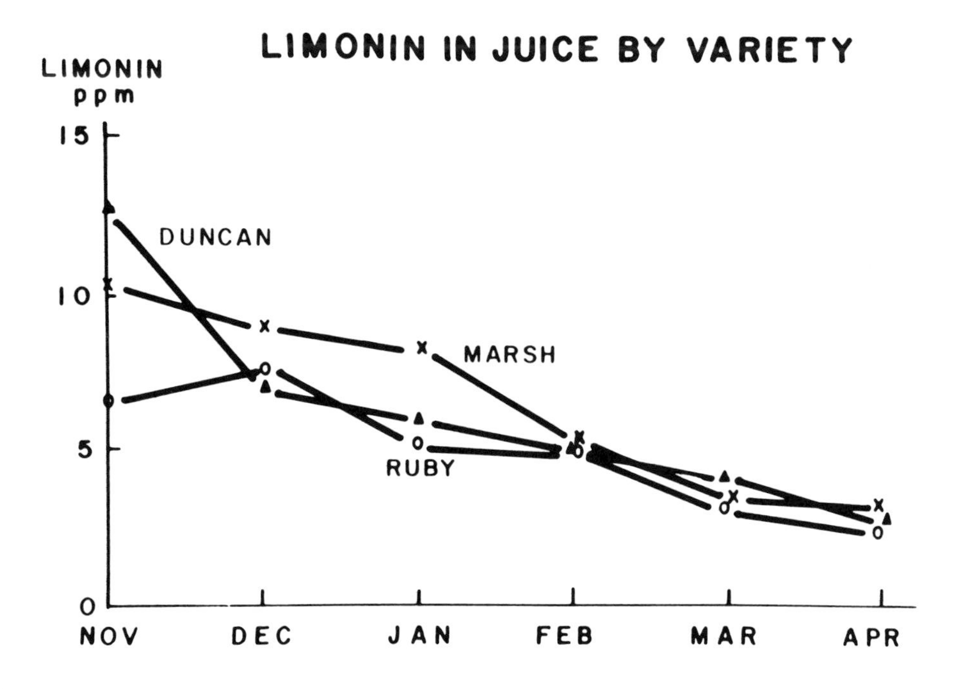

Figure 5. Limonin content of Florida grapefruit juice by variety, average of three locations. (From Proc. Fla. State Hortic. Soc. 83, 270 (1970). Copyright by Fla. State Hortic. Soc.).

there is a substantial decrease in limonin levels as the fruit matures. It should also be noted that Marsh "seedless" grapefruit juice generally contained more limonin than either the Ruby or Duncan juices. It has been observed in oranges (76), as well as grapefruit, that seedy varieties generally have lower juice limonin values than "seedless" varieties, such as the navel orange. Limonin bitterness in grapefruit juice can thus be reduced by using a low limonin (seedy) variety or simply waiting for the fruit to become more mature.

Rootstocks have a profound effect on the rate at which the limonoid content decreases with fruit maturity (77, 78). When Washington navel oranges were grown on grapefruit rootstock, the juice of the commercially mature fruit was essentially free of bitterness during the early part of the season. On other rootstocks, bitterness decreased as the season progressed, however, bitterness remained high throughout the season in fruit grown on rough lemon rootstock. Therefore, maturity, variety and rootstock all have a considerable effect on juice bitterness.

XI. PROCESSING TECHNIQUES AND JUICE BITTERNESS

While limonoids are found throughout the fruit, they are concentrated primarily in the seeds and to a lesser extent in the segment membrane (40, 79, 80). Limonin levels (or more precisely, its tasteless precursor) are lowest in the juice vesicles. Therefore, the processed juice cannot have a limonin content lower than this value. However, depending on the amount of grinding and macerating during extraction, as well as the length of time the macerated tissue (pulp) is allowed to stay in contact with the juice, considerable amounts of additional limonin may be extracted into the juice. Processing equipment has been developed (81) to minimize the amount of maceration and the pulp-juice contact time. As illustrated in Figure 6, throughout the season lower limonin values are obtained from grapefruit juice extracted under mild conditions (soft squeeze) than if the juice is extracted under heavy pressure. It is also apparent from Figure 6 that extractor pressures are most critical during the early portion of the season. As the fruit continues to mature and limonin levels fall, extractor pressure ceases to have a significant effect on limonin levels in juice. Tatum and Berry (65) found similar results in the extraction of orange juice. Soft extractor pressures yielded less juice per fruit, whereas hard squeeze conditions produced more juice but of poorer quality (greater bitterness) (82).

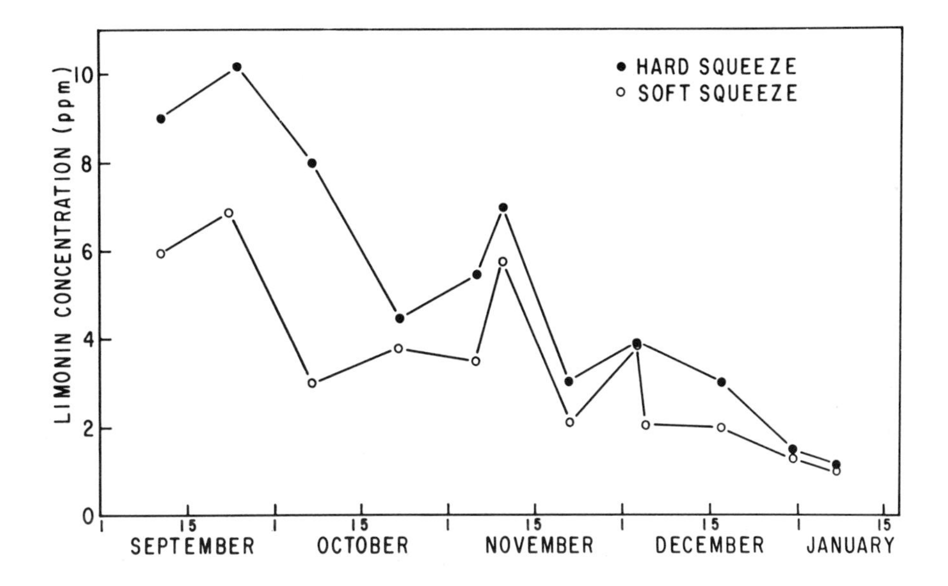

Figure 6. Effect of extractor pressure on grapefruit juice limonin content from Sept. 12, 1974 to Jan. 7, 1975. (From Proc. Int. Soc. Citriculture 3, 816 (1977)).

Naringin is concentrated in the albedo of grapefruit peel (83) with relatively small amounts found in the juice vesicles. Poncirin, another bitter flavanone glycoside, has also been found in significant concentrations in the peel (84). Juice readily solubilizes these bitter glycosides during the extraction process (85). Even if the juice is strained as it is being extracted, the naringin content will be reduced by only 20% (86). Hendrickson et al. (87) found grapefruit naringin levels to be dependent on the amount of rag and peel incorporated into the juice, and the pressure used to extract the juice. Huet (86) reported that naringin content could be reduced 10% simply by limiting the amount of pulp incorporated into the juice. The effect of extractor pressure on juice naringin content is clearly shown in Figure 7. It should be pointed out that the naringin values are from the Davis test and are considerably higher than true naringin. Extractor pressures continue to be a significant factor in juice naringin levels throughout most of the processing season.

XII. POST EXTRACTION METHODS TO REDUCE BITTERNESS

Several physical and biochemical methods have been proposed to reduce bitterness in processed citrus juices.

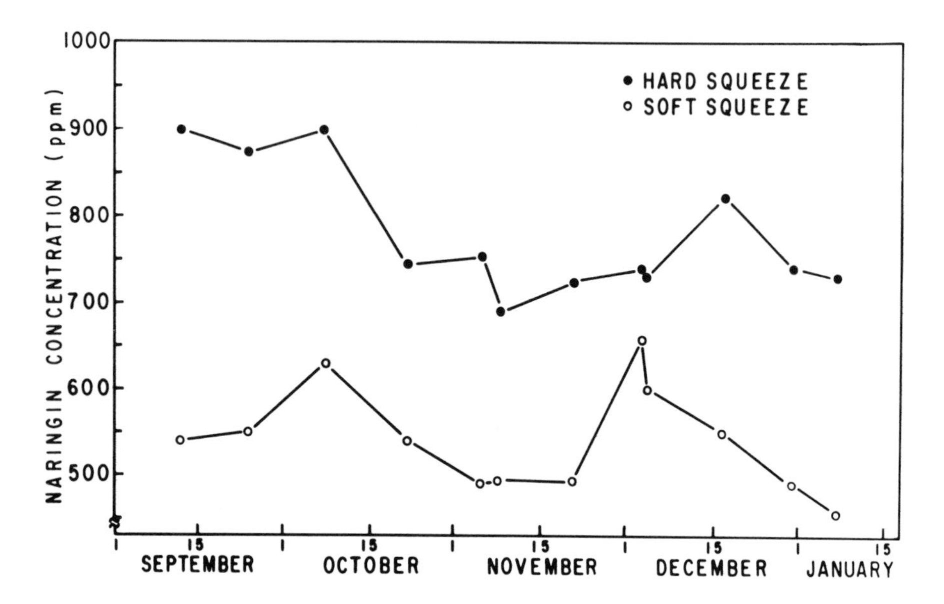

Figure 7. Effect of extractor pressure on grapefruit juice naringin content from Sept. 12, 1974 to Jan. 7, 1975. (From Proc. Int. Soc. Citriculture 3, 816 (1977)).

Processes involving precipitation, extraction or adsorption have been used with varying degrees of success. Chandler (88) found that pectin helped solubilize limonin. Thus, it was proposed that if the pectin in the juice was removed or reduced, some of the limonin might come out of the solution and the filtered or centrifuged juice would be less bitter. Juices that were treated with pectolytic enzymes became less bitter (89) but also experienced an undesirable loss of cloud. Solvent extractions using isopropanol (90) and vegetable oil (91) have been proposed but not employed commercially. Apparently desirable juice components were also extracted along with the limonin.

The most successful physical method of removing limonin has been through adsorption. Activated carbon successfully removes limonin from juice (89) but also removes some of the desirable pigments as well. Although there are some problems with carbon fines in the juice, the method has been used to a limited extent commercially on navel orange juice products (72). Polyamide powders were proposed (92, 93) to reduce limonin as well as flavanone bitterness in processed juice. The procedure had to be done on a batch basis and was multistepped. However, the primary drawback to commercialization of this process was the concurrent 30% loss of ascorbic acid. Chandler and Johnson (94) recently reported that cellulose acetate powder removed 44-70% of the juice limonin in less than one hour and removed only negligible amounts of ascorbic acid or hesperidin.

Another totally different approach has been the use of enzymes to convert limonin or its precursor to non-bitter products. Chandler reported (95) finding an enzyme in the albedo of navel and Valencia oranges capable of reducing limonin bitterness. However, the enzyme was not isolated so it is not known if the enzyme acted directly on the limonin or in some indirect manner to reduce juice bitterness. Two limonin dehydrogenases have been isolated from bacteria (96, 97). Of the two, the enzyme isolated from Pseudomonas appears to be more active at the pH of orange juices (98). This latter enzyme and the cellulose acetate process appear to hold the most commercial promise.

Naringin bitterness in grapefruit juice can also be removed or reduced with the use of enzymes. Kishi (99) isolated naringinase enzymes from strains of Aspergillus niger grown on citrus media. Thomas et al. (100) isolated a particularly active naringinase and also showed that the naringin was converted to the less bitter prunin and not to the tasteless naringinen. Naringinase was recently immobilized in a hollow fiber reactor and successfully used to reduce naringin bitterness in grapefruit juice. About 40% of the naringin was removed within three hours and juice bitterness was correspondingly reduced.

XIII. EFFECTS OF ABSCISSION CHEMICALS ON FLAVOR QUALITY OF PROCESSED JUICES

Mechanical harvesting of citrus fruit has necessitated a need for abscission chemicals that will reduce the force for separating the fruit from the stem. These abscission chemicals, however, have been shown to affect the chemical composition of cold-pressed orange oil (101, 102). Moshonas and co-workers (191) speculated that abscission chemicals enhance the aging process of the fruit and, thereby, cause adverse effects on the flavor quality of the juice extracted from these fruit. In a latter study, Moshonas and Shaw (103) showed that those abscission agents which caused injury to the peel also caused the formation of six unknown phenolic ethers, namely engenol, cis-methylisoeugenol, trans-methylisoeugenol, methyleugenol, elemicin and isoelemicin. Although none of the ethers was present at their respective flavor threshold levels (103), their flavor effects were additive and the combined effect was apparently sufficient to account for the off-flavor of the single-strength orange juice processed from these treated fruit.

REFERENCES

1. Boyd, J. M., and Peterson, G. T., Ind. Eng. Chem. 37: 370 (1945).
2. Mottern, H. H., and Von Loesecke, H. W., Fruit Prod. J. 12: 325 (1933).
3. Curl, A. L., Fruit Prod. J. 25: 356 (1946).
4. Curl, A. L., and Veldhuis, M. K., Fruit Prod. J. 26: 329 (1947).
5. Swift, L. J., Proc. Fla. State Hortic. Soc. 64: 181 (1951).
6. Nagy, S., and Nordby, H. E., J. Agric. Food Chem. 18: 593 (1970).
7. Riester, D. W., Brown, O. G., and Pearce, W. E., Food Ind. 17: 742 (1945).
8. Blair, J. S., Godar, E. M., Masters, J. E., and Riester, D. W., "Exploratory Experiments to Identify Some of the Chemical Reactions Causing Flavor Deterioration during Storage of Canned Orange Juice", American Can Co., Maywood, Ill., 1950.
9. Tatum, J. H., Nagy, S., and Berry, R. E., J. Food Sci. 40: 707 (1975).
10. Pulley, G. N., and Von Loesecke, H. W., Ind. Eng. Chem. 31: 1275 (1939).
11. Proctor, B. E., and Kenyon, E. M., Food Technol. 3: 387 (1949).
12. Henry, R. E., and Clifcorn, L. E., Canning Trade 70 (No. 31): 7 (1948).
13. Nordby, H. E., and Nagy, S., in "Fruit and Vegetable Juice Processing Technology" (P. E. Nelson, ed.). Avi Publishing Co., Westport, Conn., 1980.
14. Peterson, G. T., Continental Can Co. Bull. 18 (1949).
15. Kefford, J. F., McKenzie, H. A., and Thompson, P. C. O., J. Sci. Food Agric. 10: 51 (1959).
16. Continental Can Company, "Quality of Canned Orange Juice", Bull. 2 (1945).
17. Kefford, J. F., McKenzie, H. A., and Thompson, P. C. O., Food Preserv. Q. 10: 44 (1950).
18. Nolte, A. J., and Von Loesecke, H. W., Food Res. 5: 457 (1940).
19. Braddock, R. J., and Kesterson, J. W., J. Agric. Food Chem. 21: 318 (1973).
20. Askar, A., Bielig, H. J., and Treptow, H., Dtsch. Lebenem.-Rundsch. 69: 162 (1973).
21. Nagy, S., and Dinsmore, H. L., J. Food Sci. 39: 1116 (1974).
22. Nagy, S., in "Citrus Science and Technology" (S. Nagy, P. E. Shaw and M. K. Veldhuis, eds.), Vol. 1, p. 266. Avi Publishing Co., Westport, Conn., 1977.

23. Blair, J. S., Godar, E. M., Masters, J. E., and Riester, D. W., Food Res. 17: 235 (1952).
24. Continental Can Company, "Flavor Studies of Canned Single Strength Orange Juice", Continental Can Company, Maywood, Ill., 1954.
25. Kirchner, J. S., and Miller, J. M., J. Agric. Food Chem. 5: 283 (1957).
26. Senn, V. J., J. Food Sci. 28: 531 (1963).
27. Rymal, K. S., Wolford, R. W., Ahmed, E. M., and Dennison, R. A., Food Technol. 22: 1592 (1968).
28. Dinsmore, H. L., and Nagy, S., J. Agric. Food Chem. 19: 517 (1971).
29. Murdock, D. I., in "Citrus Science and Technology" (S. Nagy, P. E. Shaw and M. K. Veldhuis, eds.), Vol. 2, p. 474. Avi Publishing Co., Westport, Conn., 1977.
30. Vandercook, C. E., J. Food Sci. 35: 517 (1970).
31. Dinsmore, H. L., and Nagy, S., J. Food Sci. 37: 768 (1972).
32. Dinsmore, H. L., and Nagy, S., J. Assoc. Off. Anal. Chem. 57: 332 (1974).
33. Nagy, S., and Randall, V., J. Agric. Food Chem. 21: 272 (1973).
34. Maraulja, M. D., Blair, J. S., Olsen, R. W., and Wenzel, F. W., Proc. Fla. State Hortic. Soc. 86: 270 (1973)
35. Nagy, S., Randall, V., and Dinsmore, H. L., Proc. Fla. State Hortic. Soc. 85: 222 (1972).
36. Dougherty, M. H., Ting, S. V., Attaway, J. A., and Moore, E. L., Proc. Fla. State Hortic. Soc. 90: 165 (1977).
37. Horowitz, R. M., in "Biochemistry of Phenolic Compounds" (J. B. Harborne, ed.). Academic Press, New York, 1964.
38. Horowitz, R. M., and Gentili, B., Tetrahedron 19: 773 (1963).
39. Zemplen, G., Tettamanti, A. K., and Terago, S., Chem. Ber. 71B: 2511 (1938).
40. Dreyer, D. L., Phytochemistry 5: 367 (1966).
41. Rouseff, R. L., Personal observation (1979).
42. Maier, V. P., and Beverly, G. D., J. Food Sci. 33: 488 (1968).
43. Shinoda, J., Pharm. Soc. (Japan) 48: 214 (1928).
44. Kwietny, A., and Braverman, J. B. S., Bull. Res. Council Israel C7: 187 (1959).
45. Horowitz, R. M., J. Org. Chem. 22: 1733 (1957).
46. Rowell, K. M., and Winter, D. H., J. Am. Pharm. Assoc. Sci. Ed. 48: 746 (1959).
47. Davis, W. B., Anal. Chem. 19: 476 (1947).
48. Horowitz, R. M., and Gentili, B. Food Res. 24: 757 (1959).

49. Hendrickson, R., and Kesterson, J. W., Proc. Fla. State Hortic. Soc. 75: 289 (1962).
50. Hagen, R. E., Dunlap, W. J., Mizelle, J. W., Wender, S. H., Lime, B. J., Albach, R. F., and Griffiths, F. P., Anal. Biochem. 12: 472 (1965).
51. Horhammer, L., and Wagner, H., Dtsch. Apoth. Ztg. 102: 759 (1962).
52. Hagen, R. E., Dunlap, W. J., and Wender, S. H., J. Food Sci. 31: 542 (1966).
53. Fisher, J. F., Nordby, H. E., and Kew, T. J., J. Food Sci. 31: 947 (1966).
54. Tatum, J. H., and Berry, R. E., J. Food Sci. 38: 340 (1973).
55. Fisher, J. F., and Wheaton, T. A., J. Agric. Food Chem. 24: 888 (1976).
56. Fisher, J. F., J. Agric. Food Chem. 26: 1459 (1978).
57. Galensa, R., and Herrmann, K., Z. Lebensm Unters. Forsch. 166: 355 (1978).
58. Rouseff, R. L., Unpublished data (1980).
59. Wilson, K. W., and Crutchfield, C. A., J. Agric. Food Chem. 16: 118 (1968).
60. Fisher, J. F., J. Agric. Food Chem. 21: 1109 (1973).
61. Dreyer, D. L., J. Org. Chem. 30: 749 (1965).
62. Chandler, B. V., and Kefford, J. F., J. Sci. Food Agric. 17: 193 (1966).
63. Maier, V. P., and Grant, E. R., J. Agric. Food Chem. 18: 250 (1970).
64. Chandler, B. V., J. Sci. Food Agric. 22: 474 (1971).
65. Tatum, J. H., and Berry, R. E., J. Food Sci. 38: 1244 (1973).
66. Fisher, J. F., J. Agric. Food Chem. 23: 1199 (1975).
67. Fisher, J. F., J. Agric. Food Chem. 26: 497 (1978).
68. Rouseff, R. L., and Fisher, J. F., Anal. Chem., Submitted (1980).
69. Fellers, P. J., Florida Dept. of Citrus, Unpublished data (1975).
70. Guadagni, D. G., Maier, V. P., and Turnbaugh, J. F., J. Sci. Food Agric. 24: 1277 (1973).
71. Carter, R. D., Buslig, B. S., and Cornell, J. A., Proc. Fla. State Hortic. Soc. 88: 358 (1975).
72. Maier, V. P., in "Citrus Science and Technology" (S. Nagy, P. E. Shaw and M. K. Veldhuis, eds.), Vol. 1, p. 355. AVI Publishing Co., Westport, Conn., 1977.
73. Guadagni, D. G., Maier, V. P., and Turnbaugh, J. G., J. Sci. Food Agric. 25: 1349 (1974).
74. Guadagni, D. G., Maier, V. P., and Turnbaugh, J. G., J. Sci Food Sgric. 25: 1199 (1974).
75. Albach, R. F., and Redman, G. H., Phytochemistry 8: 127 (1969).

76. Kefford, J. F. and Chandler, B. V., "Chemical Constituents of Citrus Fruits", p. 149. Academic Press, New York, 1970.
77. Marsh, G. L., Food Technol. 7: 145 (1953).
78. Kefford, J. F., and Chandler, B. V., J. Agric. Res. 12: 56 (1961).
79. Scott, W. C., Proc. Fla. State Hortic. Soc. 83: 270 (1970).
80. Levi, A., Flavian, S., Harel, S., Ben-Gera, I., Stern, F., and Berkovitz, S., Publication No. 29. Div. of Food Tech., Volcani Cent., Israel (1974).
81. Higby, R. H., Calif. Citrogr. 26: 360 (1941).
82. Attaway, J. A., Proc. Int. Soc. Citriculture 3: 816 (1977).
83. Kefford, J. F., Adv. Food Res. 9: 285 (1959).
84. Horowitz, R. M., and Gentili, B., Arch. Biochem. Biophys. 92: 191 (1961).
85. Manabe, R., Bessho, Y., Kodama, M., and Kubo, S., Nippon Shokukin Kogyo Gallaishi 11: 389 (1964).
86. Huet, R., Fruits (Paris) 16: 327 (1961).
87. Hendrickson, R., Kesterson, J. A., and Edwards, G. J., Proc. Fla. State Hortic. Soc. 71: 190 (1958).
88. Chandler, B. V., CSIRO Food Res. Q. 31: 36 (1971).
89. McColloch, R. J., Calif. Citrogr. 35: 290 (1950).
90. Pritchett, D. E., U. S. Pat. 2,816,033, Dec. 10 (1957).
91. Swisher, H. E., U. S. Pat., 3,647,476, Mar. 7 (1972).
92. Chandler, B. V., Kefford, J. F., and Ziemelis, G., J. Sci. Food Agric. 19: 83 (1968).
93. Griffiths, F. P., U. S. Pat. 3,463,763, Aug. 26 (1969).
94. Chandler, B. V., and Johnson, R. L., J. Sci. Food Agric. 28: 875 (1977).
95. Chandler, B. V., J. Sci. Food Agric. 22: 634 (1971).
96. Hasegawa, S., Bennett, R. D., Maier, V. P., and King, A. D., Jr., J. Agric. Food Chem. 20: 1031 (1972).
97. Hasegawa, S., Maier, V. P., and King, A. D., Jr., J. Agric. Food Chem. 22: 523 (1974).
98. Brewster, L. C., Hasegawa, S., and Maier, V. P., J. Agric. Food Chem. 24: 21 (1976).
99. Kishi, K., Kagaku To Kogyo (Osaka) 28: 140 (1955).
100. Thomas, D. W., Smythe, C. V., and Labbee, M. D., Food Res. 23: 591 (1958).
101. Moshonas, M. G., Shaw, P. E., and Sims, D. A., J. Food Sci. 41: 809 (1976).
102. Moshonas, M. G., and Shaw, P. E., J. Agric. Food Chem. 25: 1151 (1977).
103. Moshonas, M. G., and Shaw, P. E., J. Agric. Food Chem. 26: 1288 (1978).

ANALYSIS AND CONTROL OF MILK FLAVOR

W.F. Shipe

Department of Food Science
Cornell University
Ithaca, New York

Flavor is a major factor affecting the consumption of milk and therefore dairymen are concerned about its control. As is the case with any control program there are three essential aspects. First, it is necessary to characterize the standard (e.g. establish flavor specifications). Secondly, methods must be available or developed for determining if specifications have been met. Finally, it is necessary to determine how to prevent significant deviations from the standard. The ultimate objective of milk flavor control is satisfying the consumers who base their judgements on sensory characteristics. Therefore one must use sensory evaluations to establish standards and monitor quality. Because of the limitations of sensory methods, objective methods are being developed for routine quality evaluations. Objective methods are also useful in determining the causes of off-flavors. This report deals with the sensory and objective measurements and the factors affecting flavor.

SENSORY STANDARDS AND PROCEDURES

In the beginning flavors were described in sensory terms since the chemical nature was not known. It was assumed that fresh clean raw milk from healthy animals had the ideal milk flavor. This ideal flavor was characterized as being bland and pleasantly sweet. Any deviation from this was considered undesirable. Because of the bland flavors of the fresh product it was difficult to describe it in precise terms whereas the undesirable flavors often seemed easier to characterize.

ISBN 0-12-169065-2

Consequently, in practice quality control programs are designed primarily to eliminate undesirable flavors. This approach is valid if the quality control personnel and the consumers agree on what is desirable and undesirable.

The American Dairy Science Association (ADSA) developed (1) scorecards for the sensory evaluation of a number of dairy products, including milk. These scorecards are used to rate the overall quality of samples and identify any defects. To simplify and standardize the evaluations the scorecards contain lists of defects which have been encountered in the various products. In rating the samples, panelists check off any detectable defects. Space is provided to write in any additional defects that are perceived. The list of milk flavor defects is shown in Table I along with a scoring guideline.

TABLE I. ADSA Milk Flavor and Scoring Guide (2)

	Flavor intensity		
Flavors	Slight	Definite	Pronounced
Astringent	8	7	5
Barny	5	3	1
Bitter	5	3	1
Cooked	9	8	6[a]
Cowy	6	4	1
Feed	9	8	5
Fermented/Fruity	5	3	1
Flat	9	8	7
Foreign	5	3	1
Garlic/Onion	5	3	1
High acid	3	1	...[b]
Lacks freshness	8	7	6
Malty	5	3	1
Metallic	5	3	1
Oxidized	6	4	1
Rancid	4	1	...[b]
Salty	8	6	4
Unclean	3	1	...[b]

[a] A slight cooked flavor that is not objectionable may be scored 10.

[b] ... Denotes unsalable

This list constitutes a flavor profile - of undesirable flavors. This list has evolved over many years and includes a mixture of associative terms such as barny and cowy with more scientific terms such as high acid and oxidized. Recently, the flavor nomenclature committee of the American Dairy Science Association (2) recommended the adaption of a classification of undesirable flavors on the basis of their causes. Because this classification system (Table II) focuses attention on causes it is particularly useful in quality control work.

TABLE II. Categories of Off-Flavors in Milk

Causes	Descriptive or Associative Terms
1. HEATED	Cooked, carmelized, scorched
2. LIGHT-INDUCED	Light, sunlight, activated
3. LIPOLYZED	Rancid, butyric, bitter[a], goaty
4. MICROBIAL	Acid, bitter[a], fruity, malty, putrid, unclean
5. OXIDIZED	Papery, cardboard, metallic, oily, fishy
6. TRANSMITTED	Feed, weed, cowy, barny
7. MISCELLANEOUS	Absorbed, astringent, bitter[a], chalky, chemical, flat, foreign, lacks freshness, salty

[a]*Bitter flavor may arise from a number of different causes. If the specific cause is not known it should be classified under miscellaneous.*

A numerical score is used to rate the overall flavor quality. Originally, 45 points was the perfect score for milk flavor, but samples were normally scored between 30 and 40. Currently, milk flavor is being scored on a 10 point scale. A score of 10 is given to a sample that has a "typical fresh" taste and is free of undesirable flavors. Samples lacking in flavor or having off-flavors are scored lower. The magnitude of the deduction depends on the intensity and nature of the defect as is indicated in the scoring guideline (Table I). For example, a pronounced cooked flavor is not considered very objectionable and therefore would be scored 6 whereas even a slight oxidized would be scored 6. The scoring system appears somewhat arbitrary but it is based on years of experience and is supposed to reflect consumers' reaction to the various flavors. Although there has not been a systematic study of consumer reactions to all flavors there have been separate studies on individual flavors. In spite of the limitations of trying to estimate consumer acceptance, reports (4,5,6,7,8) indicate that the scores of trained panelists can be used to predict consumer ratings. However, periodic studies should be made to determine if the scoring system reflects current consumer reaction.

Of course, the reliability of sensory data is dependent on the ability and training of panelists. Panelists are trained by having them repeatedly evaluate samples with the various off-flavors. Procedures have been developed (3) for preparing some of the off-flavors in the laboratory but in other cases this is not possible. In these cases the instructor must try to obtain samples in the marketplace. This approach has two limitations - first it is difficult to find the variety of flavors at the right time and the instructor may have to rely solely on his judgement for identification of the flavor. Additional research is needed to develop objective methods for confirming the identification of flavors that have not yet been chemically characterized.

In a critical review, O'Mahony (9) claims that the ADSA system does not follow general psychophysical procedures and that the results cannot be analyzed statistically. He points out the limitations of simultaneously trying to identify the nature and intensity of flavors and to predict consumer response. Certainly, it would be more conventional to use a trained panel for descriptive evaluations and a consumer panel for preference ratings. However, if the available evidence indicates that a given intensity of a particular flavor produces a strong negative consumer reaction it seems appropriate to give it a low score. Of course, there are three potential mistakes in this evaluation - namely, improper flavor identification, erroneous estimate of intensity and incorrect evidence concerning consumer response. But the same potential errors would be involved even if three separate evaluations had been conducted. In conclusion, the ADSA sensory evaluation approach provides a useful quality control tool for products with established consumer acceptance standards. Conventional descriptive analysis, intensity ratings, and hedonic evaluations should be used in establishing standards or specifications for new or modified products.

CAUSES AND MEASUREMENT OF UNDESIRABLE FLAVORS

Since research on causes and measurements of undesirable flavors has been interwoven discussion of the two topics will be combined. The flavor classification scheme shown in Table II will be used for this discussion.

Heated Flavors

This term is used to include all of the flavors which

are produced by thermal processing of fluid milk. The nature and intensity of the flavor depends on the time and temperature of heating. These heated induced flavors have been classified (3) into the following four types: Cooked or sulfurous, heated or rich, caramelized, and scorched. Normal pasteurization treatment imports a slight cooked or sulfurous note. If the thermal treatment is the minimum required for legal pasteurized many people cannot detect this flavor. As the temperature increases the flavor becomes more obvious. The evidence (9) indicates that cooked flavor is due to volatile sulfides derived from the sulfhydryl groups of β lactoglobulin. Hydrogen sulfide appears to be the major component but other volatile sulfides contribute to the flavor (12). The cooked or sulfurous note dissipates upon storage and usually has disappeared within 2 to 3 days.

Milk exposed to 135 to 150 C for several seconds has a strong sulfurous or cooked note immediately after the heat treatment (13). Reports (14,15) indicate that after several days of refrigerated storage, the sulfurous note disappear and a rich or heated note appears. The chemical nature of the rich, heated note has not been confirmed but recent evidence (16) indicates that diacetyl is involved. Lactones, methyl ketones, maltol, vanillin, benzaldehyde, and acetophenone may also contribute to this flavor (16).

Research workers in England (17,18) have reported a "cabbagey" flavor in milk immediately after ultra high temperature (UHT) treatment. They reported that this flavor decreased within a few days. The rate of decrease was more rapid when the initial oxygen level was high (17). However, high oxygen levels caused rapid decreases in ascorbic acid and folic acid content. The length of heating time at 140 C also affects the intensity and rate of disappearance of the cabbagey flavor (18). This flavor disappeared more rapidly in milk that was heated for 90 seconds than in milk that was only heated for 3 seconds. The authors suggested that the cabbagey flavor might be the result of an interaction between S-bearing compounds, especially H_2S and CH_3SH, and one or more carbonyls. They speculated that there might be competing reactions for the S compounds in the presence of Maillard reaction intermediate compounds. In the 3 second treatment, compared to 90 seconds, there might be less Maillard reaction intermediates and therefore more S compounds would be available for formation of the cabbagey complex. They did find a good correlation between the total volatile sulphur content and the cabbagey flavor. The volatile S compounds included H_2S, COS, CH_3SH, CS and $(CH_3)_2S$. A decline in H_2S, CH_3SH and CS_2

parallel the decrease in cabbagey flavor. The authors (18) reported that as the cabbagey flavor disappeared a cooked, rice pudding like flavor became apparent. It appears that these authors use cabbagey for the flavor described as cooked or sulfurous by others and cooked and rice pudding like in lieu of rich or heated.

Caramelized flavor has been observed in milk which has been retorted or autoclaved (19). It has been suggested that nonenzymatic browning is the cause. A sweet caramel flavor was observed (20) in milk that had been processed at 135 or 143 C for 10 seconds and stored for 9 and 16 days. Scorched flavor is produced by localized overheating such as that created by excessive "burn on" in a heat exchanger. Fortunately caramelized and scorched flavors rarely occur in fluid milk.

In addition to the flavor changes mentioned above, heat treatment of milk produces a tactual effect. This effect has been characterized by such terms as chalky, rough, powdery, and astringent and was attributed to heat coagulable substances from the whey portion of milk (21). Subsequent research (22) led to the conclusion that astringency was primarily caused by heat altered whey proteins and to some extent to milk salts. Astringent components have also been extracted from acidified milk (23). These components appeared to be polypeptides associated with the casein fraction, particularly γ casein and were low in minerals. This suggests that there may be more than one combination of components that elicit an astringent response. This may explain the observation (18) that astringency decreased on storage of milk heated to 140 C for 3 seconds but increased in milk that had been held for 90 seconds at this temperature.

Although cooked flavor is probably the most common "defect" in commercial pasteurized milk it does not appear to have a significant impact on consumer acceptance. In some cases cooked flavor may mask some less desirable flavors. Furthermore, the sulfhydryl groups that are activated by heating may contribute to the oxidative stability of milk. However, the more pronounced heated flavor produced by high temperature treatment may adversely affect milk consumption. The higher temperatures are supposed to reduce microbial and/or enzymatic degradation but this advantage could be offset by an intense heated flavor. Therefore, various techniques have been used to minimize the development of this flavor. Less intense heated flavor is produced by using direct heating (e.g. steam injection) instead of indirect heating (e.g. plate heat exchangers). The steam injection process has the disadvantage of requiring more energy.

Ferretti (24) has shown that cooked flavor development could be inhibited by compounds that react with sulfhydryl compounds. Specifically he observed inhibition by the following 7 compounds: 2-aminoethyl 2-aminoethanethiosulfonate dihydrochloride; 5-aminopentyl 5-aminopentanethiosulfonate dihydrochloride; 2-acetamidoethyl 2-acetamidoethanethiosulfonate; cystine S-dioxide; 2-aminoethanethiosulfuric acid; S-sulfocysteine; and S-sulfoglutathione. Although these compounds were effective at concentrations from 0.003 to 0.05%, they are not legally approved additives. Badings (25) reported that the addition of 30-70 mg L-cystine (/kg milk) reduced the H_2S and cooked flavor. Since cystine is a natural constituent of milk it would be permissible to use it in some countries and perhaps might be approved in this country.

Cooked flavor can also be removed by the use of sulfhydryl oxidase. This enzyme catalyzes the following reaction (26):

$$2RSH + O_2 \rightarrow RSSR + H_2O_2$$

removing the sulfhydryl precursors of cooked flavor. This enzyme has been immobilized on porous glass beads (27). It is questionable whether a sulfhydryl oxidase treatment would be practical (28). It would certainly add to the cost of processing and would create a potential microbial problem.

Sensory evaluation provides the most practical method for monitoring the type and intensity of heated flavors. Standardized procedures (3) have been developed for producing these flavors to provide a means of training taste panel personnel. At the present time there is no simple objective test for measuring heated flavors. The nitroprusside test (29) provides an indication of the sulfhydryl content of milk and cooked flavor intensity but it does not measure the other heated flavors. One could use chromatographic techniques for monitoring the concentration of volatile compounds associated with heated flavors. However, existing techniques are not suitable for routine quality control work.

Light-Induced Flavors

Sensory and chemical characterization. Exposure of milk to radiant energy produces two distinct types of flavors (3). One type arises from a degradation of proteins the other from an oxidation of lipids. The flavor of the protein degradation products is described as resembling burnt hairs or feathers or cooked cabbage. This flavor is sometimes referred to as the sunlight or activated flavor. Photoinduced lipid oxidation produces a flavor that is similar to the autoxidized or metallic induced oxidized flavor. These two flavors develop simultaneously, but the rates of development are different. Consequently, the nature of the flavor changes with time. Protein degradation is more rapid so that initially the "burnt" note predominates but after 2 or 3 days the oxidized note becomes more pronounced. Several days after light exposure it is difficult to differentiate between light and metal induced oxidation.

Methional appears to be the major component of the burnt or cooked cabbage flavor produced by exposure of milk to sunlight (30,31). Other S containing compounds are also involved (32). Methional is believed to be a product of the Strecker degradation of methionine, as illustrated below:

$$CH_3SCH_2CH_2CHNH_2COOH \rightarrow CH_3SCH_2CH_2CHO + NH_3 + CO_2$$

It has been reported (33) that free radicals derived from methionine, methional and cysteine combine to form the following compounds:

$CH_3\text{-}S + H \rightarrow CH_3\text{-}SH$	methanethiol
$2CH_3\text{-}S \rightarrow CH_3\text{-}S\text{-}S\text{-}CH_3$	dimethyl disulphide
$CH_3\text{-}S + CH_3 \rightarrow CH_3\text{-}S\text{-}CH_3$	dimethyl sulphide

High molecular weight immunoglobulins have been identified (34) as the primary source of the photosensitive amino acids. Lipoproteins have also been implicated (35). In model systems purified α lactalbumin, β lactoglobulin and acid whey proteins underwent photodegradation (36).

A number of workers (30, 37-40) have shown that riboflavin plays an essential role in the development of light-induced flavor in milk. Aurand and coworkers (41) have postulated the following mechanisms:

$$Sen + h\nu \rightarrow {}^1Sen \rightarrow {}^3Sen$$
$${}^3Sen + {}^3O_2 \rightarrow {}^1Sen + {}^1O_2$$
$${}^1O_2 + RH \rightarrow RO_2 \rightarrow ROOH \rightarrow \text{photoxidation products}$$

(Riboflavin = Sensitizer = Sen)

They believed that riboflavin was capable of generating singlet oxygen via the excited triplet state. However, they indicated that riboflavin could also produce singlet oxygen via superoxide anion. They observed that lipid oxidation was inhibited by a singlet oxygen trapper (1,3 diphenylisobenzofuran) or a singlet oxygen quencher (1,4-diazabycyclo-[2-2-2] octane. Superoxide dismutase (SOD) did not inhibit the oxidation which suggested that superoxide anion was not involved. As illustrated below superoxide anion (O_2^{-1}) spontaneously yields singlet oxygen ($^1O_2$) whereas superoxide dismutase action would yield the less reactive triplet oxygen ($^3O_2$).

$$\text{Superoxide anion} \xrightarrow[\text{dismutation}]{\text{spontaneous}} \text{Singlet oxygen}$$

$$\text{Superoxide anion} \xrightarrow[\text{dismutase}]{\text{superoxide}} \text{Triplet oxygen} + H_2O_2$$

Korycka-Dahl and Richardson (42) suggested that the failure of SOD to inhibit oxidation might have been due to the low concentration of SOD used in this study. These scientists presented evidence to indicate that superoxide anion generation is a primary step in the photooxidation process in milk. However, they do not rule out the possibility that a direct sinlet oxygen pathway is also operative. They suggested that both low and high molecular weight serum protein components served as oxidizable substrates to photoreduce riboflavin which then reduces oxygen to superoxide anion. In a model system containing riboflavin they found that of the 23 amino acids studies only cysteine, methionine, histidine, tyrosine, and tryptophan supported superoxide anion photogeneration. In studies with methionine derivatives it was found that blocking the carboxyl group reduced the production of superoxide anion by about 50% whereas blocking the amino group had a negligible effect. Since blocking both groups did not eliminate totally the production of anion they speculated that polypeptides or proteins also may contribute to riboflavin-catalyzed photoreactions.

On the basis of studies with a model system containing methyl linoleate and methyl oleate Chan (43) postulated two pathways for photosensitized oxidations. He reported that photosensitized riboflavin reacted with the substrate to produce intermediates that were capable of reacting with triplet oxygen. This pathway gave the same positional isomers of hydroperoxides as autoxidation. By contrast a different pathway was observed when erythrosine was used as a sensitizer. In this case, the photosensitized erythrosine reacted with molecular oxygen to produce singlet oxygen. Different positional isomers of hydroperoxides were generated by this pathway. Research is needed to determine if there is more than one photoxidation pathway in milk.

There is evidence that photoxidation of lipids is different from autoxidation or metal-induced oxidation. Analysis of the volatile carbonyl from light exposed samples, revealed the presence of alkanals (44, 45) and 2 enals (46). These analyses did not reveal any 2,4dienals which are present in autoxidized samples. It has been postulated (46) that photoxidation involved the monoene fatty acids of the triglycerides while autoxidation involved the polyenes of the phospholipids. This postulated difference in substrates could also explain the reports (37, 44, 47) that the thiobarbituric acid (TBA) test is a useful method for detecting metal-induced and spontaneous oxidation but not for light-induced oxidation. Oxidation of monoenes would not be expected to yield malonaldehyde which is believed to be the principle TBA reacting component in milk. The author has observed that light induced oxidation is accelerated by traces of copper which does produce TBA reacting components. Traces of metals may be responsible for the increases in TBA values that are some time observed in light exposed milk.

Light-induced volatile compounds are produced constantly only when milk is exposed to light but not when it is stored in the dark (45). This indicates that the reaction is not maintained by autoxidation. This theory is supported by the observation (48) that α tocopherol, which is an effective inhibitor of autoxidation is virtually ineffective in photochemical oxidations.

Factors Affecting Photoxidation of Milk. Obviously, photoxidation can be prevented by avoiding exposure of the milk to light, but complete protection is not possible. Some current methods of handling milk, such as the use of glass pipelines for milk transport, fluorescent lights in milk display cases and transparent, or translucent containers contribute to the flavor problem. Efforts are being made to lower the incidence of this defect by reducing the light intensity, the length of exposure and the amount of light of shorter wave length (48, 49). Exposure of milk to light in the 400 to 460 nm seems to produce the most rapid photodegradation. Eliminating light with wavelengths below 500 nm greatly reduces the problem but blocking all light below 600 nm gives more complete control. The numerous studies that have been conducted to determine the best method for blocking the short wavelengths have been reviewed (46, 48, 49, 50). These methods include using tinted or opaque containers; using special lights, light shields or light filters in refrigerators or display cases. Milk packaged in brown, amber or ruby glass is more protected from photoxidation than when it is stored in clear glass (50). Similarily, opaque or tinted plastic containers offer more protection than clear plastic (48,49). Fiberboard containers offer considerable protection from photoxidation and this protection can be increased by using plastic coating instead of wax coating (51). Cartons with laminated aluminum foil offer even more protection but they are more expensive. Pigmented cartons offered more protection than plain cartons (51,52). Of the colors tested, yellow and red offered the least protection. Studies of different fluorescent lights and light filters revealed that milk stored in polyethylene containers did not develop light-induced flavor for 30 to 40 hours when yellow lamps or yellow and green filters were used (53). By contrast the flavor developed within 2 to 4 hours when a cool white lamp was used. The rate of development of light-induced flavor also increased with increases in the temperature of exposure and the closeness of the lamp to the milk (54).

Milk varies in its susceptibility to photoxidation (48, 49, 50). Milk from cows on green feed or pasture is reported to be more resistant than milk from cows on dry feed (51, 54). It is conceivable that the green feeds are sources of singlet oxygen quenchers such as β carotene. Milk exposed to light immediately after processing was more susceptible than milk stored in the dark for 2 days before exposure. Homogenization greatly increases the susceptibility, but the cause for this effect has not been elucidated. Skimmed milks are more susceptible than whole milk, presumably because of the greater transparency of the low fat milks.

Detection of Light-Induced Flavor. At the present time there are no suitable objective methods for routine examination of milk for light-induced flavors. Theoretically measurements of either the methional or pentanal and hexanal concentrations should provide an index of photoxidation. More research is needed to develop suitable methods and to determine the critical concentration of these compounds (i.e. the concentrations that produce adverse consumer response). Consequently, sensory evaluations are still the best means of detecting photoxidation. Standardized procedures (3) have been developed for producing light-induced flavors to provide samples for training taste panel personnel.

Lipolyzed Flavor

Sensory and Chemical Characterization. Lipolyzed is used

to describe the flavor produced by lipase catalyzed hydrolysis of milk fat triglycerides. The term "rancid" is most frequently used by dairymen, but it is ambiguous because it is often used by other food scientists and processors to describe oxidized flavor. Butyric, goaty, and bitter have also been used. Bitter is another ambiguous term since bitter flavors may also be due to protein degradation or to some alkaloids.

Chemically, lipolyzed flavor is the best defined of the milk flavor categories. It is due primarily, if not exclusively, to the C-4 to C-12 volatile free fatty acids (i.e. butyric, caproic, caprylic, capric, and lauric). Studies (55, 56) on the contribution of each of these acids to lipolyzed flavor led to the conclusion that no single acid was of predominant importance. It was noted that both capric and lauric acids had persistent soapy aftertastes (56). In spite of the simplicity of lipolyzed flavor, attempts to duplicate it by the addition of blends of these five acids to milk, were unsuccessful. This failure was believed to be due to the inability to achieve the natural physical chemical state. For example, the distribution of the added acids between the aqueous and nonaqueous phases and the extent of flavor binding may have been abnormal. Even in skim milk it has been shown (57) that free fatty acids fall into at least three categories: a) free dissociated and undissociated acids, b) fatty acids associated with the membrane material in skim milk and c) fatty acids of unknown origin requiring acidification to pH 1.5 for extraction. The impact of distribution between the aqueous and nonaqueous phases on the flavor thresholds of the fatty acids is illustrated in Table III. This data show that as the fatty acid chain length increases the threshold values decreases in water but increases in oil. In other words, the flavor threshold decreases as the affinity for the dispensing media decreases.

TABLE III. Flavor Thresholds for Volatile Fatty Acids[a]

Fatty acids	Threshold (ppm) Water	Oil
Butyric (4:0)	6.8	0.6
Caproic (6:0)	5.4	2.5
Caprylic (8:0)	5.8	350.0
Capric (10:0)	3.5	200.0
Lauric (12:0)	-	700.0

[a]Taken from Patton (58).

Comparison of the flavor threshold values for fatty acids in homogenized versus non-homogenized milk provides another illustration of the effect of physical factors on sensory impact. It was observed (59) that the same concentration of free fatty acids was perceived as being more intense in non-homogenized than in homogenized milk.

Lipase and Substrate Characteristics. There is considerable evidence (61-65) to indicate that there is more than one lipase in milk. It has been suggested that milk contains one lipase that is inhibited by formaldehyde and one that is not. However, the differential sensitivity to formaldehyde may depend on whether the lipase is free or complexed with some other milk constituent. In another study, two lipase containing fractions were obtained by high speed centrifugation. One component, designated lipase A, was obtained from the sediment and had a pH optimum of 8.6. The other component, designated lipase B was obtained from the supernatant and had a pH optimum of 7.0. Some scientists believe there is a membrane lipase and a plasma or serum lipase. It has been claimed that the membrane lipase is bound to the membrane in such a way that it becomes activated by simple cooling of the milk. However, it has not been proven that the enzyme moiety in the so called membrane and plasma lipases are different. Several articles have been written about the presence of lipoprotein lipase in milk. Some of the authors (61-63) believe that the lipoprotein lipase was distinctly different from regular milk lipase whereas some do not (64,65). The recent evidence seem to suggest that the normal milk lipase and lipoprotein lipase is the same enzyme. The observed difference in activity in the presence of serum activators is believed to be due to the effect of the activator rather than the enzyme *per se*. The presence of other activators or inhibitors may also alter the enzymic behavior. Different complexes between lipase and other milk constituents may be responsible for some of the observed differences between other lipase preparations.

The lipolytic action that produces the characteristic lipolyzed flavor involves the cleavage of short chain fatty acids from the triglycerides. Milk lipase has been shown to be highly specific for the primary ester groups of both milk and synthetic triglycerides. The lipase does not differentiate between short and long chain fatty acids attached to the primary positions of the same glycerol. However, since butyrate appears to be mostly a primary ester an apparent preferential release of butyrate occurs. The amount of volatile fatty acids required to produce an intense lipolyzed flavor is only a fraction of the total volatile acids in whole milk. Consequently, the concentration of substrate is not limiting, but the amount of available substrate may be.

Factors Affecting Lipolysis. This discussion focuses on the highlights of the numerous studies on the factors affecting lipolysis. A more comprehensive summary of the findings can be found in the review articles on this subject (60-65). At the time milk is secreted, the triglycerides which are located in the fat globules are surrounded by a protective membrane. Lipolysis does not occur until active lipase has come in contact with the triglycerides. Consequently, lipolysis is dependent on the activity of the lipase and the availability of substrate.

a. Thermal and mechanical effects. The protective effect of the fat globule membrane can be reduced by either mechanical or thermal abuse. The mere cooling of milk from 37 to 5 C is enough to cause lipolysis in some milk, presumably by lipase located in the membrane. Milk that is so easily activated has been called sponatenous milk. Fortunately, only a small percentage of the milk falls in this category. Heating cooled milk up to 30 C and recooling to 5 C causes even greater increases in lipolysis. These temperature changes apparently disrupt the membrane, but they may also alter the association of the lipase with other milk constituents. It has also been reported that very fast cooling of milk produces more rapid lipolysis than slow cooling.

Mechanical agitation such as simple stirring and pumping can increase lipolysis. The magnitude of this effect is increased by raising the temperature. Because of the bulk handling of milk excessive agitation is becoming a greater problem. Vigorous agitation such as homogenization produces even more pronounced lipolysis. The increase produced by homogenization exceeds the amount of increase in fat globule surface area. Apparently, part of the increase is due to changes in the nature of the fat globule membrane. Fortunately, most of the lipase in milk is inactivated by pasteurization which is done either just before or after homogenization. If milk is pasteurized at the minimum time and temperature (e.g. 15 seconds at 71.7 C) required by law the lipase may not be completely inactivated (66, 67). Some variability in the thermal resistance of lipase in milk from different dairy herds has been observed. It has been shown (66) that thermal resistance increases as the fat and solids-not-fat content increases. Some of the active lipase in pasteurized milk may be of microbial origin since some microbial lipases are more heat resistant than milk lipase (68).

b. Chemical effects. The activity of lipase can be affected by a number of chemicals (61). Aureomycin, penicillin, streptomycin, and terramycin inhibited lipase activity by 7.6

to 49.8%. Hammarsten casein, acid casein, α casein, β casein, ν casein, α lactalbumin, and β lactoglobulin inhibited lipase. It was postulated that the inhibitory effect of these proteins was due to the formation of a complex between them and the lipase (69). On the other hand κ casein, lactalbumin, pseudoglobulin and euglobulin activated the enzyme to varying degrees. Lipase can be inhibited by a number of salts but they have not been shown to have a significant impact on lipolysis in milk.

Lipase contains free as well as masked sulfhydryl groups that are associated with enzymic activity. Presumably this explains why the enzyme can be inactivated by oxidation and is sensitive to metal ions such as copper and iron. Lipase activity can be reduced by hydrogen peroxide and exposure to light.

Lambda carrageenan, a polysaccharide containing primarily 1,3 D-galactose-2-sulfate and 1,4 D-galactose-2, 6-disulfate inhibits lipolysis (70). This polymer is water soluble and reacts readily with milk proteins. As little as 0.05% carrageenan can inhibit lipolysis in milk that is subjected to either thermal or mechanical activation. It has been postulated that this inhibition is due primarily to an interaction between the carrageenan and the enzyme, although adsorption on the fat globule membrane may also be involved.

Lecithin and casein protected a milk fat emulsion from attack by added bovine lipase (71). Lecithin was found to be more effective than casein. In this model system it appeared that the inhibitory effect of the compounds were due to a partial encapulation of the fat globule. Addition of trypsin to the system reduced the inhibitory effect of casein. In a subsequent study (72), it was shown that lipolysis was increased in both a model emulsion and in raw milk by the action of phospholipase C. Since this enzyme cleaves phospholipids, it is presumed that it enhanced lipolysis by reducing the protective coating provided by the phospholipids.

Hsu (73) has shown that a number of surfactants inhibit lipolysis. Triton X-155 a polyoxyethylene octyl phenol was the most effective compound tried but it did not inhibit the enzyme _per se_. It was postulated that it provided a protective coating for the substrate in a manner analogous to phospholipids.

c. Variability in fresh raw milk. Marked differences have been observed in the lipolytic activity in milk from individual cows. Some of those differences may be attributed to genetic factors, but this has not been proven. Several scientists have reported increases in the incidence of lipolyzed with advancing stage of lactation. However, this relationship has not been found in all studies. The difference between studies may merely reflect differences between cows. Since the fat content increases and the size of globules decrease with advancing lactation there is a substantial increase in the fat globule surface area. This increased surface area might be expected to increase the susceptibility to lipolysis.

The cow's feed has been shown to influence the susceptibility of milk to lipolysis (61,65). In general, green pasture decreases lipolysis whereas dry feeds tend to increase it. Poor quality rations with low energy value also contribute to increased susceptibility to lipolysis.

Control of Lipolysis. To avoid lipolysis the following steps should be taken:

a. Milk should be cooled immediately after milking to 5 C or below with minimum agitation. Freezing should be avoided since it can cause disruption of the fat globule membrane.

b. Milk should be maintained at this temperature until pasteurized.

c. Stirring and pumping of milk should be kept to a minimum. Long distance hauling of partially filled containers or tanks should be avoided.

d. Milk should be pasteurized as soon as possible at temperature high enough to inactivate lipase. A thermal treatment equivalent to at least 76.7 C for 16 seconds is recommended.

e. Precautions should be taken to avoid post-pasteurization contamination with lipase producing microorganisms.

f. Milk should be distributed as rapidly as possible and kept at 5 C or below to avoid possible development of lipolysis.

g. Samples should be tested periodically to ensure that the consumer is getting milk free of lipolyzed flavor.

Detection of Lipolysis. Numerous methods have been developed for measuring free fatty acids in milk to estimate the intensity of lipolyzed flavor. Many of the methods have been based on titrimetric measurements. The first step in these methods usually involve some type of separation, such as solvent extraction (74) silica gel extraction (75) and centrifugation after detergent solubilization of the proteins (76).

These methods do not give identical results because of variations in the recovery of the fatty acids. Research on these methods have led to a modified detergent method with improved reproducibility (77). Modification of the solvent extraction gave higher butyric acid recoveries and overcame the potential interference from lactic acids (78). This increased recovery of butyric acid should improve the correlation between objective values and sensory response. Of the titrimetric methods, the detergent method is most commonly used for quality control work because of its simplicity.

Considerable attention has been given to replacing titrimetric methods with colorimetric methods since they are more sensitive and are easier to automate. One method (79) utilizes the dye Rhodamine B to form soluble complexes with fatty acids. An automated method (80) has been developed that measures the intensity of color of a phenol red solution. Fatty acids alter the pH and hence the color of the indicator. Another colorimetric method (81) involves a conversion of the fatty acids to copper soaps. These copper soaps are then extracted and reacted with sodium diethyl dithiocarbanate to form a color complex. A semiautomated modification of this method has been developed in our laboratory (59). Results were obtained with this method that were comparable to the detergent titrimetric method. It is possible to test about 200 samples per hour by this colorimetric method compared to about 12 per hour by the detergent method. Comparison of the flavor intensity ratings given by a trained panel with the corresponding colorimetric values for free fatty acids revealed a correlation of 0.82. Thus the objective values can be used to predict sensory response.

Microbial Flavors

A variety of flavor defects may be caused in both raw and pasteurized milk by an accumulation of the products of bacterial metabolism (3,82,83). The nature and intensity of these flavor defects depends on the types and numbers of contaminating organisms and the temperature history of the milk.

Acid Sour Flavor. The term sour has frequently been used to describe the flavors of microbial origin. Prior to the use of mechanical refrigeration and pasteurization this was an appropriate term since the predominant contaminating organism in milk was *Streptococcus lactis* which is a fast acid producer. Consequently, microbial contamination usually led to the production of lactic acid plus traces of acetic and propionic acids. Titratable acidity development in milk of .07 to .10%, calculated as lactic acid, is commonly detected by most individuals. The odor of the volatile acids may be detected by experienced individuals when the titratable acidity has increased by as little as 0.01%. As a result of improved sanitary practices, better cooling and pasteurization, high acid or sour flavor is rarely encountered in milk. However, the term sour is still commonly applied to flavors of microbial origin.

Malty Flavor. A flavor that has been described as burnt, caramel, "grapenuts", and malty may develop in raw milk as the result of the metabolism of *Streptococcus lactic* subsp. *maltigenes* (51). This organism has been encountered occasionally in milk in Northeastern U.S. but it has rarely been found in the Pacific Coast states. Contamination of milk with this organism usually results from improperly sanitized equipment. Growth of this organism is favored by delaying cooling of milk and storage at 10 C or above. Recently a new organism has been isolated from milk which produces a malty flavor (84). This organism has been named *Lactobacillus maltaromicus*.

S. lactis subsp. *maltigenes* produces a number of volatile aldehydes and alcohols which are derived from amino acids. However, the characteristic flavor is due principally to the production of 3-methylbutanal from leucine. The addition of as little as 0.5 ppm of 3-methylbutanal to milk simulates the malty flavor. The mechanism for the formation of aldehydes and alcohols from amino acids is shown below.

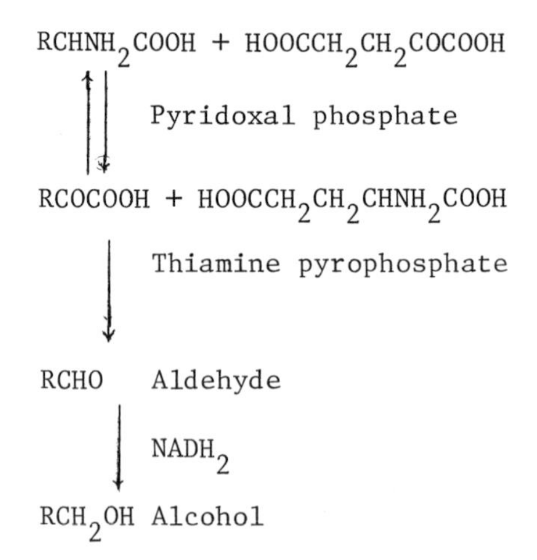

Fruity Flavor. A flavor which has been described as strawberry-like, ester like and fruity may develop in past eurized milk as a result of the metabolism of *Pseudomonas*

fragi. These organisms are destroyed by pasteurization so its presence in the pasteurized product indicates post-pasteurization contamination. At 5 to 7 C this organism will outgrow most other species in milk. *P. fragi.* produces a lipase which liberates butyric and caproic acids from triglycerides. An esterase then esterifies these acids with ethanol to form ethylbutyrate and ethylhexanoate, the primary components of fruity flavors. This flavor can be simulated by "fortifying" with 3.5 ppm of ethylbutyrate and .50 ppm of ethylhexanoate (3).

Fermented, Unclean, Bitter, and Putrid Flavors. In addition to sour, malty, and fruity flavors there are a variety of other flavors of microbial origin. Most of these flavors are produced by psychrotrophic bacteria (i.e. bacteria that can multiply at or below 7 C irrespective of their optimum growth temperature). Present handling and processing of milk have favored the proliferation of these organisms (68). Many of these organisms produce proteinases and lipases (including phospholipases) which survive pasteurization. The combination of enzymes in these organisms give rise to mixtures of flavor components that are difficult to characterize. Different mixtures of lipid and protein degradation products are probably responsible for fermented, unclean and putrid flavors. Bitter flavor is generally attributed to certain amino acids and peptides but there are undoubtedly other bitter flavor components.

Control of Microbial Flavors. Rigorous sanitary practices are required to prevent microbial contamination of milk. Milk should be cooled rapidly to 4.4 C or below and maintained at this temperature. Milk should be pasteurized as soon as possible to reduce microbial growth and the production of heat resistant enzymes. Care should be exercised in the packaging and storage of milk to minimize post-pasteurization contamination and growth of organisms. It has been shown (85) that raw milk contains heat resistant psychrotropes that may contribute to spoilage of milk. A recent study (8) revealed that 12 of 24 commercial milk samples had developed fruity-fermented flavors by the time they had reached the "sell-by" date. The 12 fruity-fermented samples had high bacterial counts. Obviously, these samples had been contaminated and consequently had a short shelf-life. In these cases the quality had not been adequately monitored, partly because there are no simple tests for these flavors except taste tests. Measurement of titratable acidity can be used to detect sour samples but there are no other direct quality control tests for microbial flavors.

Oxidized Flavor

Sensory and Chemical Characterization. Oxidized flavor is the general term applied to the flavors produced by the oxidation of unsaturated fatty acids, such as arachidonic, lenolenic, linoleic, and oleic. In fluid milk the oxidized flavor arises primarily from the oxidation of the fatty acids in the phospholipids of the fat globule membrane. Oxidation of these polar lipids generate a variety of aldehydes and ketones. The types and quantity of the carbonyl compounds is dependent on the extent of oxidation and the specific fatty acid precursors. Some of the aldehydes that can be formed from the different fatty acids are illustrated in Table IV.

TABLE IV. Possible Origin to Some Aldehydes Obtained from the Oxidation of Four Fatty Acids (86)

Fatty Acid	Hyd roxide Position at	Aldehyde Obtained
Oleic	C11	Octanal
	C8	2-undecenal
	C9	2-decenal
	C10	Nonanal
Linoleic	C13	Hexanal
	C9	2,4-decadienal
	C11	2-octenal
Linolenic	C16	Propanal
	C14	2-pentenal
	C12	2,4-heptadienal
	C13	3-hexenal
	C11	2,5-octadienal
	C9	2,4,7-decatrienal
Arachidonic	C15	Hexanal
	C13	2-octenal
	C12	3-nonenal
	C11	2,4-decadienal
	C10	2,5-undecadienal
	C7	2,5,8-tridecatrienal

As oxidation progresses the sensory characteristics may be altered by changes in the relative and absolute concentration of the various carbonyls. The sensory characteristics are also affected by additive, synergistic, antagonistic and masking interactions between the carbonyls (87). It has been shown (88,89) that the combination of subthreshold levels of carbonyls can be additive and thereby give rise to detectable flavors. An increase in the saturated carbonyls of a fishy flavored butterfat caused it to acquire a painty tallowy taste (90). Antagonistic and masking interactions was noted among a number of flavorful aldehydes (91). For example, decadienal and 2-nonenal completely masked the perception of suprathreshold levels of cis-3-hexenal. Specific compounds have been found to produce specific flavors (Table V). Since oxidized milk samples often contain more than one of these flavor groups they may elicit more than one flavor response. The flavor analyst should be aware of the heterogeneous nature of oxidized flavor.

TABLE V. Some Descriptive Flavors and Associated Compounds Identified in Oxidized Milk Fat (86)

Flavor	Compounds
Oxidized	oct-1-ene-3-one, octanal, hept-2-enal, 2,4-heptadienal, n-alkanols (C2-C9)
Cardboard, tallowy	n-octanol, n-alkanals (C9-C11), alk-2-enals (C8,C9), 2,4-dienals (C7,C10), 2,6-dienal (C9)
Oily	n-alkanals (C5,C6,C7), hex-2-enal, 2, 4-dienals (C5,C10)
Painty	n-alkanals (C5-C10), alk-2-enals (C5-C10), 2-4,dienal (C7), 2-alkanone (C7)
Fishy	n-alkanals (C5-C10) alk-2-enals (C5-C10), 2,4-dienal (C7), 2-alkanones (C3-C11), oct-1-ene-3-one
Grassy	alk-2-enal (C6), 2,6-dienal (C9)
Metallic	oct-1-en-3-one
Beany	alkanals, non-2-enal
Mushroom	oct-1-en-3-ol
Cucumber	2,6-dienal (C9)
Nutmeg	octadienal; 2,4-dienals
Creamy	4-cis-heptenal
Fruity	n-alkanals (C5,C6,C8,C10)

Mechanism of Lipid Oxidation. The oxidation of lipids involves an autocatalytic reaction, in that the oxidation products themselves catalyze the reaction. The autooxidation mechanism is a chain reaction involving the following steps:

Initiation

$RH + O_2 \rightarrow ROOH$

Propagation

$2ROOH \rightarrow ROO\bullet + RO\bullet + H_2O$

$ROO\bullet + RH \rightarrow R\bullet + ROOH$

$R\bullet + O_2 \rightarrow ROO\bullet$etc

Termination

$ROO\cdot + R\cdot \rightarrow ROOR$

$R\cdot + R\cdot \rightarrow R\text{-}R$

$ROO\cdot + ROO\cdot \rightarrow ROOR + O_2$

where RH = unsaturated lipid, R· lipid radical

ROO· = lipid peroxy radical.

Although the free radical mechanism of lipid oxidation has been well established the initial reaction or the formation of the first hydroperoxide has not been explained satisfactorily. A direct reaction between an unsaturated fatty acid and oxygen is unlikely. A change in the electron spin would be required since the hydroperoxides and lipids are in a singlet state while oxygen is in a triplet state. Spin conservation could be preserved by converting triplet oxygen to singlet oxygen. Aurand (41) postulated that in milk a copper-ascorbic acid complex catalyzes the conversion of triplet oxygen to superoxide anion. The superoxide anion then generates singlet oxygen by spontaneous dismutation. The involvement of the anion was indicated by the observation that superoxide dismutase inhibited metallic-induced oxidation. Inhibition by a singlet oxygen trapper and a quencher confirmed the involvement of singlet oxygen. Lillard (92) has questioned whether singlet oxygen is involved in all oxidation systems.

The hydroperoxides formed in the initial step are flavorless and unstable and break down to produce flavorful compounds. The hydroperoxide decomposition has been illustrated (92) by the following reactions:

$$R\text{-}\underset{\substack{|\\ OOH}}{CH}\text{-}R \rightarrow R\underset{\substack{|\\ O\cdot}}{CH}\text{-}R + \cdot OH$$

$$R\underset{\substack{|\\ O\cdot}}{CH}\text{-}R \rightarrow R\underset{\substack{\|\\ O}}{CH} + R\cdot$$

$$R\text{-}\underset{\substack{|\\ O\cdot}}{CH}\text{-}R + R^1H \rightarrow R\underset{\substack{|\\ OH}}{C}R + R^1\cdot$$

$$R\text{-}\underset{\substack{|\\ O\cdot}}{CH}\text{-}R + R^1\cdot \rightarrow R\underset{\substack{\|\\ O}}{C}R + R^1\text{-}H$$

$$R\underset{\substack{|\\ O\cdot}}{CH}\text{-}R + R^1O\cdot \rightarrow R\text{-}\underset{\substack{\|\\ O}}{C}\text{-}R + ROH$$

Most of the oxidative products are formed by the above reactions. Other compounds are produced by secondary reactions such as isomerization, e.g. non-3 enal to non-2 enal. Linoleic, linolenic, and arachidonic acids contain only cis double bonds separated by methylene groups. Forss (33) observed that double bond conjugation may precede hydroperoxide formation and the moving double bond changes to the trans form. For example, deca-trans-2, cis 4 dienal is one of the products of linoleic oxidation.

The observation that oct-1-en-3-one had a very intense metallic taste has stimulated considerable research. Wilkinson and Stark (93) report that the compound can be generated by the oxidation of either linoleic or arachidonic acids or their esters. They postulated the following scheme for the formation of oct-1-en-3-one.

$$CH_3(CH_2)_4\text{-}CH{=}CHCH_2{=}CH[OOH] \rightarrow CH_3(CH_2)_4\text{-}(CH{\cdots}CH{\cdots}CH_2)$$

$$\downarrow RH \mid O_2$$

$$CH_3(CH_2)_4\text{-}\underset{O}{\overset{\|}{C}}\text{-}CH{=}CH_2 \xleftarrow[RO_2\cdot]{RO\cdot \text{ or}} CH_3(CH_2)_4\underset{O\cdot}{CH}\text{-}CH{=}CH_2$$

Swoboda and Peers (94) have reported an even more potent metallic flavored compound, namely octa-1, cis-5-dien-3-one. They reported that it was formed from the C_{20} and C_{22} (n-3) pentaenoic acids in a manner analogous to oct-1-en-3-one.

Factors Affecting Oxidative Stability. The oxidative stability of the lipids in milk is dependent on the composition of the freshly secreted raw milk and the subsequent treatment that it receives.

a. Variability in fresh raw milk. On the basis of differences in oxidative stability, milk has been classified (95) into three categories:

1. Spontaneous milk i.e. milk capable of developing oxidized flavor without the addition of iron or copper.
2. Susceptible milk i.e. milk which does not develop oxidized flavor spontaneously but is susceptible if copper or iron is added.
3. Nonsusceptible i.e. milk which does not even become oxidized in the presence of copper or iron.

Several studies have indicated differences in the susceptibility of milk from different breeds (95). However, the available evidence does not indicate that heredity has a major impact on susceptibility.

Feed can influence oxidative stability by affecting the composition of milk. The stability is usually higher when cows are fed a predominantly green feed ration, but there are significant differences between green feeds. It has been reported (96,97) that pasture, silage or hay made up primarily of blade grasses or good quality corn silage produce more stable milk than legume hay or silage. The fact that legumes generally contain more copper and less manganese (98) may be partly responsible for the differences in the stability of milk produced on these types of forages. Variations in the amounts of antioxidants, such as tocopherols and carotenes, in feed also contribute to the differences in oxidative stability produced by different rations (95,96). To increase the antioxidant content of rations, they have been supplemented with tocopherols (99,100) ethyoxyquin (101) N,N' diphenyl paraphenylenediamin (102) and cacao shells which contain antioxidants (103). Although these supplements help to protect the lipids, direct addition of antioxidants to milk is more reliable. Only a small fraction of the added antioxidant may show up in the milk e.g. only 2% of added α tocopherol was secreted in the milk (99). Intermittent intravenous or intramuscular administration of tocopherol is more effective but too costly (104).

Conventional feeding practices do not appear to have a significant effect on the amount of unsaturated fatty acids. However, it is possible to increase the unsaturated fatty acid content of milk by feeding protected lipids i.e. unsaturated lipids encapsulated to protect against hydrogenation by rumen microflora (105). This increase in the unsaturated fatty acid content of milk may be beneficial from a nutritional standpoint but the oxidative stability is lowered (106).

Although there is some disagreement about the effect of stage of lactation on oxidative stability, the bulk of the evidence (95,107) appears to indicate that milk produced during the first 60 to 90 days after calving is the most susceptible to oxidation. Normally the ratio of concentrate to roughage in the ration is higher during this period. It has been reported (108) that the copper content in early lactation milk is higher than in late lactation milk.

The concentration of ascorbic acid in milk affects oxidative stability (95). The level normally found in fresh raw milk (i.e. 10-20 mg/L) has a prooxidant effect whereas rapid and complete destruction of ascorbic acid inhibits oxidized flavor development. As previously noted (41) the prooxidant effect can be attributed to the ability of ascorbic acid to react with copper to generate active oxygen. However, the ability of high levels (50-200 mg/L) of ascorbic acid to inhibit oxidation has not been explained.

b. Milk handling and storage. In the past metallic contamination of milk was the major factor affecting oxidation.

This problem has practically been eliminated now that most milk contact surfaces are stainless steel, glass or plastic. Occasionally, equipment gets contaminated from cleaning solutions that contain copper or iron. Added copper is a stronger catalyst than iron and ferrous iron is more effective than ferric iron. It has been observed (109) that added iron does not become associated with the fat globule whereas added copper does. Some of the natural iron is located in the fat globule membrane in the form of heme proteins. These heme proteins may contribute to lipid peroxidation (110).

The incidence of oxidized flavor appears to be higher in milk stored at low temperatures (e.g. 5 C vs. 10 C). This effect may be due to reduced bacterial growth at the lower temperature. Lower bacterial growth could reduce the competition for oxygen and/or the production of reducing subtances. The control of bacterial growth has led to prolonged storage of milk which provides more time for chemical and enzymatic reactions and therefore may increase flavor problems.

C. Milk processing. Pasteurization under mild conditions may enhance oxidation possibly by causing a migration of copper to the cream phase (111). Such a shift would bring the copper in closer proximity to the phospholipid substrate. Higher heat treatments inhibit oxidation, presumably by releasing sulfhydryl groups which can act as antioxidants. It is also possible that heated flavor may mask a slight oxidized flavor or heating may cause a change in the availability of substrate or activity of metal ions.

Homogenization is a very effective inhibitor of spontaneous or metallic induced oxidation. Since homogenization does not alter the chemical composition the inhibitory effect must be due to physical changes such as reduction in the availability of the substrate or the activity of the catalysts. For example, Corbett (112) suggested that homogenization resulted in the adsorption of a protective casein layer on the surface of the fat globule membrane. On the other hand, Smith and Dunkley (113) postulated that catalytic activity was reduced by a change in the copper-protein complex.

Oxidation is retarded by using a steam injection vacuum treatment process which releases sulfhydryl groups and removes oxygen (114).

A mild trypsin treatment of milk can inhibit oxidation (28) presumably by either exposing antioxidants such as sulfhydryl groups or reducing the activity of prooxidants (115).

Oxidation can be inhibited by the action of phospholipase C which splits off the polar end of phospholipids to yield a diglyceride and a phosphoryl fraction. It is postulated that this cleavage reduces the exposure of the unsaturated lipid (28). The addition of phenolic type antioxidants which inhibit free radical formation or chelating agents which bind metallic catalysts retards oxidation. However, it is not legal to use any of these antioxygenic treatments in processing milk.

Control of Oxidized Flavor. In most cases oxidized flavor can be prevented by avoiding contamination of milk with copper or iron and by homogenizing milk. If the milk is highly susceptible it may be necessary to pasteurize at higher temperatures or use a steam injection vacuum process.

Detection of Oxidized Flavor. The thiobarbituric acid (TBA) test is the commonly used objective method for measuring oxidized flavor in milk. In oxidized samples, malonic dialdehyde is believed to be the principal compound that reacts with TBA although other compounds do react with TBA. Dunkley and Jennings (116) and King (117) have developed TBA procedures for testing milk. The King procedure avoids the use of an isoamyl alcohol-pyridine extraction step and is easier to perform but it may be slightly more sensitive to interfering substances. A more specific and reproducible method would be desirable but the TBA test does provide a useful index of spontaneous and metallic-induced lipid oxidation. Chromatographic methods can be used to measure the carbonyl compounds responsible for oxidized flavor, but current methods are not suitable for routine quality control.

Transmitted Flavors

Transmitted flavors arise from the passage of substances in the cow's feed or surroundings to milk while it is in the udder. This transfer may be via the respiratory and/or digestive system and blood stream. The majority of the flavor variations in fresh raw milk are due to these transmitted substances. Some of these flavors are mild and may not be considered objectionable by some consumers.

A series of experiments at Cornell (118,119) demonstrated the mechanism of flavor transmission. Two cows were fitted with tracheal and ruminal fistulae so that it was possible to introduce flavored substances directly into the rumen or lungs. Odors were introduced into the lungs by pumping air up through a chamber containing odoriferous material into tubing connected to the tracheal catheter. The top of the chamber contained a T-tube to permit the introduction of oxygen or air to supplement the air passing through the odoriferous material. The rate of flow of volatile materials into the lungs was regulated by adjusting the air input at the bottom and top of the chamber.

To determine if flavors could enter the milk via the digestive tract, test substances were introduced through the ruminal cannula while the lungs were being supplied with fresh air through the tracheal catheter. To determine if the eructated gases from the rumen contributed to off-flavors, the tracheal catheter was removed thus re-establishing the normal passage of air.

Samples of milk were taken before, and after the introduction of the test substances. Samples were coded and judged by an experienced flavor panel.

Results of a study with onions are shown in Table VI.

TABLE VI. Transmission of Flavor From Onions to Milk (118)

Expt. No.	Sample[a]	Transmission Route	Off-flavor Intensity
1.	Odor from fresh onion slurry	Lungs	None
2.	Odor from onion slurry after incubation with rumen ingesta	Lungs	Strong
3.	Odor from rumen ingesta	Lungs	None
4.	Onion slurry (Fresh air in tracheal cannula)	Rumen	Mild
5.	Onion slurry (tracheal cannula removed)	Rumen & Lungs	Strong

[a] *Odors were introduced via the tracheal cannula; slurries were introduced into the ruminal cannula.*

These experiments indicates that the odors from a fresh onion slurry did not contain volatiles that were transmitted by the lungs. Following incubation of the slurry with rumen ingesta, flavor was transmitted to the milk. This indicated that the microflora of the rumen ingesta liberated flavor substances from the onion. Even though the ingesta alone was odoriferous, no odors were transmitted from it to the milk.

Apparently, the cows were able to absorb or block these odors. When the slurry was placed directly in the rumen with the lungs connected to the fresh air supply some flavor was transmitted to the milk (Experiment 4). This indicated that some flavor substances were transferred through the rumen wall to the blood stream. More flavor was transferred when the cow was permitted to breath normally (Experiment 5). Under these conditions the eructated gases entered the lungs and these odors supplemented the flavor substances passing through the rumen wall.

TABLE VII. Transmission of a Variety of Flavors to Milk (119)

Flavor substance	Terms used to describe flavor
Ethanol[a]	Sweet, vanilla-like, ester-like
Propanol[a]	Alcohol-like, vanilla-like, Duco-cement
2-Butanol	Sweet, ester-like, xylene
Acetone[a]	Feedy, cowy, sweet, silage
2-Butanone[a]	Hay-like, sweet, aromatic, cowy
Propanal	No detectable effect
n-Butanal[a]	Malty, chemical, butanal
n-Pentanal	No detectable effect
Butyric acid	No detectable effect
Propionic acid	No detectable effect
Methyl acetate	Sweet, ester, grassy
Ethyl acetate	Sweet, fruity, ester, odd
Propyl acetate	Sweet, vanilla-like, feed, odd
Butyl acetate	Banana, ester, malty, odd
Dimethyl sulfide[a]	Weedy, cowy, unclean, onion
cis-3-hexenol[a]	Grassy, weedy, musty, grass juice
Green pasture	Grassy, cowy, barny, feedy
Green corn silage	Barny, cowy, feedy, sweet
Grass silage distillate[b]	Fruity, sweet, ketone
Corn silage distillate[b]	Fruity, fermented, aromatic, sweet

[a]*These substances were introduced by both the lung and rumen routes. The pasture and silage were consumed in the normal manner. The other substances were introduced by the lung route only.*

[b]*The silage distillates represent the neutral carbonyl-free fraction boiling between 36 and 100 C.*

Subsequent experiments demonstrated the cow's ability to transmit a variety of substances commonly found in feeds (Table VII). Fortunately, cows are able to prevent some substances, such as butyric acid from being transmitted directly into the milk.

Feed Flavors. When cows consume and/or inhale the strong odor of many feeds (e.g. silage, green forage, lush green pastures, etc.) within 2 to 4 hours before milking the milk will have a sweet and aromatic flavor that may be characteristic of the feed. If these feeds are fed immediately after milking and are withheld for 4 to 5 hours before milking they may not produce a feed flavor. Since the flavor components are transported through the blood to the mammary gland, the time it takes to impart a detectable off-flavor depends on the concentration of the flavor components in the blood. The concentration in the blood depends on the amount being supplied to the cow's lungs or rumen. As the supply is removed it is assumed that there is some reversal of flow of flavor components with a consequent reduction of flavor in the milk.

Although feeds contain a variety of flavorful compounds or precursors of them, many of the components are filtered out. Consequently, milk produced on quite different rations may have flavors that are quite similar. Some feeds contain suprathreshold levels of transferable components that impart distinctive flavors. For example, fresh cut alfalfa hay contains high concentrations of *trans*-2-hexenal, and 3-hexenals and 3-hexenols which have a green grassy flavor (120). When cows were fed protected sunflower seed oil in a basal diet of chopped lucerne hay and crushed oat the milk fat had a sweet raspberry-like flavor due to a mixture of γ-dodec-*cis*-6-enolactone and γ-dodecanolactone (33). When the basal diet was fed alone there was a sweet flavor due almost entirely to the saturated lactone. In this incidence it was postulated that the lactones were produced in the rumen from precursors in the feed. Trimethylamine amine has shown to be responsible for the fishy flavor in milk produced by cows on wheat pasture (121). Methylamine was also found to be present but at levels below the threshold value. If the flavors of amines are additive like the carbonyls it is possible that the methylamine did contribute to the flavor.

Weed Flavors. A number of weeds impart serious off-flavors to milk. One of the most common and readily recognized weed flavors is that caused by the consumption of wild garlic. The flavor components from some weeds are relatively nonvolatile and are not exhausted rapidly from the cow's body via the lungs. Therefore, they may affect the flavor of the milk until they are excreted or otherwise metabolized, a process which may take as long as 12 hours.

The off-flavor produced when cows eat plants of the *Lepidium* species has been attributed to skatole and indole with skatole being the major component (122). Benzyl methyl sulfide has been shown to be the primary cause of the flavor produced by cows eating *Coronopus* or land cress (123). It was postulated that this compound is a metabolite of benzyl thiocyanate. The land cress tainted milk also contained traces of benzyl isothiocyanate, benzyl cyanide, indole and skatole. Off-flavors have been attributed to many other weeds (124) but the chemical identity of the offending compounds has not been established.

Cowy Flavor. It has been suggested (125) that cows suffering from ketosis or acetonemia produce milk having a cowy-like odor. The odor of the breath of affected cows is similar to that of the milk and may be so strong that it is transmitted to the milk of neighboring cows if the ventilation of the area is inadequate. The cowy odor at one time reputed to be due to the acetone bodies released into the blood stream from incomplete metabolism of fat. The acetone level in milk can also be affected by the amount of acetone in feeds, such as silage (126). However, the amount of acetone in the milk did not appear to be related to the intensity of off-flavor. It has also been reported that suprathreshold amounts of methyl sulfide impart a cowy flavor to milk (127).

Barny Flavor. Flavor panelists have used the term barny to describe the flavor that they associate with a poorly ventilated barn. The nature of the barny flavor has not been characterized or distinguished from cowy flavor.

Control of Transmitted Flavors. Most of these flavors can be avoided if strong flavored feeds are not fed within 4 hours of milking time and the milking environment is kept free of strong odors. The cows and barns should be kept clean and the barns properly ventilated. In cases where the pastures or forages are contaminated with strong flavored feeds or weeds it may be necessary to subject the milk to vacuum treatment.

Detection of Transmitted Flavors. Because of the heterogeneous nature of these flavors, sensory evaluation provides the best method for identifying and quantifying the flavor. A simple test has been developed for measuring the trimethylamine (i.e. fishy flavor) associated with feeding wheat pasture (128). In this method formaldehyde was added to milk to bind primary and secondary amines. The sample was made alkaline and air bubbled through it for 6 minutes. The exhaust vapors were forced through a narrow tube which contained a pH indicator strip. The strip was made from a orlon acrylic fiber which had been saturated with bromcresol green. The amount of trimethylamine was estimated from the length of the fiber that had undergone a color change. This is a simple quick test which was reported to be more sensitive than the sensory panel. Reliable tests of this type are needed for the other flavors.

Miscellaneous Flavors

Flavors that have not been attributed to a specific cause or specifically defined in sensory terms have been placed in the miscellaneous category.

Absorbed Flavors. Until recently it was assumed that most feed and environmental odors were absorbed by milk directly from the air. Now it is known that such odors are transmitted frequently through the cow. However, some volatile substances may be absorbed directly from the air. The evidence indicates that fat-soluble substances, such as turpentine and other volatile solvents are absorbed readily, particularly if the cream has risen. (Homogenized milk and skim milk do not readily absorb these odors). The milking and milk storage areas should be kept free of these odors. Milk should not be stored in an open container in a refrigerator.

Astringent Flavors. Astringent has been used to describe a dry puckery, oral sensation which involves the sense of touch or feel rather than taste. The terms rough, chalky, or powdery also have been used to describe this sensation. It was noted that in the section on heated flavors that astringency has been associated most frequently with milk products that have been processed at high temperatures. Astringency has been observed in milk fortified with iron salts, especially ferrous salts (129). Astringency occurs occasionally in fresh raw milk but the cause has not been identified.

Bitter Flavor. This flavor is frequently caused by proteolysis since some of the liberated peptides and amino acids are bitter. In fluid milk microbial proteases are usually responsible for proteolysis because the activity of natural milk proteases is low. Bitter flavor may also be caused by lipolysis or certain weeds e.g. bitterweed.

Chalky Flavor. The term chalky is used to describe a tactual effect which is similar if not identical to astringent. Chalky has been described as a sensation suggesting finely-divided insoluble powder particles. It has been suggested that homogenization and/or high heat treatments are responsible.

Chemical Flavors. This designation is given to flavors that are believed to be due to contamination of milk with chemicals associated with cleaners, sanitizers, and disinfectants or pesticides. Chlorine and iodine compounds are probably the most frequent contaminants. Occasionally milk has been contaminated with traces of phenolic compounds from disinfectants or pesticides. A chorophenol flavor has been found in milk which was attributed to a reaction between a chlorine sterilizing agent and phenols in the water supply (130).

Flat. This defect is characterized by a lack of flavor and a thin or watery consistency. It can be simulated by the addition of 3 to 5% water. Some people believe that vacuum treatment causes a flat taste.

Foreign Flavor. This term is generally applied to samples with an alien or abnormal flavor. The so called "catty" odor might be placed in this category. This odor is produced by the reaction between mesityl oxide and H_2S or sulfhydryl groups (33). The mesityl oxide is an impurity in ketone containing paint solvents and therefore such solvents should be kept out of milk handling and processing areas.

Lacks Freshness. This term is used when the panelist perceives a flavor deterioration that cannot be described more precisely. It could be caused by concentrations of substances that fall between the absolute and recognition threshold levels of two or more off-flavors could be responsible.

Salty Flavor. This defect is identified easily by tasting and can be objectively measured by a silver nitrate titration. It is most commonly found in milk from cows in late lactation and occasionally from cows with mastitis.

CONCLUSIONS

Milk quality control is a process that starts on the farm before milk is secreted and continues until it is consumed. Steps necessary to maintain quality is summarized below.

1. Use feeds and feed practices that avoid the transfer of offensive flavors from the feed to the milk.
2. House and milk cows in well ventilated buildings.
3. Cool milk as soon as possible with the minimum agitation.
4. Use milk transporting system that keeps milk cool and minimizes agitation.

5. Pasteurize milk as soon as possible in properly sanitized equipment. (Thermal treatments that are at least equivalent to 76.7 C for 16 seconds is recommended).
6. Homogenize milk and package milk in containers that minimize exposure of milk to light, particularly to wavelengths below 500 nm.
7. Store milk below 4.4 C and minimize light exposure.
8. Use a milk distribution system designed to get the milk to the consumer as soon as possible.
9. Use proper sanitary practices during all stages of milk handling.
10. Do not use equipment that would lead to copper or iron contamination of the milk.

Although these steps are relatively simple they are not always followed. The consolidation of farms and processing plants have reduced the variability in the milk supply, but the consequent bulk handling of raw milk has led to increased agitation on the farm, in transit, and in the plants. The risk of temperature fluctuations is increased in this handling system. These practices have increased the danger of lipolyzed and microbial flavor development. The length of time between processing and consumption has increased so that the risks of chemical, enzymatic and microbial deterioration has increased. The packaging of milk in clear plastic containers and storing it under fluorescent light have increased the incidence of light-induced flavor. Some of the flavor problems are due to the failure of some retail stores to properly refrigerate and monitor the age of milk.

Some of the difficulty in maintaining quality is due to a lack of suitable analytical tests for monitoring and predicting quality. Data showing flavor deterioration would enable the analyst to convince the management that a problem exists. Furthermore, data on the rate of deterioration would be useful in establishing a realistic milk dating system. Although objective methods will facilitate analysis good quality control can be achieved with properly trained flavor panelists.

REFERENCES

1. Nelson, J. A., and Trout, G. M., Judging Dairy Products 4th ed., Olsen Publ. Co., Milwaukee, WI (1965).
2. ADSA Committee on Dairy Products Evaluation, Mimeographed Guidelines for Dairy Products Evaluation, Revised 1979.

3. Shipe, W. F., Bassette, R., Deane, D. D., Dunkley, W. L., Hammond, E. G., Harper, W. J., Kleyn, D. A., Morgan, M. E., Nelson, J. H., Scanlan, R. A., J. Dairy Sci. 61, 855 (1978).
4. Colson, T. J. and Bassette, R., J. Dairy Sci. 45, 182 (1962).
5. Sather, L. A., Calvin, L. D., and Liming, N. E., J. Dairy Sci., 48, 877 (1965).
6. Sather, L. A., Calvin, L. D., and Tamsma, A., J. Dairy Sci., 46, 1054 (1963).
7. Shipe, W. F., 38th Annual Report NY State Assoc. Sanitarians, pp. 41-48 (1964).
8. Shipe, W. F., Senyk, G. F., Ledford, R. A., Bandler, D. K., and Wolff, E. T., J. Dairy Sci, 63 (Suppl 1), 43 (1980).
9. Hutton, J. T., and Patton, S., J. Dairy Sci, 35, 699 (1952).
10. Boyd, E. N., and Gould, I. A., J. Dairy Sci, 40:1294 (1957).
11. Blankenagel, G., and Humbert, E. S., J. Dairy Sci, 46, 614 (1963).
12. Shankaranarayana, M. L., Raghanan, B., Abraham, K. O., and Natarajan, C. P., CRC Crit. Rev. in Food Technol., 4, 395 (1973).
13. Burton, H., Dairy Sci Abst., 31, 287 (1969).
14. Thomas, E. L., Burton, H., Ford, J. E., and Perkin, A. G., J. Dairy Res. 42, 285 (1975).
15. Zadow, J. G., and Birtwistle, R., J. Dairy Res. 40, 169 (1963).
16. Scanlan, R. A., Lindsay, R. C., Libbey, L. M., and Day, E. A., J. Dairy Sci., 51, 1001 (1968).
17. Thomas, E. L., Burton, H., Ford, J. E., and Perkin, A. G., J. Dairy Res, 42, 285 (1975).
18. Jannou, H. A., Paney, J. A., and Manning, D. J., J. Dairy Res, 45, 391 (1978).
19. Cobb, W. Y., Diss. Abstr., 24, 1132 (1963).
20. Hansen, A. P., Turner, L. G., and Jones, V. A., J. Dairy Sci., 57, 280 (1974).
21. Patton, S., and Josephson, D. V., J. Dairy Sci, 35, 161 (1952).
22. Josephson, R. V., Thomas, E. L., Morr, C. V., and Coulter, S. T., J. Dairy Sci., 50, 1376 (1967).
23. Harwalkar, V., J. Dairy Sci., 55, 1400 (1972).
24. Ferretti, A., J. Agric and Food Chem, 21, 939 (1973).
25. Badings, H. T., In "Progress in Flavour Research", (Land, D. G., and Nursten, H. E., ed) Appl. Sci. Publ., London, p. 263 (1979).
26. Janolino, V. G., and Swaisgood, H. E., J. Biol. Chem., 250, 2532 (1975).

27. Janolino, V. G., and Swaisgood, H. E., J. Dairy Sci., 61, 393 (1978).
28. Shipe, W. F., In "Enzymes in Food and Beverage Processing", (R. F. Gould, ed) ACS, Washington, DC.
29. Patton, S., and Josephson, D. V., J. Dairy Sci., 32, 398 (1949).
30. Patton, S., J. Dairy Sci.,37, 446 (1954).
31. Allen, C. and Parks, O. W., J. Dairy Sci., 58, 1609, (1975).
32. Samuelsson, E. G., Milchwissenschaft 17, 401 (1962).
33. Forss, D. A., J. Dairy Res., 46, 691 (1979).
34. Dimick, P. S., J. Dairy Sci., 59, 305 (1976).
35. Finley, J. W., and Shipe, W. F., J. Dairy Sci., 54, 15 (1971).
36. Gilmore, T. M., and Dimick, P. S., J. Dairy Sci., 62, 189 (1979).
37. Singleton, J. A., Aurand, L. W., and Lancaster, L. W., J. Dairy Sci., 46, 1050 (1963).
38. Aurand, L. W., Singleton, J. A., and Matrone, G., J. Dairy Sci., 47, 827 (1964).
39. Aurand, L. W., Singleton, J. A., and Noble, B. W., J. Dairy Sci., 49, 138 (1966).
40. Wishner, L. A., J. Dairy Sci., 47, 216 (1964).
41. Aurand, L. W., Boone, N. H., and Giddings, G. G., J. Dairy Sci., 60, 363 (1977).
42. Korycka-Dahl, M., and Richardson, T., J. Dairy Sci., 61, 400)1978).
43. Chan, Henry W. S., J. Am. Oil Chem. Soc., 54, 100, (1977).
44. Bassette, R., J. Milk Food Technol, 39, 10 (1976).
45. Mehta, R. S., and Bassette, R., J. Food Protection, 42, 256 (1979).
46. Wishner, L. A., J. Dairy Sci., 47, 216 (1964).
47. Dunkley, W. L., Franklin, J. D., and Pangborn, R. M., J. Dairy Sci., 45, 1040 (1962).
48. Sattar, A., and deMan, J. M., CRC Critical Rev. Food Sci and Nutri, 7, 13 (1975).
49. Bradley, R. L., Jr., J. Food Protection, 43, 314 (1980).
50. Stull, J. W., J. Dairy Sci., 36, 1153 (1953).
51. Bradfield, A., and Duthie, A. H., Vt. Agri. Expt. Sta. Bull, 645, 19pp. (1966).
52. Coleman, W. W., Watrous, G. H., Jr., and Dimick, P. S., J. Milk and Food Technol., 39, 551 (1976).
53. Hansen, A. P., Turner, L. G., and Aurand, L. W., J. Milk Food Technol., 38, 388 (1975).
54. Dunkley, W. L., Franklin, J. D., and Pangborn, R. M., Food Technol., 16(9), 112 (1962).

55. Scanlan, R. A., Sather, L. A., and Day, E. A., J. Dairy Sci., 48, 1582 (1965).
56. Bills, D. D., Scanlan, R. A., Lindsay, R. C., and Sather, L., J. Dairy Sci., 52, 1340 (1969).
57. Parks, O. W., and Allen, C., J. Dairy Sci., 62, 1045 (1979).
58. Patton, S., J. Food Sci., 29, 679 (1964).
59. Shipe, W. F., Senyk, G. F., and Fountain, K. B., J. Dairy Scil, 63, 193 (1980).
60. Herrington, B. L., J. Dairy Sci., 37, 775 (1954).
61. Chandran, R. C., and Shahani, K. M., J. Dairy Sci., 47, 471 (1964).
62. Jensen, R. G., J. Dairy Sci., 47, 210 (1964).
63. Jensen, R. G., and Pitas, R. E., J. Dairy Sci., 59, 1203 (1976).
64. Intern. Dairy Federation Lipolysis Symposium. Intern. Dairy Federation Annual Bulletin, Document 82, 40 pp. (1974).
65. Intern. Dairy Federation Lipolysis Symposium. Intern. Dairy Federation Annual Bulletin, Document 86, 199 pp. (1975).
66. Harper, W. J., and Gould, I. A., Proc. 15th Intl. Dairy Congr. 1, 455 (1959).
67. Nilsson, R., and Willart, S., Milk and Dairy Research (Alnarp, Sweden), Rept. No. 64 (1964).
68. Law, B. A., J. Dairy Res., 46, 573 (1979).
69. Shahani, K. M., and Chandran, R. C., Arch. Biochem. and Biophys., 111, 257 (1965).
70. Shipe, W. F., Senyk, G. F., and Boor, K. J., J. Dairy Sci., 63 (Suppl 1) 46 (1980).
71. Marshall, R. T., and Charoen, C., J. Dairy Sci., 59 (Suppl 1) 51 (1976).
72. Chrisope, G. L., and Marshall, R. T., J. Dairy Sci., 59, 2024 (1976).
73. Hsu, Hsien-Yeh, "The Effect of Surfactants on the Enzymatic Hydrolysis of Triglycerides and Phospholipids in Milk". M.S. Thesis, Cornell University, Ithaca, NY (1978).
74. Frankel, E. M., and Tarassuk, N. P., J. Dairy Sci., 39, 1506 (1956).
75. Harper, W. J., Schwartz, D. P., and El-Hgarawy, I. S., J. Dairy Sci., 39, 46 (1956).
76. Thomas, E. L., Nielsen, A. J., and Olson, J. C., Jr., Am. Milk Rev. 77(1), 50 (1955).
77. Driessen, F. M., Jellema, A., vanLuin, F. J. P., Stadhouders', J., and Walkers, G. J. M., Neth. Milk Dairy J., 31, 40 (1977).
78. Salib, A. M. A., Anderson, M., and Tuckley, B., J. Dairy Res., 44, 601 (1977).

79. Mackenzie, R. P., Blohm, T. R., Auxier, E. M., Luther, A. C., J. Lipid Res., 8, 589 (1967).
80. Lindquist, B., Roos, T., and Fujita, H., Milchwissenschaft 30, 12 (1975).
81. Koops, J., and Klomp, H., Neth. Milk Dairy J., 31, 56 (1977).
82. Morgan, M. E., J. Dairy Sci., 53, 270 (1970).
83. Morgan, M. E., J. Dairy Sci., 53, 273 (1970).
84. Morgan, M. E., Biotechnol. and Bioeng. 18, 953 (1976).
85. Mikolajcik, E. M., and Simon, N. T., J. Food Protect. 41, 93 (1978).
86. Kinsella, J. E., In "Frontiers in Food Research", Cornell Food Sci. Symposium, p. 94 (1968).
87. Kinsella, J. E., Patton, S., and Dimick, P. S., J. Am. Oil Chem. Soc., 44, 449 (1967).
88. Lillard, D. A., and Day, E. A., J. Dairy Sci., 44, 623 (1961).
89. Day, E. A., Lillard, D. A., and Montgomery, M. W., J. Dairy Sci., 46 291 (1963).
90. Forss, D. A., Dunstone, E. A., and Stark, W., J. Dairy Res.,27, 381 (1960).
91. Meijboom, P. W., J. Am. Oil Chem. Soc., 41, 326 (1964).
92. Lillard, D. A., In "Lipids as a Source of Flavor" (F. F. Gould, ed.) p. 68, ACS, Washington,DC (1978).
93. Wilkinson, R. A., and Stark, W., J. Dairy Res. 34, 89 (1967).
94. Swoboda, P. A. T., and Peers, K. E., J. Sci. of Food and Agric., 28, 1019 (1977).
95. Shipe, W. F., J. Dairy Sci., 47, 221 (1964).
96. Krukovsky, V. N., Agri. and Food Chem., 9, 439 (1961).
97. Dunkley, W. L., Smith, L. M., and Ronning, M., J. Dairy Sci., 43, 1766 (1960).
98. Price, W. O., Linkous, W. N., and Engel, R. W., J. Agri., Food Chem., 3, 226 (1955).
99. Tikriti, H. H., Burrows, F. A., Weisshaar, A., and King, R. L., J. Dairy Sci., 51, 979 (1968).
100. Dunkley, W. L., Franke, A. A., Robb, J., J. Dairy Sci., 51, 531 (1968).
101. Dunkley, W. L., Ronning, M., Franke, A. A., and Robb, J., J. Dairy Sci., 50, 492 (1967).
102. Teichman, R., Morgan, M. E., Eaton, H. D., and MacLeod, P., J. Dairy Sci., 38, 693 (1955).
103. Mueller, W. F., and Blazys, K., J. Dairy Sci., 38, 695 (1955).
104. Dunkley, W. L., Franke, A. A., Ronning, M., and Robb, J., J. Dairy Sci., 50, 100 (1967).
105. Wrenn, T. R., Bitman, J., Weyant, J. R., Wood, D. L., Wiggens, K. D., and Edmondson, L. F., J. Dairy Sci. 60, 521 (1977).

106. Goering, H. K., Gordon, C. H., Wrenn, T. R., Bitman, J., King, R. L., and Douglas, F. W., Jr., J. Dairy Sci., 59, 416 (1976).
107. Bruhn, J. C., Franke, A. A., and Goble, G. S., J. Dairy Sci., 59, 828 (1976).
108. King, R. L., and Dunkley, W. L., J. Dairy Sci., 42, 420 (1959).
109. King, R. L., Luick, J. R., Litman, I. I., Jennings, W. L., and Dunkley, W. L., J. Dairy Sci., 42, 780 (1959).
110. Gregory, J. F., Babish, J. G., and Shipe, W. F., J. Dairy Sci., 59, 364 (1976).
111. Pork, O. W., In "Fundamental of Dairy Chemistry" (B. H. Webb, A. H. Johnson, and J. A. Alford, ed), Avi Publ. Co. Inc., Westport, CT, p. 240 (1974).
112. Corbett, W. J., Study of Oxidized Flavor in Milk, PhD Thesis, University of Illinois, Urbana, IL (1942).
113. Smith, G. J., and Dunkley, W. L., Proc. XVIth Intern. Dairy Congress A, 649 (1962).
114. Kleyn, D. H., and Shipe, W. F., J. Dairy Sci., 44, 1603 (1961).
115. Yee, J. J., Shipe, W. F., and Kinsella, J. E., J. Food Sci., 45, 1082 (1980).
116. Dunkley, W. L., and Jennings, W. G., J. Dairy Sci., 34, 1064 (1951).
117. King, R. L., J. Dairy Sci., 45, 1165 (1962).
118. Dougherty, R. W., Shipe, W. F., Gudnason, G. U., Ledford, R. A., Peterson, R. D., and Scarpellino, R., J. Dairy Sci., 45, 472 (1962).
119. Shipe, W. F., Ledford, R. A., Peterson, R. D., Scanlan, R. A., Geerken, H. F., Dougherty, R. W., and Morgan, M. E., J. Dairy Sci. 45, 477(1962).
120. Morgan, M. E., and Pereira, R. L., J. Dairy Sci., 46, 1420 (1963).
121. Mehta, R. S., Bassette, R., and Ward, G., J. Dairy Sci., 57, 285 (1974).
122. Park, R. J., J. Dairy Res., 36, 31 (1969).
123. Park, R. J., Armitt, J. D., and Stark, W., J. Dairy Res., 36, 37 (1969).
124. Strobel, D. R., Bryan, W. C., and Babcock, C. J., "Flavor of Milk A Review", USDA Nov (1953).
125. Josephson, D. V. and Keeney, P. G., Milk Dlr 36 (10) 40 (1947).
126. Potts, R. B., and Kester, E. M., J. Dairy Sci., 40, 1466 (1957).
127. Patton, S., Forss, D. A., and Day, E. A., J. Dairy Sci., 39, 1469 (1956).
128. Kim, H. S., Gilliland, S. E., Von Gunten, R. L., and Morrison, R. D., J. Dairy Sci., 63, 368 (1980).

129. Demott, B. J., J. Dairy Sci., 54, 1609 (1971).
130. Schlegel, J. A., and Babel, F. J., J. Dairy Sci., 46, 190 (1963).

FLAVOR VOLATILES FORMED BY HEATED MILK

Takayuki Shibamoto[1]

Department of Environmental Toxicology
University of California
Davis, California 95616

Satoru Mihara
Osamu Nishimura
Yoko Kamiya
Akiyoshi Aitoku
Jun Hayashi

Ogawa & Co., Ltd.,
6-32-9 Akabanenishi, Kita-Ku
Tokyo, Japan

SUMMARY

The effect of heat treatment on volatile compounds produced from milk was studied in order to correlate changes in flavor and chemical composition of heated milk samples. Milk was heated in closed containers under various heating conditions. Volatile compounds were extracted with dichloromethane from the vacuum steam distillates of raw and heated milk and identified using gas chromatography/mass spectrometry. Forty-five compounds were positively identified and sixteen others tentatively identified. The compounds identified included acids, alcohols, esters, ketones, lactones, phenoles, furans, pyrazines, pyrroles, thiophenes, and sulfur compounds. The flavor threshold values of twenty-eight of the compounds identified in raw and heated milk samples were also determined. Furans and certain alcohols were found to be the important contributors of off-flavors in heated milk.

[1]To whom all inquiries should be addressed.

ISBN 0-12-169065-2

I. INTRODUCTION

The stability of heated milk has been of interest to researchers since the beginning of this century. Most studies aimed at discovering the physical stability of milk during heat treatment (the production of condensed milk and milk powder, or the sterilization of milk). The physical problems associated with heat treated milk, such as coagulation, were gradually solved (Rose 1963).

Within the last two decades, the problems of off-flavor in heat treated milk caused by trace amounts of chemicals which may form during heat treatment have begun to receive attention. Virtually all milk and milk products are heat processed to some degree. It is generally recognized that heat treatment affects the flavor of milk (Shipe _et al_., 1978). A considerable amount of research has already been conducted to elucidate the factors in, and mechanisms of, the formation of off-flavors in milk. Milk contains many components which may be precursors of off-flavor chemicals (proteins, fat, lactose, etc.).

It is generally accepted that a major factor of flavor deterioration in dairy products, resulting in stale and other off-flavors, is due to the interaction between lactose and milk proteins (Ferretti _et al_., 1970; Ferretti anf Flanagan, 1971a, 1971b, 1971c, 1972). Other important off-flavor precursors in milk are lipid constituents, which produce a rancid flavor (Parks _et al_., 1963; Parks, 1965, 1967). Scanlan (1968) indicated that diacetyl induced by heat from a lipid gives a rich, toasted note to milk. Jaddou _et al_. (1978) studied the effect of heat treatment of milk on headspace constituents and reported that cabbagey defects in heated milk are correlated with total amounts of volatile sulfurs present. The variety and intensity of the heated flavor depends on the time and temperature of the treatment. It is not, however, clear precisely which compound contributes the toasted or caramelized flavor to heated milk. In the present study, volatile compounds in raw and heated milk samples were analyzed in order to evaluate the off-flavors of milk.

II. EXPERIMENTAL

A. Materials

Raw mixed herd milk was obtained from Tsukuba Dairy Products Co., Ltd., in the vicinity of Mt. Tsukuba. Authentic samples were obtained from reliable commercial sources and used without further treatment, or synthesized by well-established methods.

B. Sample Preparations

1. Heat Treatment of Milk. Raw milk (1.5 L) was placed in Kjeldahl flasks and the necks of the flasks were flame-sealed. The flasks were placed in an oven at various temperatures (80 - 150° C) for 5 hours.

2. Isolation of Volatiles. The distillation apparatus used for this experiment is shown in Figure 1.

The heated milk or raw milk was placed in a 2 liter round bottom flask (A) and distillation was continued for 4 hours. The temperature of the flask was kept between 28 and 30° C, and the manometer reading was held between 12 and 15 mm Hg during the distillation. Ice-cooled water was used to cool the condenser (E) and distillate collection flask (F). The distillate was collected in an ice-cooled trap (300 mL), and extracted with 300 mL of dichloromethane using a liquid-liquid continuous extractor for 24 hours. The extract was dried over anhydrous sodium sulfate for 12 hours and concentrated with a Kunderna-Danish evaporative concentrator to 0.2 mL in volume. Further concentration to 10 - 20 μl in volume was carried out with a nitrogen stream.

Immediately after the concentraion, the samples were subjected to instrumental analyses.

C. Analyses of Volatiles

1. Gas-Liquid Chromatograph (GLC). A Hewlett Packard Model 5710A gas chromatograph equipped with a flame ionization detector and a 50 m x 0.28 mm i.d. glass capillary column coated with Carbowax 20M was used for routine GLC analysis. Figures 2 - 4 show a typical gas chromatogram of volatiles obtained from raw and heated milk. The oven temperature was programmed from 80 to 200° C at 1° C/sec. Nitrogen carrier flow rate was 0.68 mL/min. The injector and detector temperatures were 250° C. The injector split ratio was 1:100.

The nitrogen and sulfur-containing compounds were analyzed by a gas chromatograph (Hitachi 163) equipped with a flame thermionic detector (Hitachi FTD). The gas chromatographic conditions were as described for the Hewlett-Packard instrument.

A Hewlett-Packard Model 3385-A reporting integrator was used to determine the peak area.

2. Gas Chromatography/Mass Spectrometry (GC/MS). A Hitachi Model RMU-6M combination mass spectrometer-gas chromatograph (Hitachi Model M-5201) equipped with a Hitachi Model M-6010 and 1011/A data system was used under the following conditions: ionization voltage, 70 eV; emission current, 80 µA; ion accel voltage, 3100 V; ion source temperature, 200° C. The gas chromatographic column and oven conditions were as described for the Hewlett-Packard instrument.

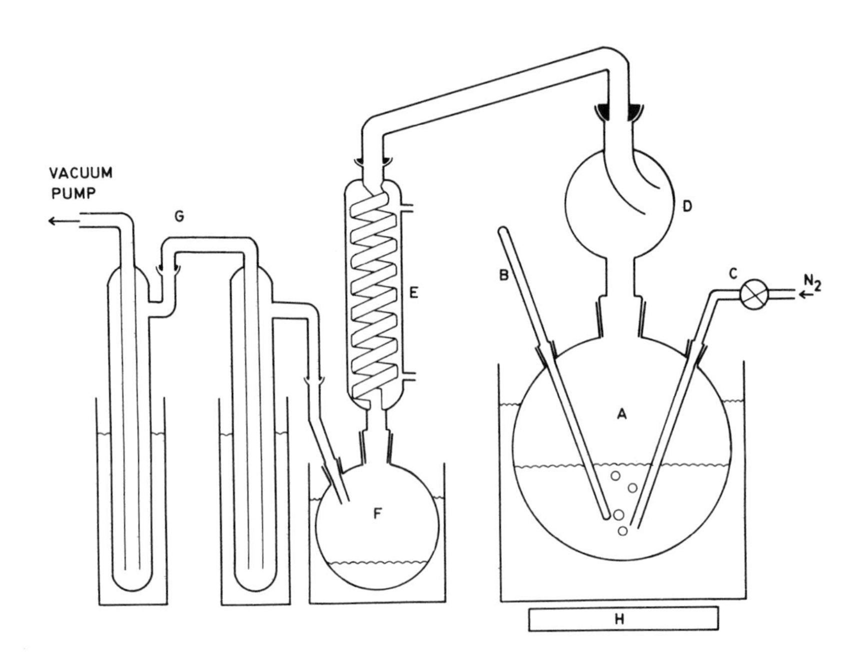

Figure 1. Distillation apparatus used to isolate volatiles. A. 2 litter flask; B. thermometer; c. nitrogen inlet; D. Wagner tube; E. condenser; F. cold trap; G. liquid nitrogen Dewer flasks; H. heater.

3. Identification of Unknowns. Identification of the volatile constituents of raw and heated milk was make by comparison of their mass spectra and Kovats' Indices to those of authentic compounds.

4. Sensory Evaluation of Milk Samples. The flavor of each sample was evaluated by three trained panels. The threshold value for each compound was determined in water, or in homogenized emulsion constructed with glycerin trioctanoate (3g), sucrose (6g), arabian gam powder (3g), and water (90 mL).

III. RESULTS AND DISCUSSION

A. General View of Analytical Results

The volatile compounds identified from raw and heated milk samples are listed in Table I. Figures 2 - 4 show a typical gas chromatogram (FID) of volatiles obtained from raw and heated milk samples. Figure 5 shows a comparison of gas chromatograms of heated (150° C) milk volatiles monitored FID and FTD.

TABLE I. Volatiles Identified in raw and heated milk

Peak[a] no.	Compound	Peak area% in raw	in heated 100° C	in heated 150° C
1.	Furan			0.01
2.	Acetone	0.11	0.78	0.26
3.	Diacetyl	0.62	1.09	0.26
4.	2-n-Pentanone		1.24	0.21
5.	2,3-Pentadione		0.32	0.55
6.	Methyl trans-propenyl ketone (tentative)			0.26
7.	1-Methyl-2-butenol (tentative)			0.08
8.	2-n-Heptanone		6.54	0.28
9.	Cyclopentanone		0.11	0.11
10.	Pyrazine		0.40	1.14
11.	Methylpyrazine		0.10	0.75
11.	2-Methyl-tetrahydrofuran-3-one (tentative)		0.25	5.74
12.	Acetoin	13.89	0.42	0.82
13.	Acetol	1.70	3.44	7.51
14.	2,5-Dimethylpyrazine			0.08
15.	2,6-Dimethylpyrazine(tentative)			0.05
16.	Ethylpyrazine(tentative)			0.21

17.	n-Propyl amine			0.46
18.	2,3-Dimethylpyrazine			0.04
19.	2-Cyclopenten-1-one(tentative)		0.08	2.67
20.	Butane-1-ol-2-one(tentative)		0.02	0.27
21.	2-Ethyl-6-methylpyrazine (tentative)			0.05
22.	2-Ethyl-5-methylpyrazine (tentative)			0.03
23.	Trimethylpyrazine			0.05
24.	2-n-Nonanone		2.66	
25.	Acetic acid			0.26
26.	2-Furaldehyde		4.28	24.35
27.	2-Acetylfuran		1.04	6.13
28.	Furfuryl methyl ketone			0.70
29.	Furfuryl acetate			0.67
30.	Dimethylsulfoxide	0.14		
31.	n-Octanol	0.42		
32.	5-Methylfurfural		0.05	3.94
33.	2-Propionylfuran		0.03	0.38
34.	(trans-1-Propenyl)-pyrazine (tentative)		0.31	0.34
35.	2-n-Undecanone		0.10	
36.	2,2'-Methylene bisfuran(tentative)		0.01	0.49
37.	n-Butanoic acid			1.35
38.	Diethylene glycol diethyl ether	1.74		
39.	Furfuryl alcohol		68.50	28.51
40.	5-Methyl-2(5H)-furanone			0.14
41.	2-Butenoic acid-γ-lactone (tentative)		1.11	1.60
42.	2-Hydroxy-3-methyl-2-cyclopenten-1-one			0.01
43.	n-Hexanoic acid	8.57		2.04
44.	Dimethylsulfone	1.29	0.01	0.09
45.	2-Hydroxy methylthiophene		0.06	0.27
46.	2,6-Di-tert-butyl-4-methyl phenol	5.32		
47.	Maltol		trace	0.01
48.	2-Acetylpyrrole(tentative)			0.02
49.	n-Dodecanol	0.58		
50.	n-Octanoic acid	17.91	0.08	0.42
51.	5-Methylpyrrole-2-carboxaldehyde(tentative)			0.05
52.	Eugenol	1.72		
53.	δ-Decalactone	3.56	0.04	
54.	n-Decanoic acid	21.24	0.04	0.03
55.	γ-Dodecalactone		0.04	trace

56.	γ-Dedecalactone		0.04	
57.	5-Hydroxymethylfurfural			0.03
58.	5-(2'-Furanylmethyl)-2-fura-methanol (tentative)			0.14
59.	n-Dodecanoic acid	3.56		
60.	n-Tetradecanoic acid	0.28		

[a]Peak numbers in Figures 2 - 5.

The compounds which were present in readily observable concentrations in raw milk were acids. The compositions (total peak area%) of each chemical group is as follows:

Chemicals	in raw milk	in heated milk 100° C	in heated milk 150° C
acids	51.6	0.1	4.1
alcohols	16.6	3.9	8.7
lactones	3.6	1.2	1.6
ketones	0.7	13.1	10.1
furans	-	74.2	65.0
pyrazines	-	0.8	2.7
others	10.2	0.1	1.1

The effct of heat treatment on various compounds in milk samples are expressed as the function of $\log(A/A_o)$ against temperature, where A is a peak area of the compound and A_o is a peak area of the internal standard (heptadecane).

Thresholds of 28 compounds evaluated by three trained panels are listed in Table II. The values of "unit flavor base" for 28 compounds were caluculated using the equation advanced by Keith and Powers (1968) and listed along with their odor thresholds in Table II.

$$\text{Unit flavor base} = \frac{\text{Concentration of the compound (ppm)}}{\text{Threshold value (ppm)}}$$

The flavor descriptions of the above 28 compounds are summarized in Table III.

TABLE II. Threshold Values and Unit Flavor Base of Flavor Volatiles Found in Milk

	Flavor Threshold Value (ppm) in		Maximum Conc. in Milk	Unit Flavor Base in	
	Water	Emulsion	(ppm)	Water	Emulsion
Acetic acid	8.000	14.000	0.079	0.010	0.006
n-Butyric acid	1.500	0.620	0.400	0.270	0.650
n-Hexanoic acid	0.200	0.200	0.630	3.200	3.200
n-Octanoic acid	0.800	0.900	0.126	0.160	0.140
n-Decanoic acid	0.700	0.500	0.063	0.090	0.130
n-Dodecanoic acid	0.500	5.000	0.010	0.200	0.002
Diacetyl	0.005	0.003	0.080	16.000	26.700
2-n-Pentanone	0.400	0.600	0.032	0.080	0.050
2-n-Heptanone	0.070	0.100	0.063	0.900	0.630
2-n-Nonanone	0.050	0.070	0.016	0.320	0.230
2-n-Undecanone	0.010	0.020	0.010	1.000	0.500
Acetol	0.200	0.800	5.010	25.100	6.260
Acetoin	0.040	0.120	1.580	39.500	13.200
Maltol	0.800	1.500	0.500	0.630	0.330
δ-Decalactone	0.001	0.015	0.005	5.000	0.330
γ-Decalactone	0.001	0.050	0.001	0.600	0.001
δ-Dodecalactone	0.010	0.700	0.003	0.300	0.004
5-Methyl-2(5H)-2-furanone	0.200	0.300	0.040	0.200	0.130
Furfuryl alcohol	0.300	0.400	10.000	33.300	25.000
Furfural	0.100	0.100	3.020	30.300	30.200
2-Methylfurfural	0.050	0.070	1.120	22.400	16.000
2-Acetylfuran	1.000	0.500	1.700	1.700	3.400
2-Propionylfuran	0.200	0.600	0.110	0.560	0.190
Pyrazine	0.160	0.300	0.650	4.060	2.200
2-Methylpyrazine	0.250	0.200	0.417	1.670	2.100
2,3-Dimethyl-pyrazine	0.100	0.500	0.005	0.050	0.010
2,5-Dimethyl-pyrazine	0.020	0.030	0.008	0.400	0.270
Trimethylpyrazine	0.010	0.022	0.005	0.500	0.230

TABLE III. Flavor Descriptions of Volatiles Identified

Compound	Flavor Descriptions In Water	In Emulsion
Acetic acid	sour	not-oily
n-Butyric acid	buttery	smooth, mild
n-Hexanoic acid	chemical	heavy
n-Octanoic acid	dry, herb	fishy, oily, thick
n-Decanoic acid	milky, creamy	milky
n-Dodecanoic acid	waxy, slightly creamy	buttery, oily
Diacetyl	roasted	burnt, fermented
2-n-Pentanone	rotten	cheesy, buttery
2-n-Heptanone	rotten	cheesy, buttery
2-n-Nonanone	fatty, creamy	fried, fatty
2-n-Undecanone	oily, waxy	heated milk-like
acetol	sweet, roasted	yogurt-like
acetoin	sweet, warm	creamy, fatty
Maltol	fruity, sweet	sweet, fishy
δ-Decalactone	milky, buttery	oily, milky
γ-Dodecalactone	fatty, milky	sweet fatty
δ-Dodecalactone	milky	fatty, milky
5-Methyl-2(5H)-furanone	sweet	roasted
Furfuryl alcohol	nutty, cooked	sweet, heated milk
Furfural	cooked	heated milk-like
5-Methylfurfural	woody	heated milk-like
2-Acetylfuran	fruity, sweet	fruity, rotten
2-Propionylfuran	creamy	rotten, creamy
pyrazine	roasted	fishy
2-Methylpyrazine	roasted, nutty	creamy
2,3-Dimethylpyrazine	peanut-like	nutty
2,5-Dimethylpyrazine	burnt, wheat-like	roasted, sweet

The lower the threshold value, the more intense the flavor. The larger the unit flavor base, the more the component contributes its flavor to milk. Based upon the definition of unit flavor base, the following chemicals seem to give the characteristic flavor to heated milk: n-hexanoic acid, diacetyl, 2-n-undecanone, acetol, acetoin, δ-decalactone, furfuryl alcohol, furfural, 2-methylfurfural, 2-acetylfuran, pyrazine, and 2-methylpyrazine.

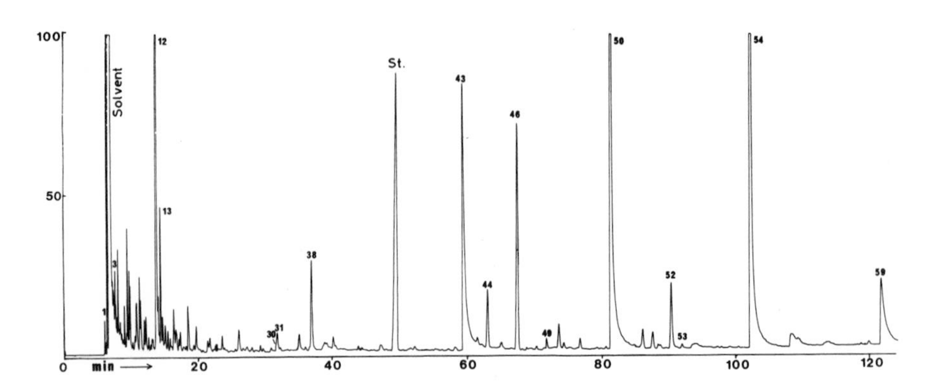

Figure 2. Gas chromatogram of flavor extract of raw milk

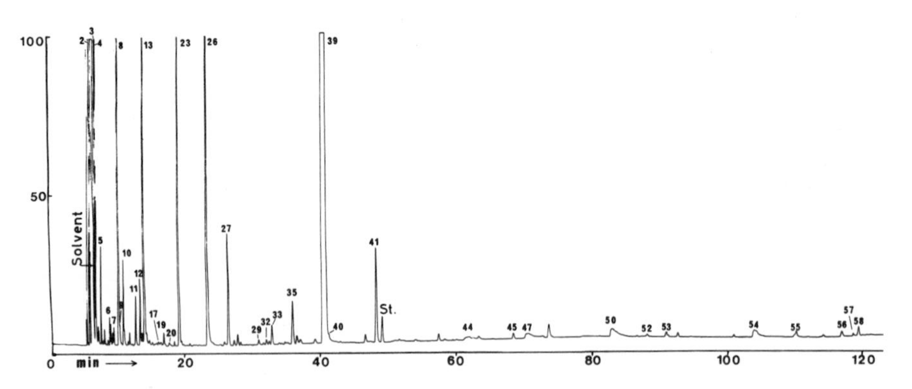

Figure 3. Gas chromatogram of flavor extract of heated milk at 100° C

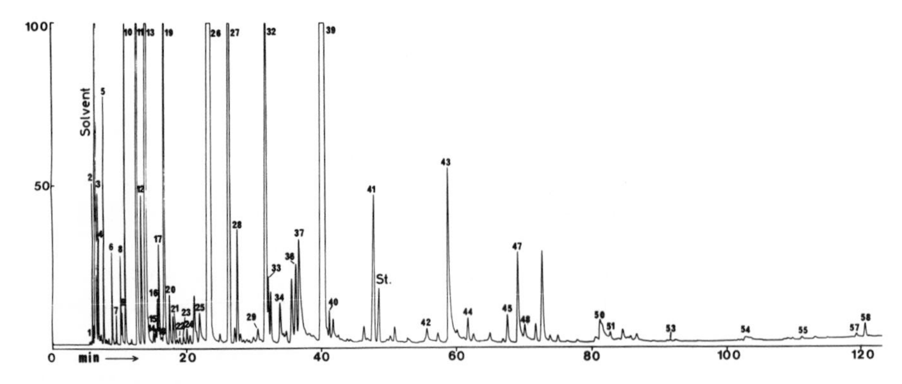

Figure 4. Gas chromatogram of flavor extract of heated milk at 150° C

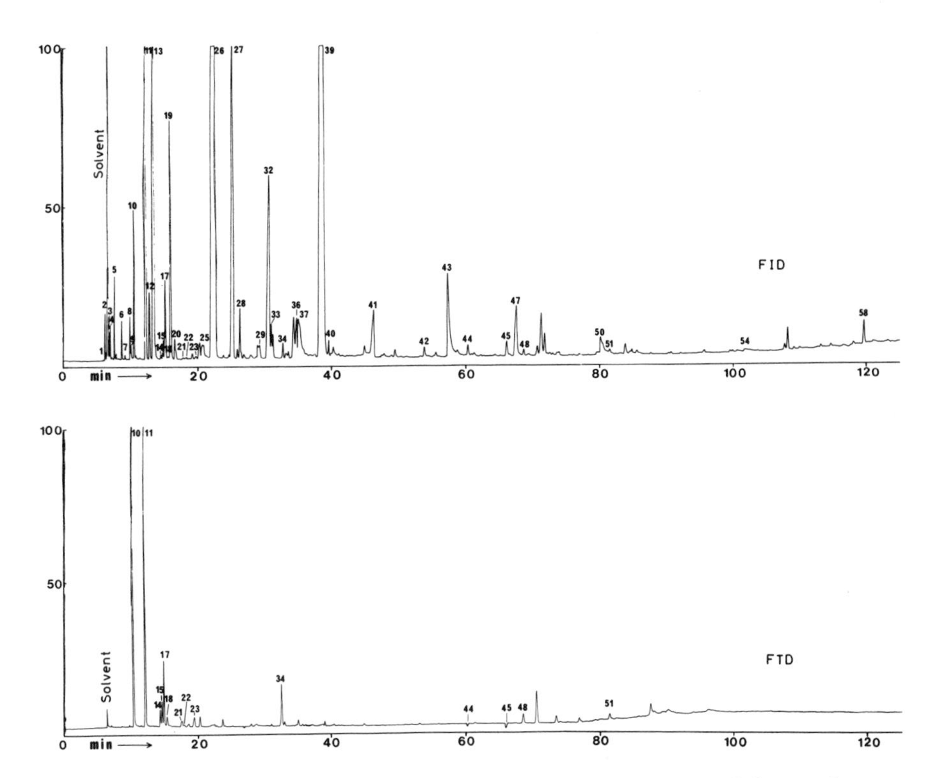

Figure 5. Gas chromatogram of flavor extract of heated milk (150° C) monitored with FID and FTD

B. FURANS

The effect of heat treatment on furan formation in milk is shown in Figure 6. Furans were not found in raw milk, but appeared suddenly in milk heated at 90° C. Furans were major volatile constituents of heated milk and their values of unit flavor base are the highest among the heated milk constituents. This suggests that furans play an important role in the flavor of heated milk.

Heyns and Klier (1968) reported that no furans were detected in their glucose pyrolysis experiment before the temperature reached 300° C. It is interesting that furans were detected in milk heated at 90° C in this study. Furfural, which is one of the main components of heated milk volatiles, is the first furan derivative to be found in a milk product (Parks and Patton, 1966). Twenty-nine furan derivatives have been reported in four milk products (dry whole milk, butter culture, whey powder, stale non-fat dry milk). Ferettis' group (1970, 1971a) have investigated the lactose-casein browning systems extensively and report that furans were formed from the casein catalyzed degradation of lactose by heat treatment at 80° C for eight days.

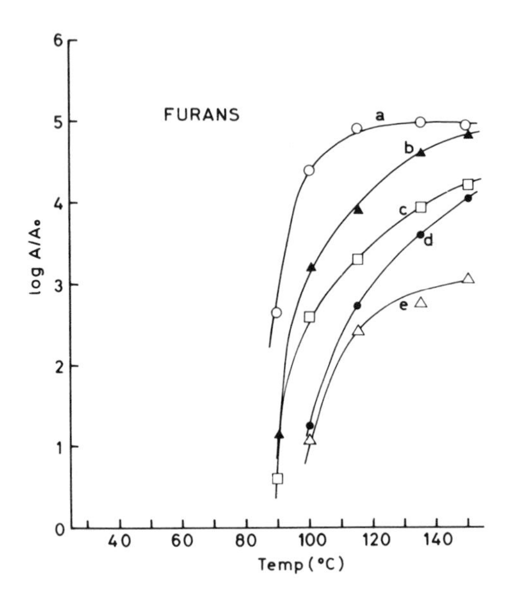

Figure 6. The effect of heat treatment on furans, a: furfuryl alcohol, b: furfural, c: 2-acetylfuran, d: 5-methylfurfural, e: 2-propionylfuran

C. Alcohols

The effect of heat treatment on alcohol formation in milk is shown in Figure 7. Acetol and acetoin were found in raw milk but butan-1-ol-2-one was not. The concentrations of acetol and acetoin decreased in milk heated to temperatures less than 80° C. The concentrations of these three compounds increased, however, with an increase of temperature from 90 to 150° C. The unit flavor bases of acetol and acetoin at their highest concentrations in milk are as high as that of furans. The taste of acetol is sweet and roasted in dilute aqueous solution and yogurt-like in emulsion. The off-flavor of heated milk is partly due to the formation of these alcohols.

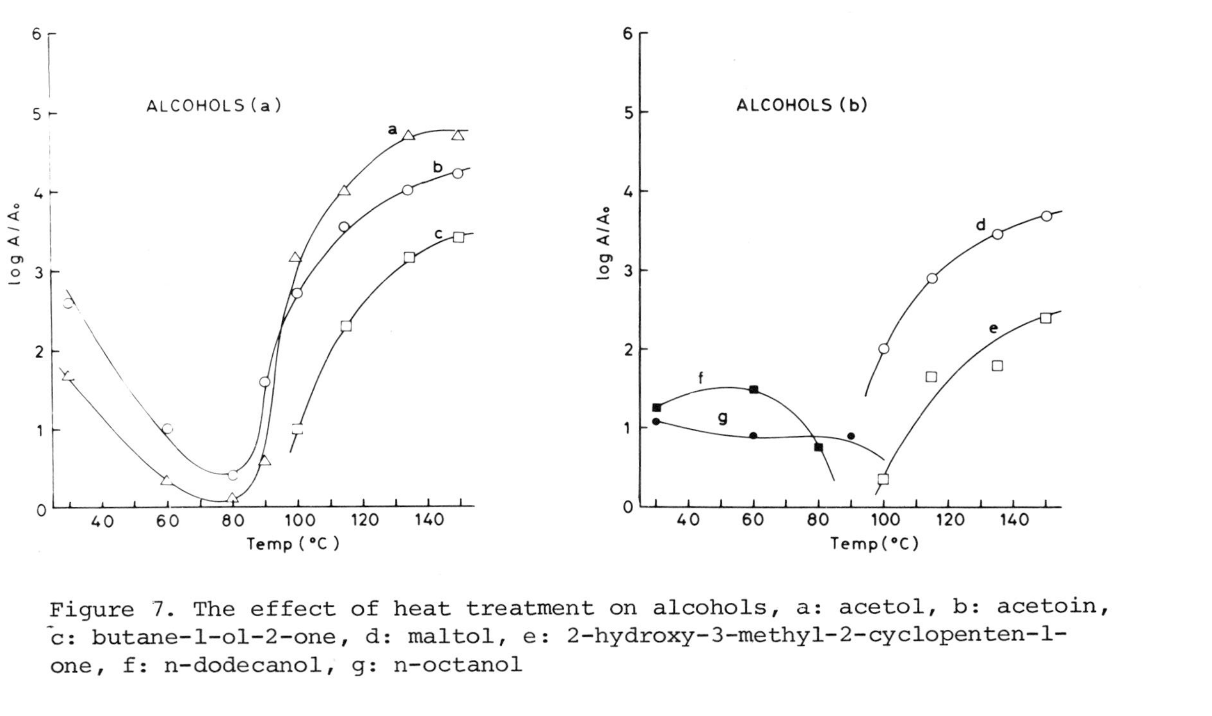

Figure 7. The effect of heat treatment on alcohols, a: acetol, b: acetoin, c: butane-1-ol-2-one, d: maltol, e: 2-hydroxy-3-methyl-2-cyclopenten-1-one, f: n-dodecanol, g: n-octanol

Acetol is a compound well known as the product of carbohydrate fragmentation or degradation which occurs during the nonenzymatic browning reactions (Heyns et al., 1966; Hodge, 1967; Heyns and Klier, 1968; Shaw et al., 1968; Ferretti et al., 1971a). Formation of acetol was also observed in butter culture (Lindsay and Day, 1965).

Maltol was also found at or above 100° C. Its unit flavor base at its highest concentration is less than 1, which indicates that maltol does not contribute significantly to heated milk flavor. Maltol has been found in overheated skim milk, condensed and dried milk, and dried whey (Hodge, 1967). it has also been formed from the caramelization of maltose and certain maltooligosaccharides in laboratory experiments, but has never been formed in sufficient quantity to isolate in the caramelization of glucose, sucrose, or pure starch (Backe, 1910a, 1910b; Diemar and Hala, 1959).

2-Hydroxy-3-methyl-2-cyclopentene-1-one formed at 100° C or above. This compound, originally found in wood distillate (Meyerfeld, 1912), also has been produced in small amounts by heating galactose in an aqueous sodium hydroxide solution (Enlevist et al., 1954). A solution of 2-Hydroxy-3-methyl-2-cyclopentene-3-one is sweet and gives a licorice-like flavor. At subthreshold concentrations, maltol and 2-hydroxy-3-methyl-2-cyclopentene-3-one showed a synergic effect on flavor. Trace amounts of octanal and dodecanal (1-2 ppb) were found in raw and low temperature-treated milk samples.

D. Ketones

1. Diketones. Figure 8 shows the variation in formation of diacetyl and 2,3-pentadione as a function of temperature. These data indicate that the production of these compounds increased as the temperature increased. The flavor threshold values for diacetyl in water and emulsion were found to be 5 and 3 ppb, respectively (Talbe II). These values are a little lower than reported values in milk (10 ppb, Bennett et al., 1965; 12 ppb, Scanlan et al., 1968; 19 ppb, Hempenius et al., 1966). It seems that the heat treatment increases the intensity of diacetyl odor from a subthreshold level to above the flavor threshold level in water and emulsion. Members of the organoleptic test panel recognized that diacetyl contributes significantly to the heated, burnt, and fermented notes in the flavor of heated milk.

Diacetyl is known to be one of the nonenzymatic browning reaction products (Hodge, 1967), and it is probably produced in heated milk in large amount.

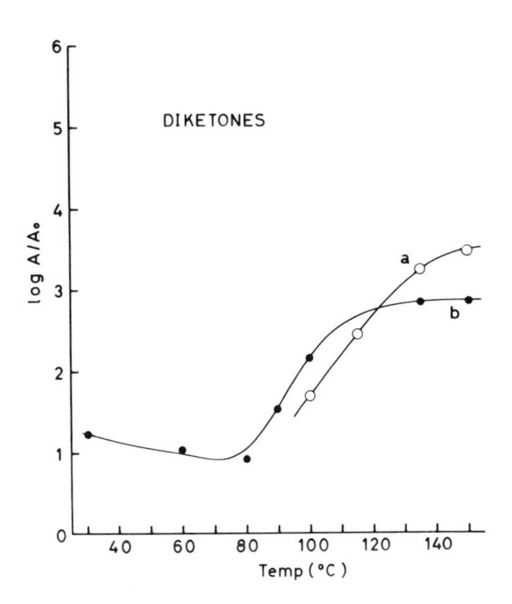

Figure 8. The effect of heat treatment on diketones, a: 2,3-pentadione, b: diacetyl

Hodge et al. (1963) proposed a formation mechanism for diacetyl in a nonenzymatic browning reaction. They suggested that diacetyl is formed from a methyl α-dicarbonyl intermediate which is in turn formed through one of the major pathways of nonenzymatic browning reactions. Heynes et al. (1966) proposed that diacetyl would form from the reaction of methyl glyoxal and glycine or its derivatives. This may be pertinent to the heat induced production of diacetyl in milk, as Jenness and Patton (1959) have indicated that methyl glyoxal is probably produced by heating lactose and trace amounts of free amino acids which are known to be present in milk. Scanlan et al. (1968) reported that a small amount of diacetyl (3-5 ppb) was observed in raw milk samples and assumed that the diacetyl in raw milk was transmitted to the milk from forage eaten by the cow.

2,3-Pentadione has been found in coffee (Stoffelsma et al., 1968) and a D-glucose/hydrogen sulfide/ammonia browning model system (Shibamoto and Russell, 1976), but has never been found in milk.

2. Cyclic Ketones. Figure 9 shows the effect of heat treatment on the formation of cyclic ketones in milk. These cyclic ketones are not easily accounted for on the basis of the known pathways of either the Maillard reaction or of sugar caramelization. Cyclopentanone and 2-methyl tetrahydrofuran-3-one have previously been found in a casein-lactose browning system (Ferretti and Flanagan, 1971a).

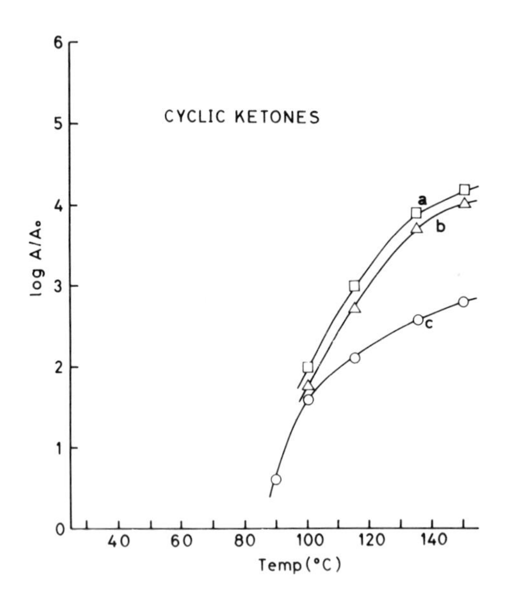

Figure 9. The effect of heat treatment on cyclic ketones, a: 2-methyl tetrahydrofuran-3-one, b: 2-cyclopentene-1-one, c: cyclopentanone

3. Methyl Ketones. Methyl ketones were not found in raw milk, but were found in heated milk (Figure 10). The highest concentration of methylketone is either a subthreshold level or a threshold level except in the case of 2-n-pentanone. Methyl ketones, at concentrations below their flavor threshold values, interacted additively to produce discernible off-flavors.

Milk which has been overheated during pasteurization may possess a stale off-flavor due to the exessive formation of methyl ketones. The precursors of these methyl ketones (β-keto acid) are biosynthesized in the bovine mammary gland from acetate (Lawrence and Hawke, 1966). These β-keto acids are apparently unfinished fatty acids, i.e., they have not undergone the final reduction, dehydration and reduction steps which normally occur during fatty acid biosynthesis (Kinsella et al., 1967). This reaction is catalyzed by heat and occurs spontaneously in stored dairy products (Schwartz et al., 1966). The proposed formation pathways of methyl ketones are as follows:

$$\begin{array}{l} CH_2\text{-}O\text{-}\overset{O}{\overset{\|}{C}}\text{-}R \\ | \\ CH\ \text{-}O\text{-}\overset{O}{\overset{\|}{C}}\text{-}R_1 \\ | \\ CH_2\text{-}O\text{-}\overset{O}{\overset{\|}{C}}\text{-}CH_2\text{-}\overset{O}{\overset{\|}{C}}\text{-}(CH_2)_N\text{-}CH_3 \end{array}$$

$$\downarrow +H_2O$$

$$HO\text{-}\overset{O}{\overset{\|}{C}}\text{-}CH_2\text{-}\overset{O}{\overset{\|}{C}}\text{-}(CH_2)_N\text{-}CH_3$$

$$\downarrow -CO_2$$

$$CH_3\text{-}\overset{O}{\overset{\|}{C}}\text{-}(CH_2)_N\text{-}CH_3$$

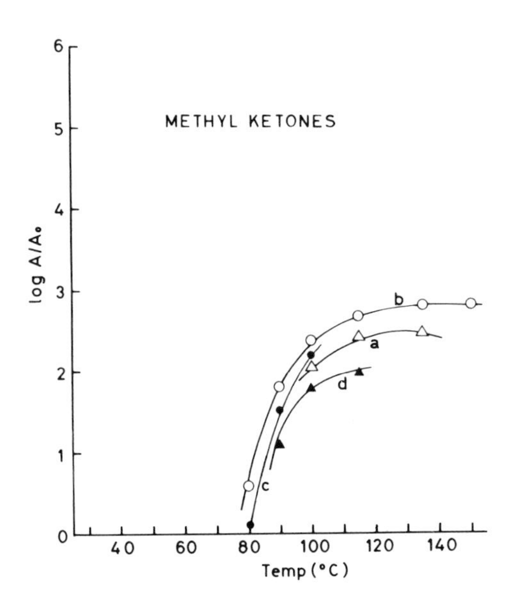

Figure 10. The effect of heat treatment on methyl ketones, a: 2-n-pentanone, b: 2-n-heptanone, c: 2-n-nonanone, d: 2-n-undecanone

E. Acids

At a temperature of 80° C, the concentrations of free fatty acids began to decrease (Figure 11). When the temperature was raised above 100° C, acetic, butyric, hexanoic, octanoic, and decanoic acids increased in concentration, in contrast to dodecanoic and tetradecanoic acids, which decreased. Butyric and hexanoic acids, previously known to be the major free fatty acids in whole and skim milk powders (Kawanishi and Saito, 1966) and in non-fat milk (Ferretti and Flanagan, 1972), are the two major constituents of the volatile oil obtained by vacuum steam distillation. Hexanoic acid is produced most abundantly and shows the lowest threshold value of the acids. It may contribute the chemical and rancid flavor to heated milk.

F. Lactones

The only lactone found in raw milk was δ-decalactone. The concentration of δ-decalactone increased in heated milk. δ-Dodecalactone, γ-dodecalactone, 5-methyl-2(5H)-furanone, and 2-butenoic acid-γ-lactone were not found in raw milk, but were found in heated milk (Figure 12). The concentration of δ-decalactone and γ-dodecalactone remained constant throughout the heat treatment from 80 to 150° C, but the concentration of 5-methyl-2(5H)-furanone and 2-butenoic acid γ-lactone increased in milk treated at higher temperatures. δ-Decalactone contributes a milky, buttery note to milk, and its flavor threshold values in water and in emulsion were 1 and 15 ppb, respectively.

It is known that when milk fat is heated or stored in the form of dairy products, δ-decalactone is formed (Keeney and Patton, 1956a, 1956b; Kinsella, 1969). The lactones are known as compounds possessing a buttery, coconut-like flavor. They account for much of the palatability derived from cooking foods in butter. On the other hand the flavor characteristics of δ-decalactone are not acceptable in dry whole milk.

When the glycerides containing these hydroxy acid esters are heated, the hydroxy acids are liberated and spontaneously form lactone upon the elimination of the water molecule.

Ferretti and Flanagan (1971a) found 4-methyl-2-butenoic acid γ-lactone in a lactose-casein browning system. They assumed that this compound would be a product of the well-known hydrolytic cleavage-rearrangement of furfuryl alcohol via the intermediate of levulic acid. 2-Butenoic acid-γ-lactone was also identified in a casein-lactose model system (Ferretti *et al.*, 1970).

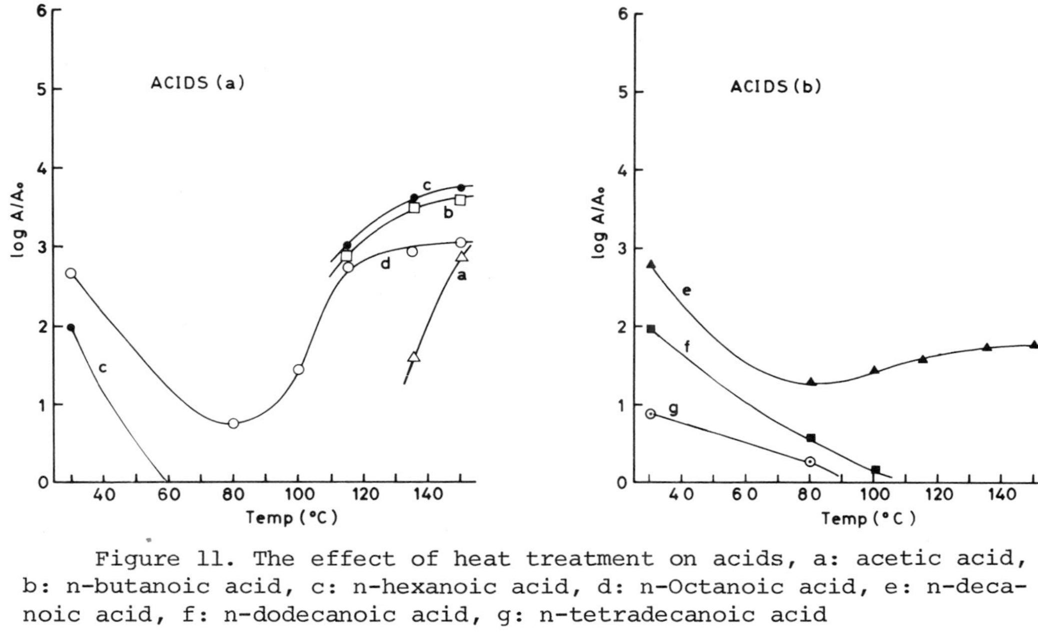

Figure 11. The effect of heat treatment on acids, a: acetic acid, b: n-butanoic acid, c: n-hexanoic acid, d: n-Octanoic acid, e: n-decanoic acid, f: n-dodecanoic acid, g: n-tetradecanoic acid

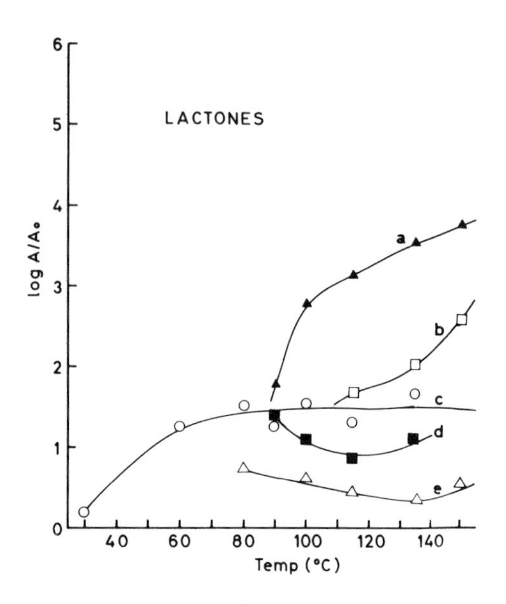

Figure 12. The effect of heat treatment on lactones, a: 2-butenoic acid-γ-lactone, b: 5-methyl-2(5H)-furanone, c: δ-decalactone, d: δ-dodecalactone, e: γ-dodecalactone

G. Sulfur- and Nitrogen-Containing Compounds

1. Pyrazines. Pyrazine and methylpyrazine formed at 90° C and trans-1-propenylpyrazine was also found above 100° C. Other alkylpyrazines were observed above 135° C (Figure 13). Figure 13 shows that the quantities of pyrazines increased with an increase in temperature. Members of the organoleptic taste panels noted that pyrazines contribute a heated, burnt, roasted flavors to heated milk. From this, one may hypothesize that pyrazines also contribute to the off-flavor of heated milk.

Alkylpyrazines were also found in spray-dried whey (Ferretti et al., 1970; Ferretti and Flanagan, 1971a), and in nonfat dry milk (Ferretti and Flanagan, 1972). The alkylpyrazine compounds found in this study have also been identified in the various Maillard type browning reactions (Shibamoto and Bernhard, 1976, 1977, 1978).

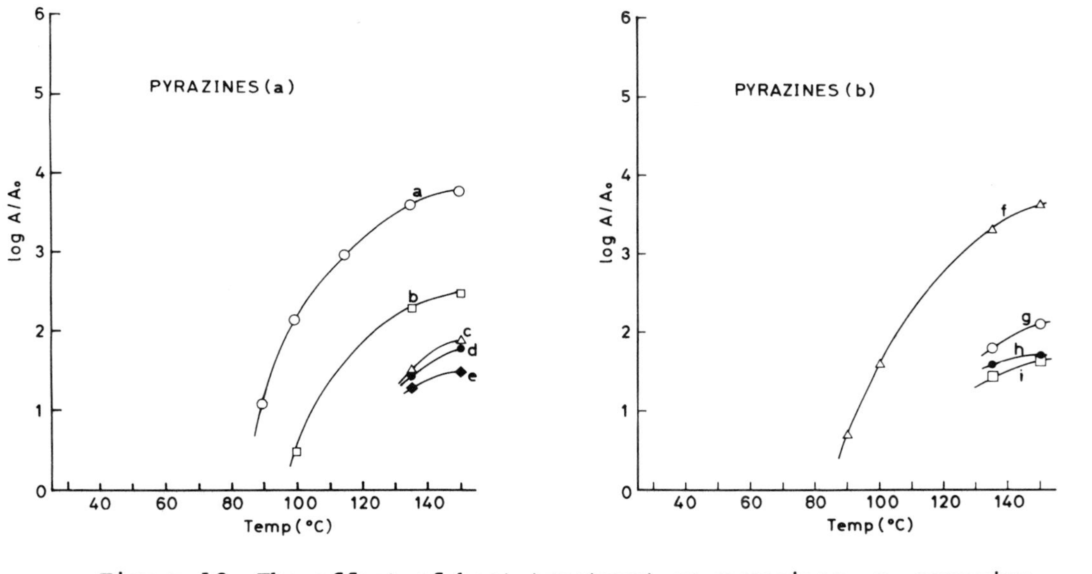

Figure 13. The effect of heat treatment on pyrazines, a: pyrazine, b: (trans-1-propenyl)-pyrazine, c: 2,5-dimethylpyrazine, d: 2-ethyl-5-methylpyrazine, e: 2-ethyl-6-methylpyrazine, f: 2-methylpyrazine, g: 2,6-dimethylpyrazine, h: 2,3-dimethylpyrazine, i: trimethylpyrazine

2. Sulfone, Thiophene, Pyrroles. Figure 14 shows the effect of heat treatment on sulfone, thiophene, and pyrroles in milk. A trace amount of dimethylsulfoxide was found in raw milk and in milk treated at 90° C. On the other hand, dimethylsulfone was found in all samples. Generally, heat treatment increased the concentration of dimethylsulfone.

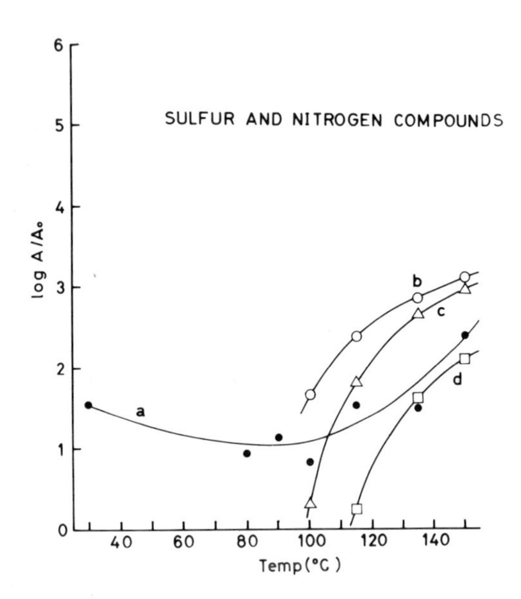

Figure 14. The effect of heat treatment on a sulfone, a thiophene, and pyrroles, a: dimethylsulfone, b: 2-hydroxymethylthiophene, c: 2-acetylpyrrole, d: 2-methylpyrrol-2-carboxaldehyde

Dimethylsulfone and dimethylsulfoxide have been reported in non-fat dry milk (Ferretti and Flanagan, 1972). Dimethylsulfide is one of the constituents of raw milk. Keenan and Lindsay (1968) related increased dimethylsulfide upon heating to a heat labile precursor which was identified as an S-methyl methionine sulfonium salt, conceivably introduced via feed from plant materials. A possible metabolic fate of dimethylsulfide is oxidation to dimethylsulfone via dimethyl sulfoxide as the intermediate.

2-Acetylpyrrole has been reported in lactose-amino acid and lactose-casein browning model systems (Langner and Tobias, 1967; Ferretti *et al.*, 1970).

2-Hydroxymethylthiophene has been found in coffee (Stoll et al., 1967) and 5-methylpyrrole-2-carboxaldehyde was isolated from roasted filberts (Kinlin et al., 1972). Pyrrole and thiophene derivatives found in this study have also been formed in the Maillard type browning reactions (Shibamoto and Russell, 1976; Shibamoto and Bernhard, 1978).

IV. CONCLUSIONS

Among the volatiles produced from milk by heat treatment, furans, alcohols, ketones, lactones, and sulfur- and nitrogen-containing compounds are those which give heated milk its characteristic flavor. The compounds which possess higher unit flavor bases, such as diacetyl, acetol, acetoin, furfuryl alcohol, furfural, and 5-methylfurfural, were recognized as the major flavor components of heated milk samples by trained panels. Furan derivatives were found to play the most important role in the heated flavor of milk.

REFERENCES

Backe, A., Compt. Rend. 150:540 (1910a).
Backe, A., Compt. Rend., 151:78 (1910b).
Bennett, G., Liska, B. J., and Hempenius, W. L., J. Food Sci. 30:35 (1965).
Diemar, W., and Hala, H., Z. Lebensm.-Undersuch.-Forsch. 110: 161 (1959).
Enkvist, T., Alfredsson, B., Merikallio, M., Pääkköneu, P., and Järvella, O., Acta Chem. Scand 8:51 (1954).
Ferretti, A., and Flanagan, V. P., J. Argic. Food Chem. 19:245 (1971a).
Ferretti, A., and Flanagan, V. P., J. Dairy Sci. 54:1764 (1971b).
Ferretti, A., and Flanagan, V. P., J. Dairy Sci. 54:1769 (1971c).
Ferretti, A., and Flanagan, V. P., J. Agric. Food Chem. 20:695 (1972).
Ferretti A., Flanagan, V. P., and Ruth, J. M., J. Agric. Food Chem. 18:13 (1970).
Hempenius, W. L., Liska, B. J., Bennett, G., and Moore, H. L., Food Technol. 20:334 (1966).
Heyns, K., and Klier, M., Carbohydr. Res. 6:436 (1968).

Heyns, K., Stute, R., and Paulsen, H., Carbohydr. Res. 2:132 (1966).
Hodge, J. E., "Symposium on Foods, Chemistry and Physiology of Flavors", Schultz, H. W., Day, E. A., and Libbey, L. M., Eds., AVI, Westrot, Conn., 1967, p465.
Hodge, J. E., Fisher, B. E., and Nelson, E. C., Am. Soc. Brew. Chem. Proc. 84 (1963).
Jaddou, H. A., Pavey, J. A., and Manning, D. H., J. Dairy Res. 45:391 (1978).
Jenness, R., and Patton, S., "Principles of Dairy Chemistry", Wiley, New York., 1959 p446.
Kawanishi, G., and Saito, K., Eiyo To Shokuryo 19:240 (1966).
Keenan, T. W., and Lindsay, R. C., J. Dairy Sci. 51:112 (1968).
Keeney, P. G., and Patton, S., J. Dairy Sci. 39:1104 (1956a).
Keeney, P. G., and Patton, S., J. Dairy Sci. 39:1114 (1956b).
Keith, E. S., and Powers, J. J., J. Food Sci. 33:213 (1968).
Kinlin, T. E., Muralidhara, R., Pittet, O., Sanderson, A., and Walradt, J. P., J. Agric. Food Chem. 20:1021 (1972).
Kinsella, J. E., Patton, S., and Kimick, P. S., J. Am. Oil Chem. Soc. 44:449 (1967).
Kinsella, J. E., Chem. Ind. 36 (1969).
Langner, E. H. and Tobias, J., J. Food Sci. 32:495 (1967).
Lawrence, R. C., and Hawke, J. C., Biochem. J. 98:25 (1966).
Lindsay, R. C., and Day, E. A., J. Dairy Sci. 48:1566 (1965).
Parks, O. W., "Fundamentals of Dairy Chemistry", Webb, B. H., and Johnson, A. H., Eds., AVI, Westport, Conn., 1965, p197.
Parks. O. W., "Symposium on Foods, Chemistry and Physiology of Flavors", Schultz, H. W., Day, E. A., and Libbey, L. M., Eds., AVI, Westport, Conn., 1967, p296
Parks, O.W., Keeney, M., and Schwartz, D. P., J. Dairy Sci. 46:295 (1963).
Parks, O. W., and Patton, S., J. Dairy Sci. 44:1 (1966).
Rose, D., Dairy Sci. Abst. 25:45 (1963).
Scanlan, R. A., Lindsay, R. C., Libbey, L. M., and Day, E. A., J. Dairy Sci. 51:1001 (1968).
Schwartz, D. P., Parks, O. W., and Yoncoskie, R. A., J. Am. Oil Chem. Soc. 43:128 (1966).
Shaw, P. E., Tatum, J. H., and Berry, R. E., J. Agric. Food Chem. 16:979 (1968).
Shibamoto, T., and Bernhard, R. A., J. Agric. Food Chem. 24: 847 (1976).
Shibamoto, T., and Bernhard, R. A., J. Agric. Food Chem. 25: 609 (1977).
Shibamoto, T., and Bernhard, R. A., J. Agric. Food Chem. 26: 183 (1978).
Shibamoto, T., and Russell, G. F., J. Agric. Food Chem. 24: 843 (1976).

Shipe, W. F., Bassette, R., Deane, D. D., Dunkley, W. L., Hammond, E. G., Harper, W. J., Kleyn, D. H., Morgan, M. E., Nelson, J. H., and Scanlan, R. A., J. Dairy Sci. 61:855 (1978).
Stoffelsma, J., Sipma, G., Kettenes, D. S., and pijpker, J., J. Agric. Food Chem. 16:1000 (1968).
Stoll, M., Winter, M., Gautschi, F., Flament, I., and Wilhalm, B., Helv. Chim Acta. 50:628 (1976).

OFF-FLAVORS OF HUMAN MILK

Mina R. McDaniel

Department of Foods and Nutrition
University of Manitoba
Winnipeg, Canada

I. INTRODUCTION

As contrasted to other topics in this symposium, human milk is generally not thought of as a "food" or "beverage" in the commercial sense. However, it does have flavor properties which may affect its acceptance or rejection by an infant. Off-flavors in human milk could have significant consequences which would be more nutritional than economic in nature. Many centers now recommend that mothers breast-feed their infants for four to six months before introducing any other type of food into the diet. This puts a tremendous responsibility on the mother to insure a nutritious and palatable food for the infant, and it may also set up a permanent flavor preference for the infant. Thus, it is surprising that so little research has been devoted to the examination of the flavor qualities of the food that constitutes infants' exclusive "flavor experience" for such a long period of time. Human milk often has been described as being sweet, perhaps watery, but until recently, no systematic study of the flavor of human milk has been available.

Adults regularly question the sensory quality of their food. Might not an infant be similarly interested? What are the taste, odor and textural properties of human milk? Does it differ from mother to mother? Does it change during the normal feeding period from one breast, or from breast to breast. Is it influenced by the diet of the mother, or by drugs, smoking, etc.? If the milk is expressed for later use, does the flavor change, and how might handling of the milk influence this? If the milk is frozen, as for use in milk banks, what is the flavor consequence? Most importantly, what can a nursing mother do to insure that the infant will not refuse the breast, a psychologically distressing phenomenon for any nursing mother, combined with a nutritional consequence for the child.

Another very important question regarding the flavor of human milk is the possibility that the infant might become conditioned to the flavor, a situation that psychologists call imprinting. One must consider that human milk, or some other formula, may be an infant's only food source for as long as six months. It is no wonder that the introduction of other foods may be somewhat difficult if their sensory make-

ISBN 0-12-169065-2

up is very different from that of the milk source or if it is more complex. The transfer of chemicals into milk is well known. Cow's milk may be highly flavored due to components in the animal's feed. What is the situation in human milk? It has been said that infants may become accustomed to the mother's diet and better accept solid foods because of flavor compounds that have been transferred during nursing. Or, infants may refuse the breast if they detect a flavor that they do not like or one that they associate with illness.

The purpose of this chapter is to pull together what is known about the flavor and off-flavors in human milk and to discuss the possible consequences. However, a problem of definition exists. What constitutes an off-flavor in human milk? Honey differs in taste depending on where the bee chooses to dine. Farmer's try to keep their cows out of fields where wild onions thrive in the spring. But then who is to say what the real flavor of cow's milk should be. Dairy scientists, through decades of research and experience, have set flavor standards for their products. It may not be inappropriate to think about human milk, a most valuable resource, in this regard. This idea will be developed in a subsequent section. Because the flavor of human milk is still largely a mystery, and because no definition of off-flavor exists at this stage of research, all aspects of the "flavor" of human milk will be presented here.

INFANT TASTE SENSITIVITY

An essential question is whether infants have the ability to perceive any aroma or taste components in human milk. Infants indeed are sensitive to taste stimuli, in fact they are probably able to taste before they are born. An infant's first exposure to taste stimuli occurs in utero since the human fetus has been reported to swallow amniotic fluid from about the twelfth week of gestation, approximately the same time as mature taste buds are present (1). Amniotic chemicals include glucose, acids, salts, proteins, urea and fecal materials providing a wide variety of taste and odor stimuli. By numerous techniques investigators have demonstrated that the taste receptors of the newborn are functional (2-10). Many researchers have been able to demonstrate what appears to be an inherent preference for sweetness in the newborn (11-15). Newborns will adjust their volume consumed according to sweetness level. Nisbett and Gurwitz (16) found that newborns consumed more of a standard milk formula when the carbohydrates were replaced by sucrose, yielding a 4.7% sucrose solution. This solution would have been many times sweeter than the original formula. Newborn's preferences for sweetness have been shown to be greater for higher sugar concentrations over lower ones (12, 14, 17) and sweeter sugars, such as fructose and sucrose, over less sweet ones, such as glucose and lactose (12, 18). Methods used to assess neonate taste discrimination included observations of facial expressions and body movements, volume of fluid ingested, and differential rates of breathing, suckling and heartbeat. Jacobs et al. (19) tested one neonate with various solutions applied with a tongue applicator and upon observing responses, set up a category scale of taste preferences based on oral behavior such as licking and spitting. Jacobs et al. (19) found consistent rejection of human milk compared to acceptance of 10% sucrose and 20% lactose, as well as when compared to water or 7% lactose. The milk had been heat sterilized in boiling water, a procedure of unknown flavor consequence. Crook and Lipsett (17) observed an increased heart rate with sweeter fluids suggesting a positive hedonic response. Johnson and Salisbury (20) observed that breathing and sucking rhythms differed between infants fed cow's versus expressed human milk. In some unknown way the cow's milk disturbed the breathing and altered the sucking patterns of the infant.

Infants have been shown to be sensitive to salt, sour and bitter stimuli (2, 3, 5-7, 9, 10, 21-24). However, the results are inconsistent from study to study, and in many cases extremely high concentrations which were far out of the range of normal foods were administered. Reasonable concentrations of the stimulants yielded little or no effect (22).

Most researchers agree that sensitivity to taste stimuli increases with age (6, 21, 22). Recent infant feeding studies have shown that infants from nine weeks to eight months of age can detect differences in and show preference for pureed vegetables (25), model infant food texture systems (26) and for model infant food taste systems (27). Harasym (25) found that preferences differed across age groups tested, as carrots and beans were liked less by older infants while corn was liked more by older infants. Canned carrots had a high residual bitterness and sourness (28), so the older infant may have been more sensitive than the younger infant to the taste components. Fabro (26) and Hogue and McDaniel (27) found that sweetness enhanced preference whereas sourness and bitterness tended to adversely affect preference for the model systems tested.

The studies mentioned here utilized infants from one day to eight months of age, covering the age range in which infants would be receiving human milk as their sole dietary food source and into the age range in which human milk would likely be supplemented with juices and semi-solid foods. There is no question that infants can perceive taste stimuli and that the stimuli have been proved to cause differences in acceptability.

INFANT ODOR SENSITIVITY

Although a suckling infant may not always "smell" its mother's milk in isolation from the other senses as an adult might "smell" a slice of apple before taking a bite, olfaction can still be expected to be an important factor in determining milk acceptability. It is well known that food aromatics reach the olfactory epithelium during normal mastication, thus allowing the perception of subtle differences in quality such as between an apple and a pear. One example of previous taste and odor experience affecting later rejection of an infant formula by odor alone does exist. Davis (29) in an experiment to determine whether a new-born child could successfully exercise a choice of formula and regulate the quantities taken at feedings, noted that differences in taste were observed early in the first week of testing (infants were about ten days old). Lactic acid milk often caused "unhappy grimaces" and was not swallowed. Davis noted that as the infant grew older the bottle of lactic acid formula might be refused without tasting, probably because the distinctive odor was recognized.

Olfactory response in newborn infants has been studied. Engen et al. (30) studied response to acetic acid, phenyl ethyl alcohol, anise oil and asafetida by 20 infants only 50 hours old. Lipsett et al. (31) investigated developmental changes in odor perception with increasing age and found that olfactory thresholds decreased drastically within the first four days of life. Engen et al. (32) found that infants habituate with repeated presentations of an odorant. Because some of the odorants used in these three studies were irritants, Self et al. (33) used purely "olfactory" odors in a subsequent experiment to study infant odor perception. They found a wide variation in response between test days and between individual infants, reflecting the often observed interindividual variation present in the adult. The above research was conducted by measuring physiological response in the sleeping infant. Two recent studies in active awake infants reflect the infants keen olfactory capabilities. Russell (34) has reported that infants six weeks of age can discriminate between odors of breast pads used by their own mother versus those of an unfamiliar mother. This breast pad odor may well reflect the odor of the milk plus the natural odor of the mother.

In another study, normal termborn neonates were shown to display differential facial expressions to food-related odor stimuli at birth (13). Facial expressions of the infants showing "acceptance" were elicited by flavors adults find pleasant, and facial expressions showing "aversion" were elicited by flavors adults find unpleasant, showing that newborns not only perceive the stimuli but also that in at least some instances, the stimuli have an innate hedonic dimension which is also perceived by the infant. However, Engen (15) suggested that taste elicits more definite hedonic reactions than odors, and cautioned against extrapolating from adult odor preferences to odor preferences of children. Attempts at predicting preferences of young children by using data from adult observers have previoulsy been unsuccessful (35, 36).

Harasym (25), using infants from about nine weeks to five months of age, found canned pureed carrots, which were very high in overall flavor intensity but not typically carrot-like (28), to be very poorly accepted by infants. It was impossible to credit the rejection response to the strong odor, because the carrots were also very bitter and sour. However, in another infant feeding study in which equivalent sour and bitter levels were tested in an otherwise bland model system the sour and bitter samples were scored higher in acceptability than the canned carrots (27).

Although the research reported here is an important first step, the importance of odor in food acceptability to infants remains to be demonstrated. If in fact the aroma of canned carrots can negatively affect the acceptability of the pro duct, it is natural to expect that aromas in human milk could also affect the milk's acceptability.

EARLY EXPOSURE - FOOD PREFERENCE

One of our main interests was to determine if an infant's early food-related sensory experience can in some way affect future food likes and dislikes or food habit development. If, according to current recommendations, an infant is maintained solely on breast milk or artificial formula for as long as six months, how might this experience influence acceptance of "novel" sensory stimuli at weaning and later in life. If an infant is maintained on commercial formula, that infant is exposed to one "flavor" which is constant within the quality control of the company manufacturing it and within the control of the person preparing it for consumption by the infant. However, human milk has a much greater opportunity to be varied in flavor due in part to its natural gross compositional variations and also due to flavor transfer from the mother's diet into her milk. Further variation may develop during the handling of expressed milk. In rats

Capretta et al. (37) found that early experience with a large variety of foods may predispose the animal to be more willing to consume novel foods. (They used black walnut, rum and vanilla flavors in training, and tested acceptance using chocolate as the novel flavor.) This effect was true for young rats and did not hold true for mature rats. Could the experiencing of a wide variety of flavors in the nursing infant cause the infant to more readily accept a wide variety of flavors upon weaning?

Evidence bearing directly on this point has developed in the psychological literature in recent years. Galef and Clark (38) found that cues transmitted from a mother rat to her pups during the nursing period were sufficient to determine the dietary preference of the young at weaning. Galef and Henderson (39) in further experiments found that the cues causing pups to select a specific diet which their mother had ingested during nursing are contained in the mother's milk. This was further tested by Galef and Sherry (40) who found that pups who had ingested milk expressed from a female eating a specific diet showed a slightly enhanced preference for that diet during the first 24 hours of weaning and a distinct preference for it after 24 hours. Bronstein and Crockett (41) found that both exposure of rat pups to the odor of garlic and consumption of mother's milk from a garlic fed dam were sufficient to cause a tolerence for this flavoring when encountered at weaning. The garlic-laced feed was always a non-preferred diet for rats not previously exposed to this flavoring. Capretta and Rawls (42) presented garlic flavored water to rats late in pregnancy and throughout nursing. A control group consumed plain water. The rat pups were weaned on either water or garlic water and later tested for preference by being presented with both water and garlic solutions. The garlic-garlic (nursed/weaned) rats drank the most garlic solution. The garlic-water and water-garlic treatments were next in garlic preference, and the rat pups not exposed to garlic showed the lowest preference. Retesting occurred after one month and the same relative relationship was observed. This is a particularly important finding as results from other tests showed only a transient preference. Tests with peppermint produced similar results.

However, Wuensch (43) noted that the preference effect was not lasting. Wuensch found that exposure to onion flavor through cues transmitted in mother's milk did enhance preference for the onion-flavored diet at weaning, but that the duration of the preference effect was short. It is very likely that the duration of the preference is a function of the different flavors, their strenghts, the length of time given, and other aspects of test design. Capretta (44) noted that early exposure to a distinctive flavor predisposes at least some rats to prefer such flavors later in life. We have noted in work with human infants (27) that some infants show distinct dislikes within a set of samples well accepted by other infants. Capretta (44) suggested that those rats without strong natural preferences may be those most affected by early exposure.

Along similar lines, while investigating the effect of prenatal or early postnatal exposure to alcohol on rat pups' ability to learn in adulthood, Phillips and Stainbrook (45) also tested to see if the preference for alcohol was affected by this early exposure. They used chablis wine, where both the ethanol plus flavor compounds could have been passed to the pups through the mother's milk. The dams were given chablis or water throughout mating, pregnancy and lactation, and the off-spring, maintained only on water and food, were tested for preference at 170 days of age. The rat pups nursed by wine-consuming mothers drank more wine over all five days tested. The effect could have been due to the physiological effect of the ethanol, or to the flavor of ethanol or to the many other flavor compounds in the wine. (The water group performed better on seven of the nine learning problems.)

Wurtman and Wurtman (46) tested the effect of giving sucrose to nursing and immature rats and their elective consumption of it later in life. They found no evidence to support the view that the high levels of sweet foods in childhood causes prolonged increase in preference for sucrose. In fact they found a significant decrease in sucrose consumption with the onset of puberty.

Results of studies such as those presented above have also interested the agricultural community. Specifically of interest was the use of flavor compounds in pig and calf rations (47, 48). Campbell (47) reported improved growth of weaners when a flavor which had been incorporated into the sow's diet during lactation was incorporated into the weaner diet. However, Morrill and Dayton (48) found no such effect with calves. This phenomenon may eventually have some commercial value if adequate time and funds are provided for its development.

TASTE AVERSION TO MOTHER'S MILK

In recent years many studies investigating taste aversions in neonates have been reported (49, 50). These studies were conducted primarily by psychologists who found this a useful paradigm in which to study the development of learning and memory. Of course, the studies also have great relevance to the study of taste and food habit development. It is of interest to know what infants can taste, if they are conditioned early on to certain flavors, and if flavor preferences or aversions early in life affect later food choices.

Much of the work on taste aversions to date has been done with rat pups. The method involves presenting the pups with a distinctive milk and then by various means causing the pups to be sick. The pups are then tested to see if an aversion was formed to that particular milk. Galef and Sherry's (40) experiment illustrated the taste aversion phenomenon. Milk, expressed from rat dams having consumed a specific diet, was placed in the months of weanling-age rat pups (21-22 days of age) who were then injected with LiCl, which caused the pups to be ill. These rats then avoided ingesting the specific diet consumed by the donor dams. No such avoidance was noted with the control groups. Campbell and Alberts (51) found that rat pups as young as ten days of age can acquire a conditioned taste aversion to grape juice. Using geraniol as the flavor, Martin and Alberts (52) found that rat pups as young as 15 days of age are capable of discriminating taste cues in cow's milk and that they express aversions to those cues. They found that the memory for the aversions was well developed in the rat, and that the magnitude of the aversion increased with age. Martin and Alberts (52) did note that in the very young rat, the suckling cues were stronger than food cues, and that in a nursing situation, an aversion will not be acquired before the rat pup is 20 days of age.

An analogy can be made to the human nursing situation. For example, alimentary upsets in the infant have been associated with the consumption of chocolate by the mother (53). It is possible that some of the flavor compounds could be transferred into the milk and perceived by the nursing infant. Subsequent illness could then be associated with the chocolate flavor. Will the effect be strong enough to cause the infant to refuse the breast the next time the mother consumes chocolate, or will the infant dislike and avoid chocolate when he or she is introduced to it as a solid food later in life. An amazingly small amount of research is available on human food habit development in relation to food preferences and food aversions (54, 55), but it is well recognized that both food preferences and food aversions seem to be long-lasting, or at least difficult to change.

COMPOSITION OF HUMAN MILK

The comparison of human milk to cow's milk may be helpful here because the sensory qualities of cow's milk are well known. The largest difference in gross composition is in protein concentration. Cow's milk is 3.3 to 3.9% protein while protein in human milk has been found to range from 1.0 to 1.6% (53). Even these numbers may be high, as researchers now suggest that, due to inaccuracies in early measurement, 0.8 to 0.9% protein may be more correct (56). The composition of the protein is also different between the two types of milk. In cow's milk, about 80% of the protein is casein and 20% is whey protein, while in human milk, 35% is casein, with the remaining 65% being whey protein. The amino acid content of human milk is unique. Human milk is high in cystine and taurine and lower in tyrosine and methionine than cow's milk.

Fat content ranges from 3.4 to 5% in cow's milk, averaging around 3.8%, while human milk contains from 4.2 to 4.8% in well nourished mothers (53). The fat content is most important as a source of calories and has been suggested to be important in appetite control (57). The higher fat content in the hind milk (the last-expressed milk out of one breast) may somehow signal the infant to stop suckling.

Lactose has been found to be the most constant of the major constituents in human milk, ranging from 7 to 8% as compared to around 4.8% in cow's milk.

Further reports of human milk composition studies are summarized by Jelliffe and Jelliffe (53).

CHEMICALS IN HUMAN MILK

It is well known that chemical substances in the diet can be transmitted into the milk (58). Anderson (59) who presented a comprehensive update on drug excretion into breast milk, noted that most substances that can get into the bloodstream can eventually be found in the milk. The time necessary for a drug to reach the milk is dependent upon the way in which the drug was administered and the method of absorption (60). Also important is the pH or pKa properties of the drug and its solubility in fat and water (61). Normal pH of milk is 6.5 - 6.8 whereas blood plasma has a pH of 7.4 (61). Goodman and Gilman (62) explain that because milk is more acidic than plasma, basic compounds are concentrated there, while the concentration of acidic compounds in the milk are lower than in the plasma. A compound such as ethanol, a nonelectrolyte, would readily enter the milk regardless of the pH relationship.

Unfortunately, due to decades of disinterest in breast feeding, little research has been done to identify ill-effects of drugs on the nursing infant. Jelliffe and Jelliffe (53) may be consulted for a list of drugs that should not be given to lactating women. Drugs are known to have very strong flavors as evidenced by the time and money pharmaceutical companies spend trying to mask the flavor of drugs given orally. It is not known whether drugs taken by the mother can affect the flavor of her milk, but the possibility exists.

Popular "social toxicants" (53) include ethanol, caffein, and nicotine. All of these compounds reach the milk and may in some way cause a flavor consequence. Nicotine, an alkaloid, is volatile, strongly alkaline in reaction, and on exposure to air acquires the odor of tobacco (62). Since high levels of the fat soluble nicotine are found in human milk (63), it seems quite possible that it provides the infant with a percievable "tobacco" aroma during nursing. However, its presence in the milk may be no stronger an influence than its presence in the air when the mother (or anyone else present) is smoking. Bronstein and Crockett (41) found that rat pups which consumed a non-flavored feed in a garlic-scented room were affected to the degree that they would tolerate the garlic flavor encountered in weaning, even though garlic was not a preferred diet to rats in general.

Smoking, of course, is a very complicated issue, and the factors causing paople to smoke are many. Although peer pressure may be a major factor influencing a teenager (or younger child) to start smoking, there is also evidence that the smoking parent "becomes a model" for the child (63). Unfortunately, parents who smoke influence their children to initiate smoking. If both parents smoke, their children are twice as likely to smoke as children of non-smoking parents. Exactly how a parent affects the decision to smoke in their children is not known, but the possible tobacco flavor in an infant's initial milk source may prove to be important. It has been estimated that tobacco smoke probably contains more than 2600 constituents (65). Readers interested in the chemical make-up of tobacco smoke and the transfer of those chemicals into human milk should consult "Smoking and Health: A report of the Surgeon General (63, Chapters 14 and 17).

With the recognition of foetal alcohol syndrome in the children of alcoholic mothers, more interest has developed in the effect of alcohol on the foetus and in the nursing infant. Although it is apparent that ethanol in human milk is an effective addictive agent in the infant, it is not known whether or not the ethanol is an effective sensory stimulant in the infant. However, the results by Phillips and Stainbrook (45) discussed previously suggest that it is possible. The concentration of alcohol in mother's milk is approximately equal to the concentration in her blood (66). A severely intoxicated mother would have a blood alcohol concentration of 400 mg%; 50% of persons are grossly intoxicated when the concentration is 150 mg% (62). These alcohol levels are probably not very significant gustatory or olfactory stimuli, but aromatic chemicals specific to the type of liquor imbibed may be of adequate strength to be perceived by the infant.

Caffeine, an intensely bitter alkaloid, reached detectable levels in the blood of nursing infants (67). However, only 1% of the total dose ingested by the mother is found in the infant (68), thus making it unlikely to have an impact on taste.

DIET RELATED INSTRUCTIONS TO NURSING MOTHERS

In order to prevent difficulties in nursing, mothers are often instructed to avoid certain foods. These foods may disrupt the nursing procedure in one of two ways. Firstly, they may cause the infant to refuse the breast (due to an immediate gustatory or olfactory cue), and secondly, they may cause the infant later distress such as gas pains. Wet nurses in the 16th century were not only selected according to strict physical and mental ideals, but they were also advised to avoid such stimuli as bad air, bad smells, salty or spiced foods, garlic, and other items (53). Although Jelliffe and Jelliffe (53) noted that prohibition of certain foods during nursing has "gone out of fashion," in fact it has not. Similar advice still exists. Pamphlets prepared by health professionals for nursing mothers suggest that strong flavored or gas producing products such as cabbage, turnip or onion should be avoided if they upset the infant (69). A pamphlet distributed by industry advises that foods such as onions, members of the cabbage family, tomatoes, chocolate, spices and seasonings have been troublesome for some nursing mothers by causing their babies to "refuse milk so flavored" (70). Recent advice to physicians included a warning that onion, garlic, tomatoes, spicy food, and other foods which cause gas or other disorders should be taken in moderation (71). Nursing women seem to be heeding this advice. A recent survey by Sims (72) found that 90% of nursing women surveyed on food avoidances have changed their way of eating. 73% avoided specific foods, many of which were the same as those mentioned above. The most common advice they received from their pediatrician was to avoid chocolate and spicy foods. The folklore associated with the diets of nursing mothers is abundant and varied (73, 74). Hopefully, research in the area will allow future mothers to be guided by scientific data rather than by folklore.

OFF-FLAVORS IN MILK

Of the many off-flavors in cow's milk described recently by Shipe et al. (75), several sources probably also contribute to human milk off-flavor. Transmitted flavors, those flavors which originate in the mother's diet and are transmitted into the milk, may be perceived by the suckling infant. In cows, the transfer may be via the digestive system, but it has also been shown that volatile compounds in the atmosphere may be transferred rapidly via the respiratory system to the udder (76, 77). Sometimes, but not always, the flavor of the milk is not characteristic of the feed (78). Honkanen et al. (79) suspect that distinct off-flavors exist in cow's milk due to an additive effect, because the percentage transfer of a compound from the diet into the milk is very low. In a study of onion flavor transfer, the off-flavor which occurred in the milk was not onion-like in nature (76).

Human milk expressed from the breast and held for later consumption may be subject to some of the same processing and storage flavors as cow's milk. Lipolized flavors may be expected due to the active lipase system present in human milk, oxidized flavors could develop dependent upon the handling of the milk, and heated flavors could develop if milk were re-heated for use, or if human milk were pasteurized prior to frozen storage in milk banks.

For detailed accounts of off-flavors in cow's milk consult Shipe (this publication) and Shipe et al. (75).

FLAVOR OF HUMAN MILK

Until recently, the few references found in the literature commenting on the flavor of human milk stated generally that human milk was sweet. Several questions remained to be asked. For example, how sweet is human milk compared to other milk sources or synthetic milk sources (infant formulas)? What is the normal flavor of human milk and what are the ranges of variability? Also, what are the texture and mouthfeel characteristics of the milk? It is actually quite amazing that up until this time no attempts have been made to qualitate and quantitate these sensory characteristics of such an important food source.

A study on the flavor of human milk has recently been completed by Elizabeth Barker (80) in my laboratory at the University of Manitoba. Since nothing is known concerning the variability of sensory properties of human milk, a limited number of factors had to be selected for study. The test involved 24 mothers who had been nursing between ten weeks and eight months. Samples were taken between 7:00 and 8:00 A.M. for three consecutive mornings. Milk was expressed by the mother, immediately refrigerated, collected by the experimentors, transferred under refrigeration to the testing laboratory and tested by an adult trained sensory panel within three hours of expressing. The samples were warmed to body temperature immediately before testing. Proper handling of the milk is critical due to the presence of a very active lipase system. The panelists pre-selected basic sensory qualities for scaling, and simply noted any other novel or unexpected sensory qualities in the milk samples. The pre-selected qualities were sweetness, viscosity, and mouth-coating.

As predicted, sweeteness was the main taste quality noted by the panel and significant differences were found between the mothers tested. Magnitude estimates ranged from a low of 0.781 to a high of 1.25, reflecting a fairly narrow range of sweetness. Magnitude estimates have ratio qualities, such that a score of 2 represents twice as much sweetness as a score of 1. The 6% lactose reference average score was 0.881. The samples were analyzed for lactose and the averages ranged from 7.03 to 8.76 gm/100ml. Differences between fore and hind milk and differences across the three-day collection period were analyzed, but no significant differences were found. These data suggest that most infants are exposed to a similar level of sweetness by way of their mother's milk.

Although viscosity and mouthcoating are not classical "flavors," they do affect the perception of tastants and deserve a brief mention here. Barker (80) found that both viscosity and mouthcoat varied from mother to mother and from for to hind milk. Some milk samples from different mothers were nearly twice as viscous as others. Smaller differences were noted between fore and hind milk, hind milk being the most viscous. Similar results were found with mouthcoating, with hind milk again being more mouthcoating than fore milk. These differences in the textural properties of the milk were probably due to differences in fat content. One mother's milk averaged 0.57 gm/100 ml across the three days of testing while another was a high of 5.58 gms/100 ml on one day of testing. Hind milk averaged 3.21 gm/100 ml and fore milk averaged 1.13 gm/100 ml. Protein content was also slightly higher in hind milk (1.15 vs. 0.98 gms/100 ml). Protein content across mothers varied from a low of 0.61 gm/100 ml to a high of 2.15 gms/100 ml.

Complex textures generally tend to diminish flavor intensities, however Barker's panelists found no difference in sweetness between fore and hind milk. Therefore, human milk was found to be sweet, but only as sweet as a 7-8% lactose solution, which is approximately as sweet as a 2% sucrose solution. Adults who consume soft drinks are tasting about a 10% sucrose solution with the sweetness being moderated by the acid blend appropriate to that beverage.

At this point it seems appropriate to compare the flavor of human milk to that of infant formula. Malcolmson and McDaniel (81) reported that infant formulas varied in odor, flavor-by-mouth, and texture qualities. The formulas tested were both commercially prepared formulas purchased in liquid concentrate form, and home-made formulas prepared by combining appropriate amounts of milk (evaporated or whole), water and sugar. Home-made formulas for young infants require larger amounts of sugar. While human milk seemed to be practically odorless, Malcolmson and McDaniel (81) found that many of the formulas had distinct but low levels of sweet, sour, or cereal type odors. The soy based samples had a relatively strong hay/beany odor. All infant formulas had an evaporated milk odor, but it was strongest in the commercially-prepared cow's milk and home-prepared evaporated milk formulas.

All of the formulas were found to be sweet, but some of the soy-based (containing sucrose instead of lactose) and home-prepared samples were particularly sweet being equivalent in sweetness to a six percent sucrose solution. The commercial formulas prepared using lactose are similar in sweetness to human milk. Although an infant on home-prepared formula would initially receive a formula of high sweetness, the infant would be exposed to increasingly smaller amounts of sweetness with time if the formula were properly adjusted. However, an infant maintained solely on a soy-based formula could be exposed to this high level of sweeteness for as long as one year.

Some of the formulas contained low levels of sour, bitter and cereal tastes, but these were low in intensity. The only significant flavor-by-mouth characteristic other than sweetness was a strong hay/beany character note present in the soy based formulas.

OFF-FLAVORS IN HUMAN MILK

In the study by Barker (80) the panelists noted any unique or novel flavors not specifically expected in the milk samples. It was suspected that these novel flavors were transferred from the mother's diet into the milk. A record of the mother's food consumption twenty four hours prior to sampling was taken so that correlations between mother's diet and the flavor profile of her milk could be analyzed.

The adult trained panel found off-flavors in 30% of the samples. However, the panelists did not always agree that an off-flavor was present, and they often used different terms to describe the flavors they perceived. Some mother's samples never were judged to have an off-flavor, while other mother's samples were repeatedly judged to contain novel flavors. Table 1 lists the descriptors used by the panel and their frequency of occurrence. These flavors could have been transmitted from the diet of the mother to her milk, or they could have developed after expression due to handling of the milk by the mother and/or by the experimentors. The nature of the flavors should be a clue to their origin.

The milk samples were heated to body temperature and maintained there for approximately 30 minutes while instrumental viscosity measurements were made. Although this cannot be considered a severe heat treatment, typical "cooked" milk descriptors were used by the panel in a few cases. Cooked milk, ice cream, vanilla, sulfur, caramel and milky, are all typical of heated milk descriptors (78, 75). Josephson and Doan (82) found that cows milk heated to 62.8°C for 30 minutes was judged to have no off-flavor, so the "heated" flavors formed here are probably due to some other factor.

In contrast to cow's milk, human milk is high in lipase, thus making lipolyzed flavors suspect in an experiment such as the one reported here. At temperatures as low as 4°C free fatty acids are liberated from the triglycerides (53) of human milk. This lipase activity of the milk is critical to the energy supply of the infant as the infant's digestive system is still immature. However, old, rotten, and soapy, are the only descriptors used here which would reflect lipolized flavor. It therefore seems unlikely that the handling of the samples caused flavors which were lipolytic in origin.

Metallic (also tinny) was the off-flavor descriptor used most often by the judges. A metallic off-flavor is known to occur when milk fats have become oxidized. Some milk is said to oxidize "spontaneously" and is catalyzed by copper ions and ascorbic acid (83, 84). Picciano and Gutherie (85) have reported that human milk contains a higher concentration of copper than cow's milk , and human milk has higher levels of ascorbic acid than does cow's milk (86). As human milk was not meant to be collected in a test tube, there would have been no reason for nature to try to prevent this type of oxidation from occurring. It seems quite likely that these metallic off-flavors were caused by the exposure of the milk to oxygen outside of the breast.

Fruity was the second most popular descriptor of an off-flavor. In cow's milk, a fruity flavor generally comes from microbial contamination with *Pseudomonas fragi*. The fruity aroma is due to ethyl butyrate and ethyl hexanoate which have been formed due to the action of two enzymes supplied by the organism; a lipase to liberate butyric and caproic acids from milk triglycerides, and an esterase which then esterifies these acids with ethanol (75). Lipase is known to be active in human milk, and perhaps a similar esterase exists as well. However, there are no short-chain fatty acids below 10:0 in human milk. Therefore, the fruity character must take some other form or be of some other origin. Perhaps this off-flavor provides the best evidence for being a transmitted flavor, as only very small amounts of esters would have to be transferred into the milk to create a fruity off-flavor in such a bland base.

Most of the remaining descriptors seem to logically fit into transmitted flavors because they describe flavors of real foods. The hot-spicy-peppery descriptors were used almost exclusively on the milk from one subject and the woman had consumed a "spicy" diet during the test period. In general, however, the mothers' diets on the day prior to expression did not provide much information on the origin of the transmitted flavors. If flavors are transmitted, they are probably often mixtures of the many components in the mother's diet, and this mixed flavor may or may not be perceived as a typical "food" flavor. All of the descriptors given in Table 1 may be transmitted flavors, but this can not be determined by this experiment alone.

A flavor problem has been shown to cause a human infant to refuse the breast. The case involved a mother whose baby nursed the right breast poorly due to chronic mastitis (87). The father, tasting the milk from each breast, found that the milk from the right breast was salty. Analysis showed that the sodium level of the milk from the normal left breast was 5 mEq/liter while milk from the mastitic right breast was 103-108 mEq/liter. This level of salinity would be equivalent to that of a reconstituted onion soup mix (80).

TABLE I. Descriptors used by Panelists to Describe Fresh Milk off-flavors.

Descriptor	# of Times Used
Metallic	29
Tinny	2
Fruity	25
Candy apples	1
Musty	2
Earthy	1
Sour	11
Old, rotten	1
Cooked milk	3
Icecream, vanilla	11
Sulfur	1
Caramel	6
"Milky"	2
Coconut	6
Salty	5
Hot, spicy, peppery	10
Limey	4
Peppermint	1
Beany	1
Soapy	1
Perfumy	1
	124 (from a total of 414 evaluations)

Barker (80).

OFF-FLAVORS IN FROZEN HUMAN MILK

Another source of off-flavor in human milk is created by the handling of the milk after expression. Milk is expressed for many reasons, often to let the mother enjoy longer times away from her infant, and to supply via milk banks infants whose mother's cannot provide the milk that is needed. Milk for milk banks was traditionally frozen, but nurses or nutritionists handling these milk samples realized that certain sensory changes occurred during freezing. For that and other reasons, milk banks are currently using only fresh milk in many areas.

Twenty-two of the twenty-four mothers in Barker's study (80) had expressed and frozen their milk for later use at least once during their nursing experience. Thirteen of the twenty-two mothers had never tasted the frozen milk. Mothers who had tasted their frozen milk described off-flavors which were soapy (n=4), oily, sour, and tin-foil like. Seven of the 18 infants given frozen breast milk rejected it. This is an alarmingly high percentage, important in the respect that the infant refused the milk and that the mother may have felt guilty giving the infant an undesirable food. The milk samples from this small survey were generally frozen within 2 3 hours of expression, stored for 1-4 weeks, and defrosted in warm water, shaking at least once to mix.

Because of the high frozen milk rejection rate, Barker (80) undertook a preliminary study to investigate the off-flavors caused by freezing human milk. She used only two mothers in this portion of the study. One sample per mother was collected for three consecutive days at the time of the first morning feed. The milk was then stored for one month. At the end of this period, fresh milk was collected and compared to the frozen, again over three consecutive days. The diets of the mothers were not controlled. on the day of testing, the frozen milk was thawed in a water bath, shaken to accelerate thawing and to mix the separated portions. During that day's testing by the panel, several panelists reported that some of the samples were nauseating. On day 2 and 3 the samples were not shaken, however off-flavors were still present. Table 2 contains the descriptors used by the sensory judges to describe the frozen and fresh milk flavor. Although similar terms were used in both frozen and fresh milks, the judges found twice as many incidences of off-flavors in the frozen milk and the intensity of the flavors were higher. Soapy was listed more often in this small test than in the initial study, and it was found to be more intense in the frozen milk. The milk was not pasteurized or in any way heat treated prior to freezing, so enzyme activity would be expected to proceed slowly during storage. Hernell et al. (88) estimated that human milk lipase is active enough to completely hydrolyze the milk dietary lipids in less than one half hour in the small intestine. Although cow's milk is higher in C10-C14 fatty acids than fresh human milk, hydrolysis would probably produce enough of these fatty acids to give the "soapy" character noted here.

Metallic, joined by cardboardy, provided over one half of the off-flavor judgments suggesting that oxidation is the key problem in the handling of frozen milk. Human milk is much higher in polyunsaturated fatty acids than cow's milk. Crawford et al. (89) found human milk triglyceride fatty acids to be 8.7% linoleic, 3.4% linolenic, 2.2% arachidonic and 1.1% docadexaenoic. The addition of antioxidants to expressed human milk would probably be an unpopular idea with nutritionists. High concentrations of ascorbic acid have been shown to inhibit oxidized flavor formation (90), and if used only occasionally, these levels may not influence the infant's need for or use of the vitamin. A solution as simple as adding ascorbic acid to expressed milk before storage will hopefully be developed as an aid to the nursing mother.

TABLE II. Descriptors used by panelists to describe frozen and fresh milk flavors (n = two mothers)

	Frozen Milk (1 mo.)		Fresh Milk	
Descriptor	No.	Intensity	No.	Intensity
Waxy	2	2 extreme	2	strong
Sour	1	1 slight	1	detectable
Soapy	5	1 strong, 1 slight, 3 moderate	3	slight
Evaporated Milk		----	1	moderate
Regurgitate	1	1 moderate		----
Cooked milk	1	1 slight		----
Peppery		----	1	slight
Metallic	12	5 strong, 2 moderate, 4 slight, 1 detectable	1	slight
Cardboardy	4	2 extreme, 1 moderate, 1 strong	1	extreme
Lime		----	1	slight
Astringent	1	slight		----
Creamy		----	2	slight
TOTALS	27 (75%)		13 (36%)	

Barker (80)

CONCLUSIONS

The data presented here represents only an initial survey of the sensory properties of human milk. Although Barker (80) was able to quantitate the major and basic sensory qualities of human milk, and to introduce the problem of off-flavors, other projects will have to be designed to delineate the origin and significance of human milk off-flavors. Hopefully, the time has arrived for the need to study human milk and that its impact on food habit development be recognized by the scientific community as the need to study cow's milk has been in the past.

It was suggested in the introduction that human milk is a valuable resource, and that there may some day be a need to standardize human milk quality, not only on a nutritional level, but on a sensory level as well. Jelliffe and Jelliffe (53) pointed out that human milk, through the use of wet-nurses, has been a saleable commodity for centuries. Also, they reported that in Britain, as recently as 1973, expressed breast milk was selling for $0.72 (U.S. funds) per pint, and it is currently probably twice that amount. One mother, who expressed milk not needed by her infant, earned $1500 one year by supplying milk to a human-milk bank. In any country, parents purchasing expressed milk for their infant would certainly be concerned about the sensory quality of the milk lest it be refused by the infant. Barker (80) has exposed the presence of serious quality defects in expressed/frozen human milk. Optimum expression, handling and storage procedures for the milk would need to be determined. Although it might remain a small industry, it could be of major importance to the health of infants from non-nursing mothers, and provide additional income to mothers who are abundant lactators.

The question of the importance of an infant's early sensory experience gained from the flavor properties of human milk must remain unanswered at this time. It is reassuring to know that human milk is not excessively sweet, as are some of the artificial alternatives. Thirty percent of the samples evaluated in Barker's (80) study were judged to have some type of off-flavor. Although it was not possible to determine exactly which character notes were transmitted flavors, it is reasonably safe to say that many flavors were transmitted from the mother's diet into the milk. These off-flavors may represent risk/benefit tradeoffs. If the infant refuses the breast due to flavors in the milk, the risk is sufficient to cause alarm. However, if it were known that certain flavors were transmitted into the milk, a mother could, via her milk, gradually introduce the novel flavor to her infant, thus circumventing the infant's refusal of the breast. If results of rat pup behavioral studies hold true for human infants, this flavor "conditioning" would enhance acceptance of the flavor when given as a semi-solid or solid food. Similarly, foods consumed by the mother which cause the infant any gastric discomfort should be avoided. The infant may form a taste aversion for this food and reject it when it is introduced as a solid food, even if at that time no gastric discomfort exists. These precautions could help create less finicky babies, and would benefit anyone whose task it is to introduce an infant to the world of flavors currently enjoyed by adults.

REFERENCES

1. Minstretta, C., and Bradley, R. (1975). Brit Med. Bull. 31, 80.
2. Kussmaul, A., (1859). Cited in Peiper, A. (1963). "Cerebral Function in Infancy and Childhood" (J. Wortis, ed.) p. 44. New York.
3. Preyer, W., (1882). Cited in Pratt, K.C., A.K. Nelson, and K. H. Sun. (1930) Ohio State University, Contributions in Psychology, 10, 105. Ohio State University Press.
4. Eckstein, A. (1927). Cited in Peiper, A. (1963). "Cerebral Function in Infancy and Childhood" (J. Wortis, ed.) p. 44. New York.
5. Kulakowskaja, E., (1930). Cited in Peiper, A. (1963). "Cerebral Function in Infancy and Childhood" (J. Wortis, ed.), p. 44. New York.
6. Pratt, K., Nelson, A., and Sun, K. (1930). Ohio State University, Contributions in Psychology 10, 105. Ohio State University Press.
7. Jensen, K. (1932). Genetic Psych. Monographs. 12, 361.
8. Martin Du Pan, R. (1955). Pediatric. 10, 169.
9. Aiyar, R. and Agarwal, S. (1969) Indian Ped. 6, 729.
10. Steiner, J. (1974). In "Fourth Symposium on Oral Sensation and Perception: Development in the Fetus and Infant." (J. Bosma, ed.) p. 254. U.S. Government Printing Office, Washington.
11. Kobre, K. and Lipsitt, L. (1972). J. Exp. Child Psych. 14, 81.
12. Desor, J., Maller, O. and Turner, R. (1973). J. Comp. Physiol. Psych. 84, 496.
13. Steiner, J. (1977). In "Taste and Development. The Genesis of Sweet Preference" (J. Weiffenbach, ed.) p. 173. U.S. Dept. of Health, Education, and Welfare, Bethesda, Maryland.

14. Weiffenbach, J. and Thach, B. (1973). In "Fourth Symposium on Oral Sensation and Perception" (J. Bosma, ed.) p. 232. U.S. Government Printing Office, Washington.
15. Engen, T. (1977). In "Taste and Development. The Genesis of Sweet Preference" (J. Weiffenbach, ed.) p. 143. U.S. Dept. of Health, Education, and Welfare, Bethesda, Maryland.
16. Nisbett, R. and Gurwitz, S. (1970). J. Comp. and Phys. Psych. 73, 245.
17. Crook, C. and Lipsitt, L. (1976). Child Dev. 47, 518.
18. Engen, T., Lipsitt, L., and Peck, M. (1974). Dev. Psych. 10, 741.
19. Jacobs, H., Smutz, E., and DuBose, C. (1977). In "Taste and Development. The Genesis of Sweet Preference" (J. Weiffenbach, ed.) p. 99. U.S. Dept. of Health, Education, and Welfare, Bethesda, Maryland.
20. Johnson, P. and Salisbury, D. (1975). In "Parent-Infant Interaction" CIBA Foundation Symposium 33. p. 119. Elsevier, Amsterdam.
21. Maller, O. and Desor, J. (1974). In "Fourth Symposium on Oral Sensation and Perception: Development in the Fetus and Infant" (J. Bosma, ed.) p. 279. U.S. Government Printing Office, Washington.
22. Desor, J., Maller, O., and Andrews, K. (1975). J. Comp. and Phys. Psych. 89, 966.
23. Fomon, S., Thomas, L. and Filer, L. (1970). J. Ped. 76, 242.
24. Osepian, V. (1958). Pavlov J. of Higher Nerv. Act. 8, 828.
25. Harasym, L. (1977) "Infant Acceptance and Trained Panel Quality Evaluation of Pureed Vegetables." Unpublished Masters Thesis, University of Manitoba, Winnipeg, Canada.
26. Fabro, T. (1979) "Infant Acceptance and Trained Panel Evaluation of Model Taste and Texture Systems." Unpublished Masters Thesis, University of Manitoba, Winnipeg, Canada.
27. Hogue, L. and McDaniel, M. (1980). Unpublished data.
28. McDaniel, M. and Harasym, L. (1979). Can. Inst. of Fd. Sci. and Technol. 12, 180.
29. Davis, C. (1935) Am. J. Dis. Child. 50, 385.
30. Engen, T., Lipsitt, L. and Kaye, H. (1963). J. of Phys. and Psych. 56, 73.
31. Lipsett, L., Engen, T., and Kaye, H. (1963). Child Dev. 34 371.
32. Engen, T., Lipsitt, L. and Kayne, H. (1963) J. of Comp. Phys. and Psych. 56, 73.

33. Self, P., Horowitz, F. and Paden, L. (1972) Dev. Psych. 7, 349.
34. Russell, M. (1976). Nature 260, 520.
35. Engen, T. and Corbit, T. (1970). Injury Control Research Laboratory, Public Health Service, U.S. Department of Helath, Education, and Welfare. Report No. ICRL-RR-69-6.
36. Engen, T. and Moskowitz, L. (1972). Injury Control Research Laboratory, Public Health Service, U.S. Department of Health, Education, and Welfare, Report No. ICRL-RR-71-4.
37. Capretta, P., Petersik, J., and Stewart, D. (1975) Nature 254, 689.
38. Galef, B. and Clark, M. (1972) J. of Comp. and Phys. Psych. 78, 220.
39. Galef, B. and Henderson, P. (1972) J. of Comp. and Phys. Psych. 78, 213.
40. Galef, B. and Sherry, D. (1973) J. of Comp. and Phys. Psych. 83, 374.
41. Bronstein, P. and Crockett, D. (1976). Behavioral Biology 18, 387.
42. Capretta, P. and Rawls, L. (1974) J. of Comp. and Phys. Psych. 86, 670.
43. Wuensch, K. (1978). J. of Gen. Psych. 99, 163.
44. Capretta, P. (1977).In "Learning Mechanisms in Food Selection"(L. Barker, M. Best, and M. Domjam eds.) p. 99. Baylor University Press.
45. Phillips, D. and Stainbrook, G. (1976) Phys. Psych. 4, 473.
46. Wurtman, J. and Wurtman, R. (1979) Science 205, 321.
47. Campbell, R. (1976) Animal Production 23, 417.
48. Morrill, J. and Dayton, A. (1978). J. Dairy Sci. 61, 229.
49. Wallace, P. (1976). Science 193, 989.
50. Garcia, J. and Hankins, W. (1977). In "Learning Mechanisms in Food Selection" (L. Barker, M. Best, and M. Domjam eds.) p. 3, Baylor University Press.
51. Campbell, B. and Alberts, J. (1979). Behaviorial and Neural Biology 25, 139.
52. Martin, L. and Alberts, J. (1979). J. of Comp. and Phys. Psych, 93, 430.
53. Jelliffe, D. and Jelliffe, E. (1978). "Human Milk in Modern World." Oxford University Press, New York.
54. Garb, J. and Stunkard, A. (1974). Am. J. of Psychiatry. 131, 1204.
55. Beauchamp, G. and Maller, O. (1977). In "The Chemical Senses and Nutrition" (M. Kare and O. Maller, eds.) p. 291. Academic Press, New York.
56. Hambreus, L. (1977). Ped. Clins. N. Amer. 24, 17.

57. Hall, B. (1975). Lancet i, 779.
58. Patton, S. and Jensen, R. (1976). "Biomedical Aspects of Lactation," Pergamon Press, New York.
59. Anderson, P. (1977). Drug Intelligence and Clinical Pharmacy 11, 208.
60. Catz, C. and Giacoia, G. (1972). Clin. N. Am. 19, 151.
61. Arena, J. (1970) Nutrition Today 5, 2.
62. Goodman, and Gilman, (1975). "The Pharmacological Basis of Therapeutics," Fifth Edition. p. 19. MacMillan Publ. Co., New York.
63. U.S. Department of Health, Education, and Welfare. Public Health Service. Smoking and Health, a report of the Surgeon General. DHEW Publication No. (PHS) 79-50066. U.S. Government Printing Office, Washington, D.C. 20402. 1979.
64. National Institutes of Health (1976) "National Patterns of Cigarette Smoking, Ages 12 through 18, in 1972 and 1974. DHEW Publ. No. (NIH) 76-931, U.S. Dept. of Health Education, and Welfare, Public Health Service, National Institutes of Health, Bethesda, Maryland.
65. Demole, E. and Berthet, D. (1972). Helvetica Chimica Acta 55, 1866.
66. Kesäniemi, Y. (1974). J. Obstet. Gynaecol. Br. Commonw. 81, 84.
67. Rivera-Calimlin, L. (1977) Drug Ther. 7, 59.
68. Illingworth, R. (1953). The Practitioner 171, 533.
69. Manitoba Department of Health and Community Services. (1979). "One to Grow On." Health and Community Services, Winnipeg, Canada.
70. Ross Laboratories. (1975). Breast Feeding Your Baby. Division of Abbott Laboratories, Montreal, Canada.
71. Smith, G., Calvert, L. and Kanto, W. (1978). Am. Fam. Physician 17, 92.
72. Sims, L. (1978). J. Am. Dietet. Assoc. 73, 139.
73. Walter, M. (1975). Bull. N.Y. Acad. Med. 51, 870.
74. Snow, L. and Johnson, S. (1978) Ecology of Food and Nutrition 7, 41.
75. Shipe, W., Bassette, R., Deane, D., Dunkley, W., Hammond, E., Harper, W., Kleyn, D., Morgan, M., Nelson, J. and Scanlan, R. (1978). J. Dairy Sci. 61, 855.
76. Dougherty, R., Shipe, W., Gudnason, G., Ledford, R., Peterson, R., and Scarpellino, R. (1962). J. of Dairy Sci. 45, 472.
77. Shipe, W., Ledford, R., Peterson, R., Scanlan, R., Geerkin, H., Dougherty, R., and Morgan, M. (1962). J. of Dairy Sci. 45, 477.

78. Parks, O. (1967). In "Chemistry and Physiology of Flavors" (H. Shultz, E. Day and L. Libbey, eds.) p. 296. AVI Publishing Co., Westport, Connecticut.
79. Honkanen, E., Karvonen, P., and Virtanen, A. (1964). Acta. Chem. Scand. 18, 612.
80. Barker, E. (1980). Sensory Evaluation of Human Milk. Unpublished Masters Thesis, University of Manitoba, Winnipeg, Canada.
81. Malcolmson L., and McDaniel, M. (1980), Can. Inst. of Fd. Sci. In press.
82. Josephson, D. and Doan, F. (1939). Milk Dealer 29, 35.
83. Smith, G. and Dunkley, W. (1962). J. Dairy Sci. 45, 170.
84. Smith, G. and Dunkley, W. (1962). J. Food Sci. 27, 127.
85. Picciano, M. and Guthrie, H. (1976). Amer. J. Clin. Nutr. 29, 242.
86. Kuratani, H. (1966). Acta. Paediat. jap. 8, 55.
87. Conner, A. (1979). Pediatrics 63, 910.
88. Hernell, O1, Gebre-Mehdin, M., and Olivecrona, T. (1977). Amer. J. Clin. Nutr. 30, 508.
89. Crawford, M., Sinclair, A., Msuya, P., and Munhambo, A. (1973). In "Dietary Lipids and Postnatal Development" (G. Calli, J. Jacini, and A. Pecile, eds.) p. 41. Raven Press, New York.
90. King. R. (1963). J. Dairy Sci. 46, 267.

FORMATION OF OFF-FLAVOR COMPONENTS IN BEER

Roland Tressl[1]
Daoud Bahri
Maria Kossa

Technische Universität Berlin

1. INTRODUCTION

Compared to other alcoholic beverages beer shows a poor flavor stability. The application of more rapid brewing systems, equipment of ever-increasing scale, long distribution distances, extended storage under unsuitable conditions, and the trend to low-gravity beers are increasing this problem. Certain beer-types are more or less susceptible to changes in aroma and flavor during fermentation, storage and aging. The oxygen content during packaging and the temperature during distribution and storage are important parameters which affect the flavor stability of beer.

Dark beers contain higher amounts of Maillard reaction products than pale beers (among them cyclic enolones like maltol, 5-hydroxymaltol, 5-hydroxy-5,6-dihydromaltol, cyclotene etc., which may act as antioxidatents) (1).

Lager beers with lower gravity are more susceptible to oxidative deterioration (staling) than high gravity beers.

[1] Present address: Technische Universität Berlin
Seestraße 13, D-1000 Berlin 65

ISBN 0-12-169065-2

2. FORMATION OF OFF-FLAVOR COMPONENTS DURING FERMENTATION AND MATURATION

During fermentation and maturation alcohols, acids, esters, lactones, and carbonyls are formed as aroma components by the yeast metabolism. Maillard reaction products of the malt as well as terpene and sesquiterpene components of hops are known as aroma contributing constituents. Depending on the yeast strain and fermentation conditions, beer can develop some off-flavors during fermentation and storage. Most of these constituents are known as common yeast metabolites and are only perceived as flavor defects when present above certain concentrations.

2.1 Diacetyl (2,3-Pentandione) = "Buttery Aroma"

Diacetyl is formed from α-acetolactate which is an important metabolite in the valine biosynthesis. Valine deficient malts may lead to high amounts of α-acetolactate which is decomposed to diacetyl and imparts to beer a buttery aroma at 0,15 ppm. Also accelerated fermentation systems produce beers with higher amounts of α-diketones and their precursors. During maturation diacetyl is reduced by the yeast to flavor-inactive acetoin and 2,3-butandiol. The formation and reduction of diacetyl in brewing have been thoroughly investigated in the last years and can now be controlled by the brewing technologist (2,3).

2,3-Pentandione is formed from α-aceto-α-hydroxybutyrate in an analogous reaction and is perceived at higher levels than diacetyl. As shown in Figure 1, α-ketobutyrate may be transformed into 2-hydroxy-2-butenolides which possess intensive "maggi"-like flavors. We characterized these constituents in some beers and could demonstrate that they are formed from α-ketobutyrate by the yeast. 3-Hydroxy-4,5-dimethyl-2,5H-furanone was characterized as a compound with a burnt flavor in aged saké. The threshold level in water was determined as 76 ppb (4). The butenolides may also be formed by aldol-type reactions from Streckeraldehydes and 2-oxo-acids.

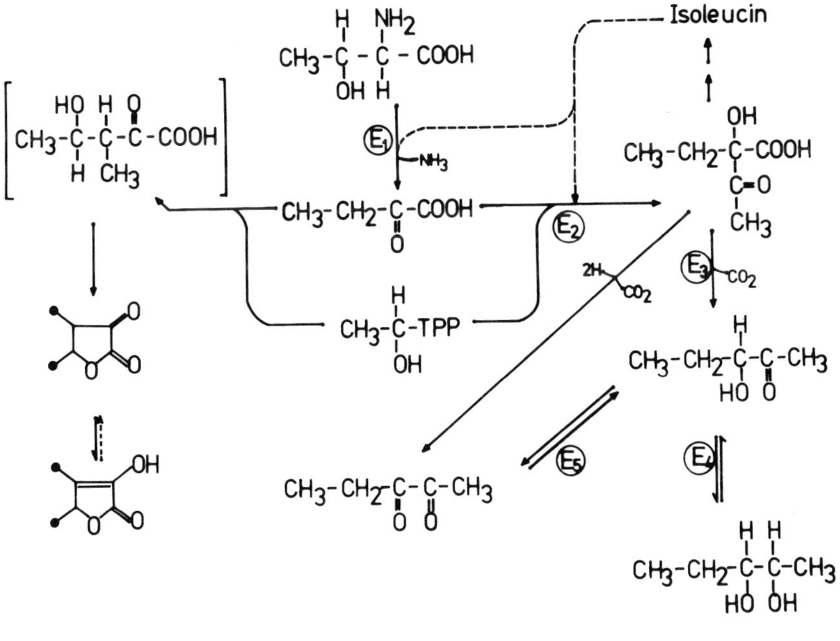

Figure 1 Formation of 2,3-pentandione and 2-hydroxy-2-butenolides

2.2 Dimethylsulfide = "Cooked Vegetable Flavor"

Dimethylsulfide (DMS) adds an aroma of "cooked vegetables" (5) to beer. The amounts of DMS in lagers are higher than in dark beers and ales. It is presumably formed from a labile precursor of the malt which is decomposed during kilning at higher temperatures. Therefore the amounts of DMS in beer may be controlled by wort boiling as shown by Zürcher et al. (6). The DMS precursor is also transformed into the aroma active principle by the yeast as shown by White and Wainwright (7).

2.3 Reduction of Carbonyls which Contribute to Cereal/Malty Aromas

Methional, a Strecker degradation product of methionine is produced during kilning of malt. During this reaction a series of flavor active Strekker aldehydes such as 3-methylbutanal, 2-methylbutanal, 2-methylpropanal (responsible for the primary malty odor) (8), and 2-phenylethanal are produced from the corresponding amino acids. Strecker aldehydes are normally reduced during fermentation as shown in Figure 2. During fermentation the flavor composition of malt is modified by the yeast. Lipidoxidation products from linoleic-, linolenic-, and oleic acids are formed by enzymatic and chemical reactions during malting and wort boiling. More than seventy volatiles of this class have been characterized in pale and dark malts (9). During fermentation the flavor-active carbonyls are reduced comparable to Streckeraldehydes and carbonyls formed by Maillard reactions (Table 1). In addition some of the corresponding alcohols are transformed into esters, and amines are acetylated to N-acetylamides which possess less desirable ("mousy") aroma notes (10).

The precursors of carbonyls are more or less transferred to beer and may be decomposed during beer aging into flavor-active carbonyls which are responsible for cardboard flavor, toffee-like, and bready aromas.

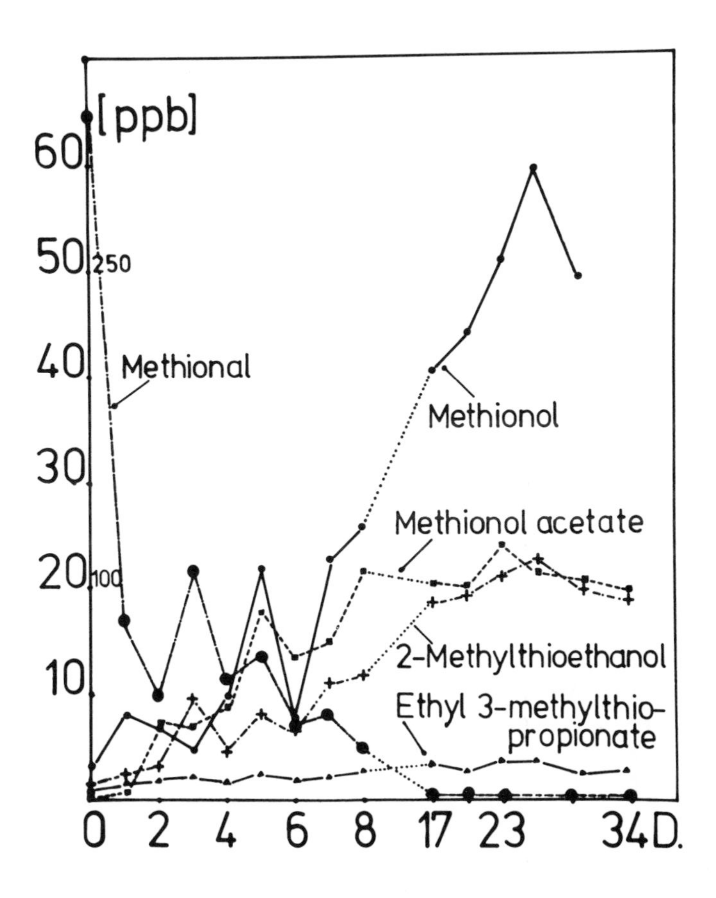

Figure 2 Reduction of methional during fermentation and maturation

Table 1 Strecker Aldehydes and Maillard Reaction Products in Pale and Dark Malts which are Reduced during Fermentation

	Components	Malt (dark) ppb	Malt (pale) ppb
1	Isobutnal	640	60
2	Isopentanal	12800	1200
3	2-Methylthioethanal	40	+
4	3-Methylthiopropanal	300	30
5	2-Phenylethanal	500	100
6	2-Phenyl-2-butenal	45	10
7	4-Methyl-2-phenyl-2-pentenal	20	+
8	4-Methyl-2-phenyl-2-hexenal	12	+
9	5-Methyl-2-phenyl-2-hexenal	65	+
10	2-Furfural	2700	450
11	2-Acetylfuran	540	
12	2-Formylpyrrole	580	50
13	2-Acetylpyrrole	1400	180
14	2-Acetyl-5-methyl-pyrrole	20	+
15	2-Formyl-5-methyl-pyrrole	400	20
16	2-Acetylthiazole	120	60
17	2-Acetylpyridine	20	10

Table 1 Strecker aldehydes and Maillard reaction products in pale and dark malts which are reduced during fermentation

2.4 Fatty Acids = "Caprylic Flavor"

During fermentation C_6-, C_8-, C_{10}-, C_{12}- fatty acids are produced by the yeast metabolism and transferred to beer according to their hydrophilic/lipophilic character. According to Clapperton (11) these straight chain fatty acids possess musty, soapy, fatty, oily, and rancid odor qualities which are described as "caprylic flavor". The threshold concentrations in beer of octanoic acid (4,5 mg/litre), decanoic acid (1,5 mg/litre), and dodecanoic acid (0,6 mg/litre) are all within the range of concentrations of the respective compounds that have been found in beer. Moreover the effects of these acids on caprylic flavor in beer are additive. People who can detect caprylic flavor in beer perceive it differently. In low concentration caprylic flavor is part of the overall flavor of a large number of commercial beers. According to Arkima (12) bottom-fermenting yeasts produce higher levels of these straight chain fatty acids than top-fermenting yeasts.

By application of the flavor profiling method to 150 commercial beers, Clapperton and Brown (13) could show that lager beers contain higher amounts of DMS and fatty acids than ales and therefore display the corresponding cooked-vegetable flavors and caprylic flavors. The yeast strain seems to be more important for caprylic flavor formation than the wort composition. This can be seen in Table 2. The formation of the corresponding ethyl esters is obviously linked to the fatty acid biosynthesis. Äyräpää and Lindström (14) showed that the oxygen content and exogenous oleic acid can increase the formation of medium chain fatty acids and ethyl esters.

2.5 Phenolic Components

Phenolic components are known as potent off-flavor components in beer and other foodstuffs. Phenolic contaminents may be chlorinated into chlorophenols in the brewing water. Some bacteria and wild yeasts are able to produce phenols (15). The investigation of a beer, which possessed a "phenolic off-flavor" (by means of GC-MS), showed that some phenols, which are known as normal beer aroma components, were present at higher levels than

Table 2 Fatty Acids of Fermenting Wort and Beer (ppm)

Fatty Acid	Hexanoic	Octanoic	Decanoic	Dodecanoic
A	3,0	8,1	2,1	0,83
B	3,9	10,5	2,1	0,55
C	2,5	8,2	2,6	0,38
D	1,7	5,3	0,9	0,35
E	3,4	12,7	1,3	0,43
Pils 1	3,0	3,8	0,18	0,06
Pils 2	2,5	7,2	0,70	0,15
Strong Lager (dark)	2,4	5,2	0,9	0,04

Beers A to E brewed with different strains of bottom-fermenting yeasts and the same conditions.

Table 3

Phenols in Commercial Beers and Model Fermentations (ppb)

Components	"Phenolic Beer"	Normal Beer	I Yeast A	II Yeast B	III Yeast C	IV Yeast A	V Yeast B	VI Yeast C
p-Cresol	12	5	-	-	-	3	-	-
4-Ethylphenol			8	6	8	15	6	20
4-Vinylphenol	170	5	85	18	13	3950	35	50
Guajacol			4	-	-	-	-	-
4-Vinylguajacol	75	7	1260	25	45	85	32	50

Yeast A = Phenolic Beer
Yeast B = Normal Beer

Experiments I to III: Model fermentations with 100 ppm ferulic acid

Experiments IV to VI: Model fermentations with 100 ppm p-coumaric acid

Table 4 Formation of 4-Vinylguajacol from Ferulic Acid by Certain Yeasts

Yeast Strains	No Decarboxylation	Decarboxylation
S. uvarum (carlsbergensis)	23	11
S. cerevisiae	21	24
S. cerv. var. ellips.	-	4
S. diastaticus	-	2
S. delbrückii	3	-
Hansenula anomala	3	4
Pichia farinosa	-	1
	49	46

Table 5 4-Vinylguajacol in Commercial Beers

Beer-type	2-Phenylethanol (ppm)	Phenylethyl acetate (ppm)	Ethyl nicotinate (ppm)	4-Vinylguajacol (ppm)
Bavarian "Weizenbier"				
1 "	8,1	1,0	1,2	1,0
2 "	15,2	1,2	1,0	2,03
3 "	20,0	0,5	0,7	0,70
4 "	22,2	1,0	0,7	1,1
5 "	11,2	0,3	1,4	0,4
6 "	11,1	1,2	0,9	0,63
7 "	18,0	1,4	2,5	6,0
8 Ale	24,9	1,4	0,7	0,4
9 Alt (German top-fermenting dark beer)	37,3	1,0	2,2	0,15
10 Pils 1	20,0	1,8	0,3	0,03
11 Pils 2	8,3	0,3	1,4	0,20
12 Strong Lager (pale)	20,0	0,7	0,6	0,20
13 Strong Lager (dark)	8,3	0,3	1,0	0,10
14 "Berliner Weiße" 1 (mixed fermentation of a top-fermenting yeast and Lactobacillus SP.)	22,0	0,5	0,5	0,01
15 "Berliner Weiße" 2	37,0	2,4	0,3	0,01

normal(15). Some of the results are summarized in Table 3. Normally these phenols are formed during kilning of malt (or wort boiling) from the corresponding phenolic acids by chemical reactions (1). In model fermentation systems it could be demonstrated, that the yeast was responsible for the phenolic off-flavor. Ferulic acid and p-coumaric acid were decarboxylated into the corresponding phenols by yeast A. Sinapic acid was not decarboxylated. Table 4 shows some results from more than 95 yeasts. It can be seen that 49 of the tested yeasts are able to decarboxylate ferulic acid. Normally, bottom-fermenting yeasts which are used in brewing do not possess this ability. Some strains of S. cerevisiae which are used in the fermentation of Bavarian "Weizenbier"produce high amounts of 4-vinylguajacol. In Table 5 the amounts of 4-vinylguajacol in some commercial beers are summarized. It can be seen, that Bavarian Weizenbier contains the highest levels of this constituent which is responsible for the typical spicy, clove-like flavor of this beer type. Clapperton could demonstrate that his flavor profiling method correlates excellently with our analytical results.

The term "phenolic" which has been used by brewers is now better described in the flavor terminology.

3. FORMATION OF OFF-FLAVOR COMPONENTS DURING BEER AGING

When a normal sound beer prepared from normal sound raw material leaves the brewery chemical changes of the flavor composition proceed. Pasteurization, provided it is carried out properly, can be regarded as a short period of accelerated aging and ensures a biological stability of 6 to 12 months. The flavor composition of beer is not a static phenomenon but changes during aging. This topic has been reviewed by Dalgliesh (16). There is a steady decrease in the sensory bitterness and a corresponding increase in sweetness, so that the sweetness-bitterness balance shifts towards a sweeter flavor. The increasing sweet flavor is accompanied by toffee-like, caramel, bready, burnt or bread-crust aromas and flavors. Maillard reactions may be responsible for these sensorial changes. In addition oxidative reactions and/or thermal fragmentation of non volatile precursors produce carbonyls which play an important role in the development of ribes, catty, papery, and cardboard flavors. These reactions may be induced and enhanced by light and metal ions. The characterization of these potent off-flavor constituents in beer is very difficult. They are perceived in the very trace range and are masked by the strongly concentrated yeast metabolites. The flavor profile method may be a useful tool for the characterization of certain beer-types, and very helpful in the practical quality control but the cooperation of an effective analytical laboratory is needed to characterize the off-flavor constituents and the reactions by which they are formed.

3.1 Formation of Maillard Reaction Products

Overpasteurization, storage at elevated temperatures, long distribution distances in extreme hot climates may induce Maillard reactions in bottled beer. The levels of furans, thiophenes, pyrroles, thiazoles, and cyclic enolones increase and some flavor-active Streckeraldehydes (already present in malt and wort) are re-formed. The amounts of furfural, 5-methylfurfural, 2-acetylfuran, and 5-hydroxymethylfurfural may be a useful index for estimation of heat damage to beer, but in our opinion they are not responsible for these flavor changes. Proline is the most abundant free amino acid in beer as shown by Masschelein (17). It can be seen, that the Strecker-active amino acids Val, Leu, Phe, and Met are also present in (finished) beer. With furfural or α-diketocomponents they can undergo degradation to aldehydes which possess malty/cereal notes and to methional, which has the aroma character of cooked potatoes and is further decomposed to methanthiol and dimethyldisulfide. Cysteine is transformed into H_2S and acetaldehyde. During these reactions sensorial relevant sulfur containing components may be formed.

L-Proline has a secondary amino group and is not degraded via Strecker as is seen by an abnormal ninhydrin reaction. In a series of model heating

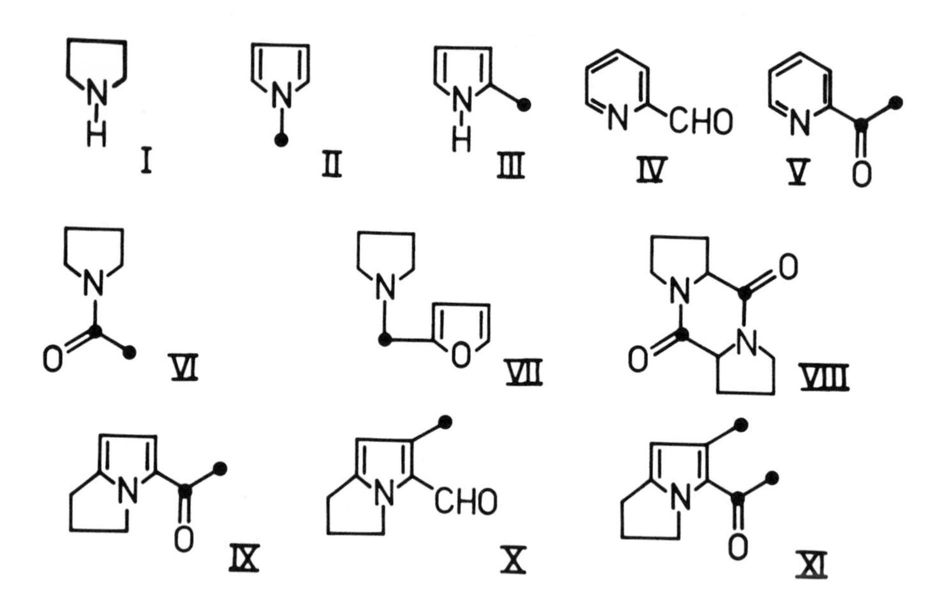

Figure 3 Proline derived components in malt and beer

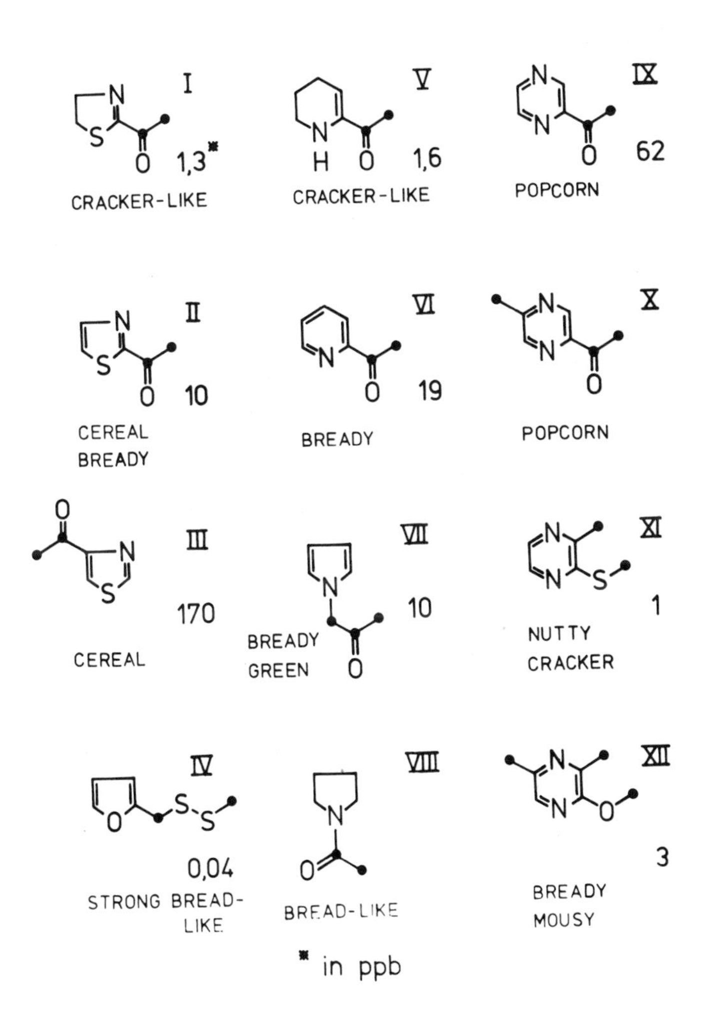

Figure 4 Maillard reaction products which possess cereal/bready aromas (Corresponding odor qualities and thresholds in water)

systems we investigated the products which are formed from proline with reducing sugars. Some of the results are shown in Figure 3. Some components possess cereal, bready, cracker-like aromas at low levels.Most of these constituents have been characterized in malt and dark beers (17). In Figure 4 some constituents which possess cereal/bready notes are summarized. Components I and II are derived from cysteine, components V to VIII are formed from proline. In heat-damaged beers the constituents I, IV, VI, VII, VIII, IX, and X were characterized by GC-MS. In our opinion component V may also play an important role, but its basic character prevents its detection in the trace range. Most of these components were determined in pale and dark malts where they contribute to the aroma. During fermentation the flavor-active carbonyls are reduced.

In the last years we investigated commercial beers which possessed ribes flavor. According to Dalgliesh (16) the formation of ribes flavor is a common flavor defect during aging of beer with high levels of oxygen. In these beers we detected dimethyldisulfide, 2-methylthioethanal, 3-methylthiopropanal, 4-mercapto-4-methyl-2-pentanone, and some unidentified sulfur components. The mercaptoketone was synthesized and proved to be responsible for the ribes flavor. It is presumably formed from mesityl oxide. The threshold value in water was determined at 0,005 ppb, in beer at 0,05 ppb. Some beers which had a strong ribes odor contained this constituent in levels of 1 to 5 ppb. But there may be other constituents which contribute to this off-flavor in beer.

3.2 Formation of Off-Flavor Components by Lipid Oxidation

Carbonyl components derived from linoleic- and linolenic acids, respectively from the corresponding oxygenated C_{18}-precursors,are the most important off-flavor constituents which are produced during beer aging. A cardboard-flavored substance produced in beer by artificial aging at elevated temperature and lowered pH was identified as 2-trans-nonenal (18). Together with 2-hexenal, 2,6-nonadienal and related components it imparts to beer a papery/cardboard flavor at very low concentration levels. Source of the carbonyls are the malt lipids. During malting linoleic-, and linolenic acids are oxygenated by lipoxygenases into 9-LOOH and 13-LOOH which are transformed to some extent into the corresponding α- and γ-ketols by an isomerase. In addition the hydroperoxides and ketols are transformed into mono-, di-,trihydroxy-, and hydroxyepoxy acids by chemical reactions (19). During malting linoleic acid is also split into hexanal, 2,4-decadienals, 9-oxononanoic acid, and caprylic acid. 2-Nonenal is formed in the trace range. During kilning the labile C_{18}-precursors are partially decomposed into carbonyls and alcohols. 2-Nonenal is now produced as a main constituent (as shown in Table 6). The hydrophilic dihydroxy-, and trihydroxy acids are transferred to beer in the ppm-range. As demonstrated in model systems the vicinal di- and vicinal trihydroxy-acids are heat labile components and split into flavor active carbonyls. This is shown in Figure 5.

In an anologous reaction the corresponding dihydroxy acids from linolenic acid are transformed into 2-hexenal and 2,6-nonadienal. Comparable to the system DMS/precursor the formation of carbonyls derived from thermal degradation of C_{18}-precursors may be decreased by modified wort boiling techniques.

Air pick-up during bottl ing increases the flavor instability e.g. oxidative deterioration of beer. Autoxidation of unsaturated fatty acids as well as photo-oxidation may be involved. The characterization of carbonyls in the sub ppb-range in beer by GC-MS is very difficult and time consuming. The flavor-active carbonyls have to be enriched by liquid-liquid extraction at low temperature, separated by means of liquid-solid chromatography from the strongly concentrated yeast aroma components, further enriched by preparative gas chromatography and investigated by capillary gas chromatography mass spectrometry. During these procedures labile precursors may be degraded and the carbonyls formed may undergo chemical reactions.

High sunlight intensities or high ultra-violet exposure on supermarket shelves cause sunstruck flavor in pale-coloured or colourless glass bottled beers (20). Riboflavin, other flavin derivatives or melanoidines may act as photoactivators. During these reactions sensorial relevant carbonyls may be formed as shown in Figure 6 and Table 7. The aroma

Table 6 Lipid Oxidation Products in Malt which are Reduced during Fermentation

	Components	Malt (dark) ppb	Malt (pale) ppb
1	Pentanal	880	900
2	Hexanal	6400	1920
3	2-Hexenal	580	760
4	cis-4-Heptenal	100	20
5	2-Heptenal	60	10
6	2,4-Hexadienal	20	+
7	2-Octenal	60	160
8	Nonanal	+	10
9	trans-2-Nonenal	400	200
10	2,4-Heptadienal	60	15
11	2,4-Nonadienal	60	10
12	2,6-Nonadienal	50	10
13	2,4-Decadienal	185	45
14	2,4,7-Decatrienal	25	10
15	2,4-Undecadienal	20	+
16	2-Heptanone	+	80
17	2-Octanone	30	40
18	3,5-Octadienone	30	+

Table 6 Lipid oxidation products in malt which are reduced during fermentation

Table 7 Formation of Carbonyls during Light Irradiation of Beer

"Sniffing" Chromatogram

LSC - F I Fresh Beer	LSC - F I Fresh Beer (2 days irradiated)	Identified Components	Threshold Level (ppb)
rancid	rancid	2,4-Undecadienal	0,01
	rancid intensive	2,4,7-Decatrienal 2,4-Decadienal	0,01 0,3
	rancid	2,4-Nonadienal 2-Decenal	0,05 1,0
	cucumber intensive	2,6-Nonadienal	0,5
woody	melon-like papery intensive	2-Nonenal	0,1
rancid	rancid	1,5-Octadien-2-one	0,002
	metallic	1-Octen-3-one	0,5
green grassy	green grassy	3-Hexenol	
	papery intensive	4-Heptenal	0,4
green intensive	green intensive	3-Hexenal 2-Hexenal	
	rancid	Hexanal	
	mercaptan-like	Prenylmercaptan	0,005

Table 7. Formation of carbonyls during light irradiation of beer.

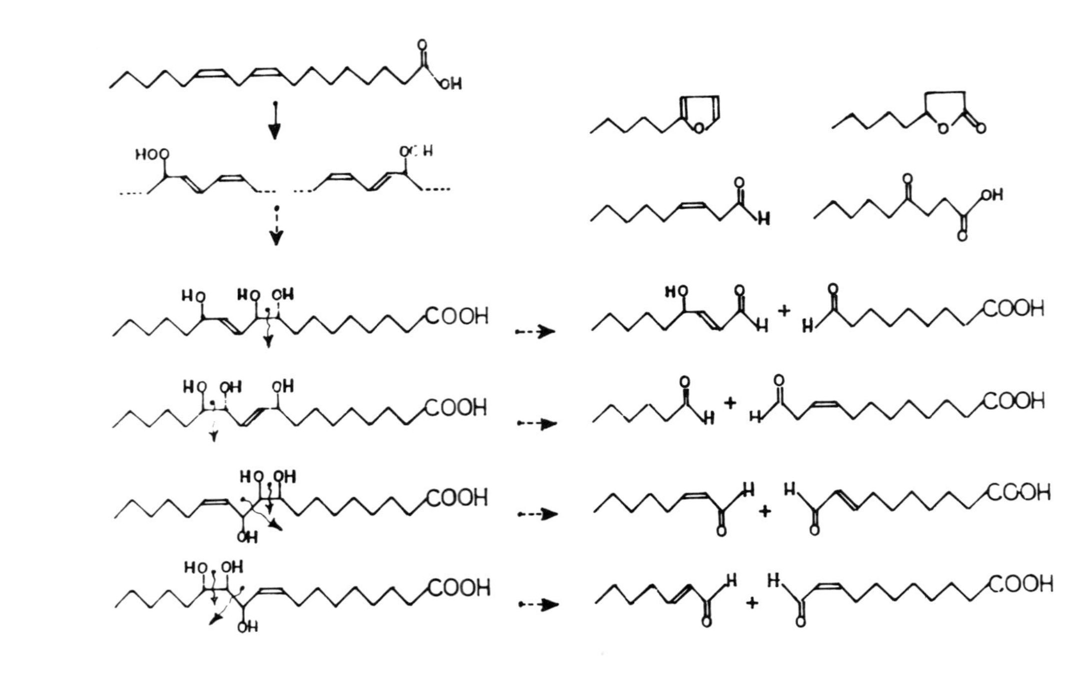

Figure 5 Possible formation of trans-2-nonenal from 5a) trihydroxy-, and

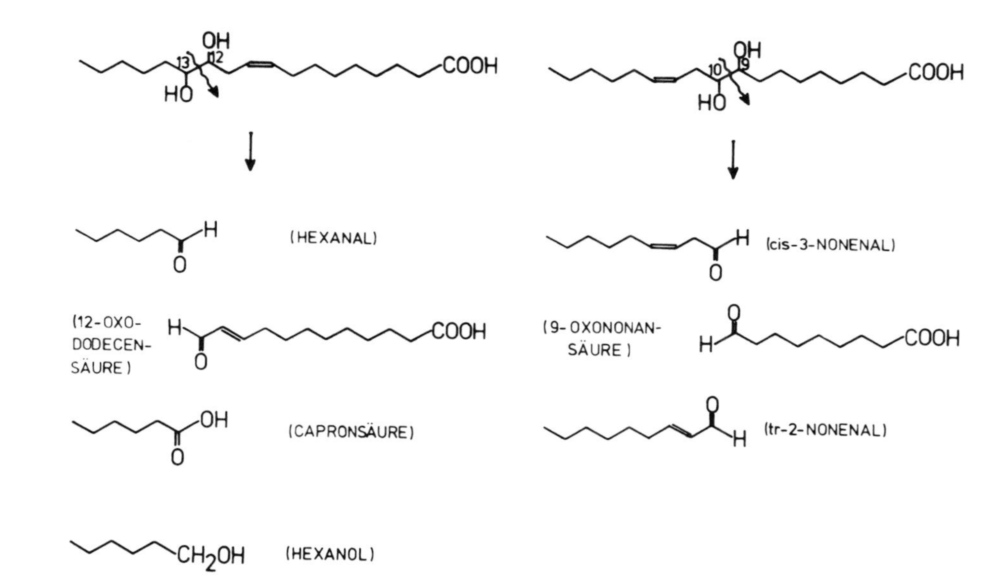

5b) dihydroxyoctadecenoic acids
in malt and beer

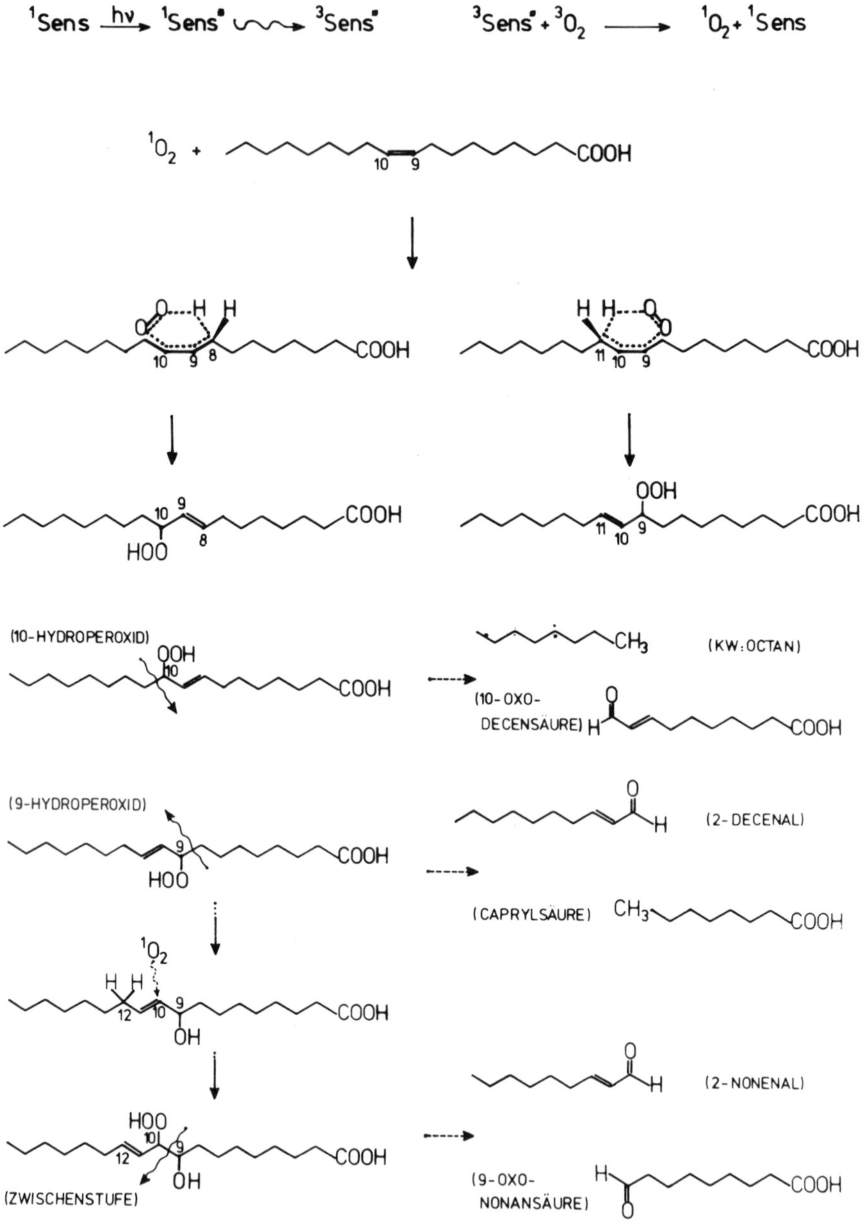

$^{1}Sens \xrightarrow{h\nu} {}^{1}Sens^{*} \rightsquigarrow {}^{3}Sens^{*}$ $^{3}Sens^{*} + {}^{3}O_2 \longrightarrow {}^{1}O_2 + {}^{1}Sens$

Figure 6 Reaction scheme which may explain the formation of
6a) hydroperoxids, and
6b) carbonyls and oxoacids
by photo-oxidation of oleic acid (19)

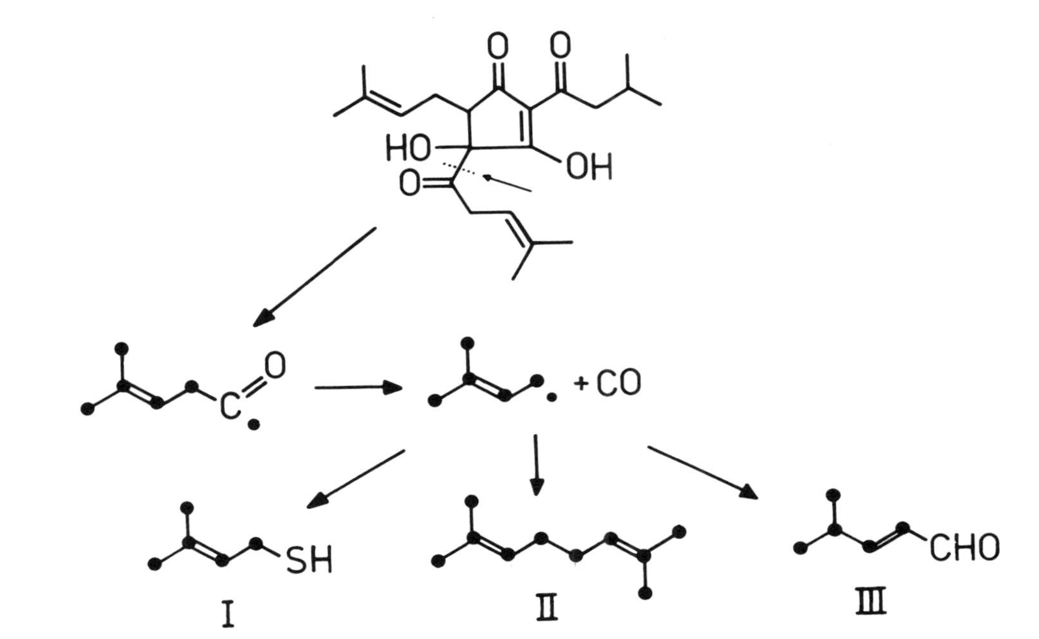

Figure 7 Photo-degradation of iso-α-humulones in a cysteine model system

extract of a German Pilsner, which contained 45 mg isohumulone, was investigated by means of GC-MS (preseparation of carbonyls from yeast metabolites by liquid-solid chromatography). No carbonyls could be detected. By means of sniffing-chromatograms (sniffing at the outlet of the inoperative detector) only some rancid and green characters were detected which correlated with the retentions of 2-hexenal, 3-hexenol, 1,5-octadien-2-one, and 2-nonenal. After irradiation of the beer with intensive light at 20°C the aroma extract showed a strong sunstruck-flavor and in addition an intensive cardboard flavor. The corresponding sniffing chromatogram showed more than 12 chromatogram sections with papery, green, rancid, metallic, and cucumber-like odor qualities which could be correlated with carbonyls by GC-MS (after enrichment of the fractions by preparative gas chromatography). In our opinion this is a very effective method for the detection of off-flavor components in beer. The results show that there are at least two pathways operative in the formation of carbonyls during beer aging: Thermal fragmentation of oxygenated C_{18}-precursors and photo-oxidation of unsaturated fatty acids.

3.3 Formation of Off-Flavor Components by Degradation of Iso-α-humulones

An early success of beer flavor research was the identification of 3-methyl-2-butene-1-thiol (prenylmercaptan) which is formed by photodegradation of iso-α-acids and sulfhydril components (20). Pilsner beers brewed with high amounts of hops, bottled in green glass tend to develop "sunstruck flavor" in the sunlight or in ultra-violet exposure on supermarket shelves. In a number of German pilsner beers which possessed sunstruck-flavor we characterized prenylmercaptan in levels from 0,1 to 1 ppb. The threshold level of the synthesized and purified component in water was determined at 0,005 ppb, in beer at 0,05 ppb. In a series of model experiments we investigated the formation of prenylmercaptan from isohumulone and cysteine. As shown in Figure 7 the 4-methyl-3-pentenoyl-side chain is released during illumination and components II (2,7-dimethyl-2,6-octadiene) and III (4-methyl-2-pentenal) are formed as main constituents. In the presence of cysteine,

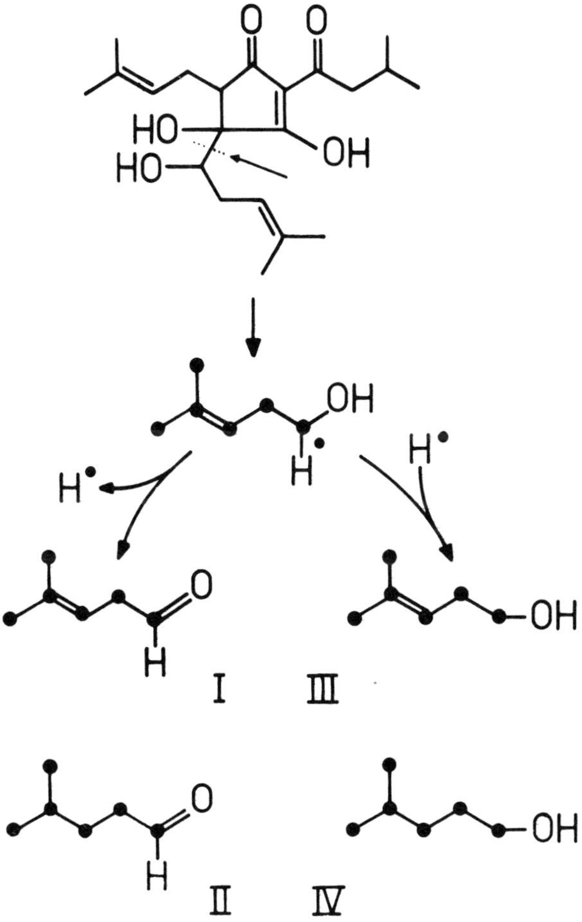

HO
HO
O O
OH
OH
H•
H•
H•
O
H
OH
I III
O
H
OH
II IV

Figure 8 Photo-degradation of reduced iso- - humulones in a cysteine model system

prenylmercaptan is detected as a minor constituent together with diprenyldisulfide and some sulfur constituents in the trace range. The prevention of sunstruck flavor in beer is claimed in a patent by bittering the beer with iso-α-acids, in which the side chains have been reduced (21). We therefore investigated reduced iso-α-acids in our model system. No prenylmercaptan could be detected and no 2,7-dimethyl-2,6-octadiene was formed. As shown in Figure 8 two carbonyls and two alcohols were characterized which displayed a green aroma character. The formation of sunstruck-flavor may be reduced or prevented by using dark coloured glass for bottles or by bittering the beer with reduced iso-α-acids. Photodegradation and thermal degradation of isohumulones lead to a series of carbonyls which may contribute to beer staling as recently shown by Hashimoto (22).

Beer flavor is changing all the time after a beer has left the brewery. The flavor instability of beer is a challenge for flavor chemists. Some progress has been made in the last years but we are far from controlling off-flavor formation in beer. Intensive future research is required.

ACKNOWLEDGMENTS

This work was supported by ARBEITSGEMEINSCHAFT INDUSTRIELLER FORSCHUNGSVEREINIGUNGEN, Köln.

REFERENCES

1. R. Tressl, T. Kossa, and R. Renner, EBC-Proc. 15th Congress, Nice 1975, p. 737
2. A. Scherrer, Dissertation No. 4585, Eidgenössische Technische Hochschule Zürich, 1971
3. T. Wainwright, Brewer 12, 638 (1974)
4. K. Takahashi, M. Tadehuma, and S. Sato, Agr. Biol. Chem. 40, 325 (1976)
5. J.F. Clapperton, C.E. Dalgliesh, and M.C. Meilgaard, MBAA Techn. Quart. 12, 273 (1975)
6. Ch. Zürcher, R. Gruss, and K. Kleber, EBC-Proc. 17th Congress, Berlin 1979, p. 175
7. F.H. White, and T. Wainwright, J. Inst. Brew. 83, 224 (1977)
8. J.E. Amoore, L.J. Forrester, and P. Pelosi, Chem. Senses and Flavour 2, 17 (1976)
9. T. Kossa, D. Bahri, and R. Tressl, Mschr. Brauerei 5, 249 (1979)
10. R. Tressl, R. Renner, T. Kossa, and H. Köppler, EBC-Proc. 16th Congress, Amsterdam 1977, p. 693
11. J.F. Clapperton, J. Inst. Brew. 84, 107 (1978)
12. J.F. Clapperton, and D.G.W. Brown, J. Inst. Brew. 84, 90 (1978)
13. V. Arkima, EBC-Proc. 12th Congress, Interlaken 1969, p. 507
14. T. Äyräpää, and I. Lindström, EBC-Proc. 17th Congress, Berlin 1979, p. 507
15. K. Wackerbauer, T. Kossa, and R. Tressl, EBC-Proc. 16th Congress, Amsterdam 1977, p. 495
16. C.E. Dalgliesh, EBC-Proc. 16th Congress, Amsterdam 1977, p. 623
17. C.A. Masschelein, and J. van de Meerssche, MBAA Techn. Quart. 13, 240 (1976)
18. B.W. Drost, P. van Eerde, S.F. Hoekstra, and J. Strating, EBC-Proc. 13th Congress, Estoril 1971, p. 451
19. R. Tressl, D. Bahri, R. Silwar, EBC-Proc. 17th Congress, Berlin 1979, p. 27

20. Y. Kuroiwa, and N. Hashimoto, ASBC-Proc. 19, 28 (1961); Rep. Res. Lab. Kirin Brewery 4, 35 (1961, ibid. 5, 57 (1962)

21. D.A. Hougen, U.S. Patent 3.079.262 (1960)

22. N. Hashimoto, Rep. Res. Lab. Kirin Brewery 19, 1 (1976)

THE WORDS USED TO DESCRIBE ABNORMAL APPEARANCE, ODOR, TASTE, AND TACTILE SENSATIONS OF WINES

Maynard A. Amerine[1]

Department of Viticulture and Enology
University of California
Davis California

and

Wine Institute
San Francisco, California

Introduction

The meaning of words used to describe wines obviously changes with time and place. Adlum's (1) definitions of wine "tasting" terms undoubtedly reflected the state of the art of wine appreciation in his time, but they would hardly apply today. As a matter of fact, his definitions reflected the French usage of his time, not that of Washington, D.C. This discussion is mainly confined to words which are used in the United States to describe abnormal sensory sensations of appearance, odor, taste and tactile stimulation. Andrews uses "taints" for such sensory impressions in beer (2). Obviously, one would prefer to describe the desirable sensory impressions of wines. But, in the format of this symposium, we are emphasizing the abnormal sensations. Philosophically, one may justify this from Murphy's Law, that if something abnormal can occur, it will. Modern technology has reduced the number and seriousness of abnormal sensations in wines. Perhaps now it is true that "the most serious 'off' odor is the wine

[1]Present address: P.O. Box 208, St. Helena, California

ISBN 0-12-169065-2

with no odor." (3)

Since most of the wines produced in the world are from western Europe, the vocabulary of many consumers naturally reflects the sensory definition useage given in the French, German, Italian and other languages. Unfortunately, there are no definitive definitions of traditional words for the attributes of wines in any of these languages, and certainly not in English.

Harper et al. (4) noted that even for English American and British subjects differed in their description of the odor qualities of some common compounds, but agreed well in their descriptions of others.

The multilingual definitions of the Office International de la Vigne et du Vin (5) are often helpful but not definitive. In some cases the definitions of different languages are mirror-images (usually basically from the French). In other cases the definitions for what appears to be the same word differ in degree from each other. In some cases the definitions of different terms in one or more languages are not clear, confusing, or contradictory. This discussion is an attempt to give some standard definitions for words used to describe the undesirable or abnormal appearance, odor, taste, and tactile sensations currently used for wine in this country.

Words have different meanings to people of varying degrees of experience. The ordinary wine consumer experiencing only a limited number and quality of wines needs only a few words to describe and appreciate the wines he consumes. The wine connoisseur exposed to a wide variety of wines of different types, qualities and ages may require and use a much wider vocabulary, even in more than one language. Depending on the experience of the connoisseur, he may have a consistent vocabulary for the wines with which he is most familiar. For example, if he is afficionado of red wines he may use a variety of terms to distinguish astringency q.v., bitterness q.v., and various combinations of compounds having astringent or bitter properties. The connoisseur of white wines requires a less extensive vocabulary for these taste sensations since they are seldom required for white wines but may have a rich vocabulary of terms having to do with fruitiness and sweetness.

Then there are the professional experts who buy, sell, blend, etc. wines. They use a vocabulary necessary for their commercial transactions. It may, or may not, be compatible with the vocabulary or meanings used by connoisseurs or ordinary consumers. It may be very specific for some sensations (acetic acid, for example) and much less specific for others (nuances of bouquet, etc.) because they deal mainly with young wines in their profession.

Enologists should try to find a single chemical compound (or a group of related compounds) as the cause of a specific off-odor. However, in fact, many off-odors are a combination of several compounds often having distinctly different undesirable odors. In beer, Moll et al. (6) used 7 to 13 physico-chemical parameters to forecast the "good" or "passable" quality reactions of a panel to beer. Discriminate analysis was used to make the correlations. Comparable studies seem not to have been done for wine. They have the value that once the correlation has been established the physico-chemical parameters can be used for predictive purposes. The accuracy of Moll et al. predictions was from 82% to 94%, not really high enough for predictive purposes, but indicative of what can be accomplished. Similar studies have been done of several other foods using both positive and negative factors.

It is not only one sensation but combinations of sensory impressions which may influence the consumer: alcohol/acid, acid/sugar, alcohol/tannin, etc. ratios. Color and appearance interactions may also be important.

Another problem is that what is desirable to one individual may be obnoxious to another. This may not be true in the professional world of commercial wine production. But small consumer groups, families, ethnic enclaves, public relations experts, etc. may have quite skewed biases for and against certain odors and tastes. In this country, wine consumers in California are not only unappreciative of the foxy q.v. odor of Concord grapes, but may reject the wine forthwith. In New York, the same odor is appreciated or at least tolerated. A further example, along the Dalmatian coast, high-alcohol, high-tannin wines are produced and appreciated. To the western European, they are considered harsh q.v., bitter q.v., hot q.v., or frankly undrinkable.

Finally, sensory stimuli may be pleasant at a low concentration and unpleasant at a higher concentration. Some of the sulfides may be in the first group. All are unpleasant at higher concentrations.

Sources of Off-Odors and Tastes

The grape vine and its fruit are subject to the vagaries of nature and man as far as the suitability of the fruit for wine making are concerned. Grapes grown in very cool climatic conditions often do not ripen. Thus the grapes arriving at the crusher may be low in sugar and high in total acidity, and often of low pH (3 or below). Such grapes produce wines of low alcohol (thin, q.v.) and green q.v. In warmer regions the grapes may become overripe, and the resulting wines acquire an overripe q.v. odor or a sunburned q.v. or raisin-like q.v. odor. In extreme cases, there may be a distinct caramel q.v. aroma in the wine from the use of partially or wholly dried grapes or from addition of grape concentrate or boiled-down grape juice.

The region where the grapes are grown may also impart undesirable odors to the resulting wines. The earthy q.v. odor is the common term used for this. It is not common. The earthy flavor is surely an odor. Flinty q.v. is probably not an odor, if it exists. It is said to be associated with siliceous soils. More likely it is another way of describing wines of low alcohol, low pH and high total acidity.

As a result of climatic conditions, a number of undesirable odors may get into the resulting wines. Hail q.v. removes leaves and fruit and bruises the fruit. The bruised fruit may turn brown and be subject to infection by bacteria and yeasts, thus compounding the off-odors produced by enzymes from the bruising of the fruit. Early frosts may also damage the grapes leading to wines with a frost q.v. odor. If the weather during the ripening period is moist and not too cool, a wide variety of cryptogamic organisms may invade the vineyard and the fruit on the vines. The risk of off-odors from the infected fruit varies, depending on the time, extent, and nature of the microbial invasion, but in extreme cases it can seriously damage the sensory quality of the wines.

If *Botrytis cinerea* develops on the green fruit, it prevents ripening, the fruit may split, and the condition known as grey mold develops. Other moldy q.v. odors develop from *Penicillium* sp., *Aspergillus* sp., *Personospera* sp., *Rhizopus* sp., *Uncinula necator*, etc. Not only are the fungi capable of being the direct source of an off-odor, they are surely involved in secondary odors from oxidation q.v., acetic acid q.v., etc. If the grapes are in poor physical condition (from whatever source) a variety of undesirable yeasts may also develop and have an influence on the quality of the fermentation. Undesirable odors, tastes, and haziness may result from their growth.

In recent years a variety of new organic fungicides and pesticides have been introduced in viticultural practices. Few of these are used in sufficient amount to transmit directly a foreign odor to the resulting wine. They may however, have an indirect effect on flavor since in some cases they can inhibit the growth of yeasts and allow undesirable microorganisms to grow in the must. The variety of off-odors thus produced has not yet been systematized since the number of compounds used is large and is constantly changing and the biochemical changes are variable and often unknown.

Normally Saccharomyces cerivisiae develops in the must, and the enzymes, coenzymes, and metals of the glycolytic cycle result in the production of ethanol and carbon dioxide. Glycerol is a constant and important by-product. Small amounts of acetic, lactic and other acids and acetaldehyde accumulate. They normally have little influence on an abnormal taste or odor of the product. Succinic acid is a by-product but even in the substantial amounts produced has little effect on the sensory properties of the product, certainly not in a negative sense.

Small amounts of hydrogen sulfide q.v. are produced during alcoholic fermentation, more with some strains of yeast than with others. Hydrogen sulfide may also be formed during storage of the wine on the lees (accompanied by yeast antolysis). Since the threshold is only 1 or 2 parts per billion, this can be an important negative quality factor. The amount of elemental sulfur on the grapes (used for mildew control) is also important. If it gets into the must it is easily reduced to hydrogen sulfide under the highly reducing conditions that occur during alcoholic fermentation.

Under conditions that are poorly understood mercaptans may also be produced during or after fermentation.

Normally sulfur dioxide q.v. is added at the time of crushing. In the amounts usually employed, 75 to 150 mg/l, little remains in the free (sulfurous acid) condition at the end of the fermentation. Some is lost with the carbon dioxide, some is oxidised, some reacts with acetaldehyde or carbonyls, etc. It may react with hydrogen sulfide, if present, to produce elemental sulfur. The fixed or combined sulfur dioxide has little of the odor of the free.

In the case of red wine fermentations the skins remain in contact with the liquid for days to weeks. The floating cap of skins may, if unattended, develop growth of Acetobacter sp., with the resulting formation of large amounts of acetic acid q.v. and ethyl acetate q.v.

In the case of white wines, if the fermentation is kept under anaerobic conditions, very little of these compounds is formed. But if the fermentation is slow and aerobic conditions exist, a number of yeasts (Pichia sp., Hansenula sp., Saccharomyces sp., Kloeckera sp.) may grow on the surface of the wine. The major product of these film yeasts is acetaldehyde, which gives the wine an aerated q.v. or oxidized q.v. odor.

The containers in which musts and wines are placed may be major sources of undesirable odors. First of all, the containers may not be made from properly prepared (aged) wood, and the wine then soon acquires a woody q.v. odor. Or the cooperage may be dirty from stagnant q.v. water, or moldy q.v.

growths, or hydrogen sulfide q.v., etc. Even if the wood is perfect, if the wine is left in the wood cooperage too long it may acquire too intense a woody q.v. character. However, connoisseurs generally tolerate more of the woody odor in red than in white table wines, and differ in how disagreeable they perceive a woody odor to be.

Heat is generated during alcoholic fermentation. If the temperature is not controlled dangerously high temperatures may ensue during fermentation. These result in pomace q.v. or heated q.v. odors. In some cases the fermentation ceases (sticks) and bacterial growths may ensue with sauerkraut q.v. odors or sweet-sour q.v. tastes.

As the fermentation subsides, the yeast cells gradually settle to form a deposit, called the lees. If the lees is left in contact with the wine for too long a period, and especially at too high a temperature, the yeasts cells may autolyse, resulting in a variety of off-odors, q.v. pomace, hydrogen sulfide, lees, mercaptans, etc.

In the production of dessert wines the spirits added may have a marked influence on the quality of the product. High-aldehyde q.v. and high fusel q.v. spirits are especially bad. Compressed air is sometimes used to aid in mixing the spirits It may contain traces of petroleum q.v.

In the United States dessert wines contain 14 to 21% ethanol, usually 17 to 19%.

During storage the wines are subject to a variety of sources of undesirable odors. Sulfur dioxide may be added in excessive amounts, especially to white table wines. The wine may be kept in containers that are not full and become oxidized. Film yeasts growing on the surface may exacerbate this. Fining with organic fining agents (casein, egg white, gelatin, etc.) is commonly practiced. If the fining agent is not free of off-odor, it may give the wine a rotten q.v. type odor (particularly true of low quality casein). Or the fining agent may remain in contact with the wine for too long a period and protein decomposition may result, tainting the wine with rotten odors. Potassium ferrocyanide is legally used as a fining agent in certain countries (not the United States) to remove excess copper and iron. When improperly used in excessive amounts the residual ferrocyanide may decompose to yield hydrogen cyanide (or bitter almond q.v.) odor.

When asbestos filters were used and not properly washed before use an asbestos q.v. odor is sometimes acquired. Asbestos filters are rarely used in the United States at present. They have been replaced with membrane and other types of filters.

In fact, most filters, if improperly prepared, may contribute a filter q.v. odor to the wine, particularly to the first wine passing through the filter. Use of excessive amounts of filter aid may also give the wine an off-odor. Likewise, excessive use of ion-exchange materials may unduly increase the sodium content, resulting in a flat q.v. or even a salty q.v. taste, or, with too much hydrogen ions from some ion exchangers, in a tart q.v. taste.

Whenever table wines are handled during storage, if too much air is present and the wine is not high in sulfur dioxide, then the wine may acquire an aerated q.v. or oxidized q.v. or even a sherry-like q.v. odor. In the case of some wines this may be intentional (sherries and rancio-type wines) in which case the resulting odor is, of course, not considered undesirable.

A major source of undesirable colors, odors, or tastes results when uncompatible wines are inadvertantly mixed: white with red or vice versa, fortified with table wines, flavored wines (vermouth or retsina) with non-flavored types, high acid with high acid wines and low acid with low acid wines, muscat-flavored wines with non-muscat-flavored wines, foxy flavored with non-foxy flavored wines, young with old wines, and even vice versa, etc. (Murphy's Law)

The hoses through which wines are pumped may be a source of off-odor. Wine left in hoses or glass lines or filters or bottling equipment may spoil and contribute an off-flavor or excessive numbers of bacteria to the succeeding wine. Later the microorganisms may grow and produce undesirable cloudiness or yeasty q.v. or acetic q.v. odors. Even the bottles may be dirty with petroleum or mold, etc., and unless thoroughly washed, may contribute moldy q.v. or petroleum q.v. odors to the wine.

In practice, most of the terms used can be roughly quantified by the use of slightly, moderately or very. In some cases, low, medium, and high may be similarly used.

Color

The color appropriate to one type of wine may be entirely inappropriate to another type. Thus dry sherries have a more or less light amber color. The same color in a young Riesling wine would indicate undue darkening owing to action of oxidising enzymes, excessive fermentation time on the skins, or other serious defects. There are so many types of wines that to list the possible inappropriate colors for all the types would require a separate and lengthy discussion; therefore, we have omitted them from this list, although many are undesirable sensations. In fact, the off-colors in many cases offer the first and best clue as to the probable quality or lack of quality of the wine.

In general, we may assume that dry white table wines with excessive amber color have been left on the skins or in contact with air too long. Young red wines with a purple color may have been produced from over-ripe grapes of low acidity (high pH), have been fermented at too high a temperature, or have undergone an early and excessive malolactic fermentation.

Appearance

There are many words used to describe undesirable appearances in wines. In the past (Amerine et al., 7; Amerine and Roessler, 8) we have used only three and see no reason to extend the list.

Clear. A wine which is nearly brilliant but which suffers from a very slight colloidal haze. Sometimes this can only be observed with a strong beam of light. Though it may not indicate much deterioration in the odor quality, it usually is a warning to be on the lookout for such.

Dull. Such a wine is obviously hazy with a visible colloidal content. However, a deposit in the bottle is usually not found. There is often an alteration of odor, depending on the cause of the haziness: bacterial q.v. or oxidized q.v.

Cloudy. Turbid wines containing large amounts of suspended colloidal and non-colloidal material, usually with a deposit of yeast lees, or of bacteria, or of precipitated tannins and coloring materials. In young wines the odor is often yeasty q.v. In older wines it may be bacterial q.v. or oxidized q.v.

Leaden. (plombe Fr.) This refers to the dull grey appearance and color of some white wines. I confess to not have seen a good example in this country.

Oily or Ropy. Rod bacteria may develop in low acid, high pH wines, resulting in wines which pour like oil. Off-odors and gassiness usually develop - very rare in California.

These words for color and appearance are not repeated in the list which follows. There are a few compounds with taste, odor, and tactile properties. Among these is ethanol. When there is too much alcohol in a table wine it is said to be alcoholic. See also hot.

Tastes

Acidity. Surely some acid (tart or sour) taste is present in all wines. It is the excess of acidity, especially at low alcohol levels, that results in an undesirable acidulous q.v. taste. For lack of acidity see flat.

Acidulous. Unpleasantly sour. The total acidity is usually above 1.0 gr/100ml (as tartaric) and the pH below 3.1 or at most 3.2. These limits will be influenced by sugar and alcohol content and by local custom.

Bitter. A true taste arising mainly from polyphenolic (tannin) compounds (Singleton and Noble (9). So many words are used in French and Italian for astringency and bitterness that it is not easy to separate them. See astringency under tactile. Flavonol phenolics and acrolein have been blamed for the bitter taste in some red wines. It is, therefore, mainly noted in young red wines; Singleton and Noble (9) attribute the bacterial-induced bitterness to be resultant "from reactions of acrolein and perhaps other aldehydes with the phenols of wine."

Flavonoids also give a bitter taste to white wines. In vermouths, it is expected and not objectionable, within limits. Bitterness is also reported to be due to compounds produced by bacterial action (amertume in French) but this has apparently not been noted in California. Ough and Berg (10) did report that grapes attacked by oidium (*Uncinula necator*) gave wines "with bitter to distinct off tastes." Probably they meant "off" odors. See moldy.

Bitter-sweet. (ammandorlato, It.) The juxtaposition of these two tastes is rare. It may be the result of bacterial action.

Flat. (plat, mou, Fr.; piatti, It.) Lack of sufficient acidity is more common in California table wines than is usually recognized: total fixed acid below 0.5 gr/100ml (as tartaric) and pH above 3.7 (for table wines). Nowadays deficiency in acidity can be and usually is made up by adding citric or other approved acids. Nevertheless, if the grapes are harvested too late in a warm year there may be so little acidity and such a high pH in the resulting wine that it is difficult to correct. Flatness may also result from excessive malo-lactic fermentations. Flat-tasting wines may arise from over-correction of excess total acidity (use of calcium carbonate, etc.) or from over-amelioration of the original musts. (In the United States, outside of California, the volume may be increased by as much as 35 percent by the use of sugar and water, providing the total acidity is not reduced below 0.55 percent. Similar regulations exist in Canada. Most other countries prevent use of water but many allow addition of sugar.) California permits neither.

Freshness. This is probably another aspect of good acidity. It is generally desirable so the undesirable aspect is the lack of freshness, most often in white table wines. So flat is probably all one needs.

Fruitiness. This appears to be still another aspect of good acidity with perhaps some odor factors, and is certainly desirable. It is the lack of fruitiness, often in low acid white and red wines, that is undesirable. Again flat is probably sufficient.

Green. (vert, verdeur, Fr.; bruschetto, It.; Grasig, Ger.) The unbalanced acidulous q.v. taste in wines, usually made from immature grapes, is most undesirable. However, in regions where all of the wines have a high acidity, and often low alcohol, consumers may learn to tolerate (or even prefer) the green taste. The wines of the Vinhos Verdes region of Portugal, of some varieties in very cool seasons in Germany, and in the Steiermarkt district of Austria have their local afficionados. For the odor see green.

Metallic. Probably not a taste. May be an odor or tactile sensation. It does seem to be associated in some cases with low-ethanol, dry, high-acid white table wines.

Salty. (salato, It.) Surely there is a true salty taste but it rarely occurs in wines. Wines made from grapes grown near the ocean may be salty, but this is rare. Very old wines may gradually acquire a salty taste but this is very rare indeed. Some wines are made from musts treated with calcium sulfate (plastering). The resulting wines are high in potassium sulfate which has a slightly salty taste. Wines that have been over-treated with sodium ion-exchange may have a sodium content of 600 or more mg/l. These may have a slightly salty taste. Wines that have been over-treated with sodium content are sold in this country as "cooking wine" without paying certain taxes. There is always the possibility of brine refrigerant leaking into the wine. Murphy's Law again!

Stemmy. This may be more of an odor than a taste. But there is a bitter q.v. taste in wines fermented on the stems. A similar taste may also arise where the musts remain in contact with the seeds and skins for too long a period during fermentation.

Berg and Webb (11) call it a taste due to too long a fermentation on the skins and "is a modified complex of the bitter and green sensations."

Sour. The acid taste. See acidity.

Sweet. No one objects to a sweet taste when it is found in appropriate wines. It is the lack of sweetness, say 3 instead of 10 percent sugar in a port, that is undesirable. Also, it is possible to have an excess of sweetness. See sweetish.

Sweetish. (douceâtre, Fr.) This undesirable balance of excess sweetness, lack of acidity and low alcohol resulting in a sort of sickly taste that is disagreeable. The French doucereaux and the Italian dolcigno appear to express the same thing. It is worth remembering the old saying that sweetness covers up a multitude of sins. Broadbent (12) and Amerine and Roessler (8) use cloying and unctuous for excessively sweet wines of low acidity. Grazzi-Soncini (13) emphasized the undesirable aspect of this taste.

Sweet-sour. The bacterial fermentation of sugars may produce mannite (a sugar alcohol) as well as fixed (non-volatile) and volatile acids. The resulting sweet-sour (aigre-doux, Fr.; agro-dolce, It.) taste is very unpleasant, expecially as it may be accompanied by mousy q.v. off-odors. Mannitic fermentations are now very rare owing to better harvesting of the grapes, use of pure yeast cultures, sulfur dioxide, and cool fermentations.

Tart. See acidulous.

Odors

The general concept is that an off-odor comes from a single compound (or from a group of related compounds). This assumes that odors are not additive, subractive, or synergistic. With beer, Moll et al. (6) found 15 or more chemical determinations that gave correlations with "good" or "possible" quality. Similar studies have been made with other foods. For a good theoretical discussion see von Sydow and Okesson (14). Hannack (15) reports that even below their threshold value various beer components together may have undesirable effects on beer quality.

The regulations on the removal of off-odors from wines

vary from country to country. France and Spain generally prohibit treatment. Italy allows activated carbon, neutral vegetable (olive) oil or mineral oil. Activated carbon, under certain limitations, is permitted in the United States.

The influence of moderate amounts of alcohol on odor detection has not been defined. See Engen and Kilduff (16) for some of the problems.

Acetic. (aigre, Fr.; Stichig, Ger.) Acetic acid is a normal by-product of alcoholic fermentation but the amounts formed are small, about 0.03 mg/l (as acetic). When wines of below 15 percent ethanol are exposed to air they may become contaminated with Acetobacter and large amounts of acetic acid and ethyl acetate may be rapidly formed. The threshold for acetic acid in ethanol solutions of about 10 percent is about 0.050 percent (as acetic), 0.07 in wines. The legal limits for acetic acid in wines varies from country to country and for different types of wines, from 0.090 to 0.25 gr/100ml according to Robertson and Rush (17). In the United States, the limits are 0.120 gr/100 ml (as acetic) for white and fortified wines and 0.140 gr/100 ml for red wines. A slight spoiled odor (pique, Fr.) is discernible at about 0.070 percent according to Amerine and Ough (18). The spoiled odor of acetic acid (aigre) is similar to that of ethyl acetate q.v. (ascescent). Possibly the odors of the two compounds are additive. Since the ethyl acetate odor is more undesirable, it has been suggested that a legal limit for ethyl acetate would be more rational than one for acetic acid. Robertson and Rush (17) suggest a limit of 150 mg/l (as ethyl acetate). As more wineries have GLC equipment, the determination could be easily made. It is of interest to note that it is legal in the United States to add a small amount of acetic acid to wines to improve their odor complexity as long as the total acetic acid content does not exceed the legal limits. It is said (Hallgarten, 19, for example) that high sugar content "conceals the acid." Most American enologists have not found this to be correct.

Aerated. The odor of a table wine exposed to air is said to be faded (goût d'éventé, Fr.) It is due to the formation of acetaldehyde which is "free." With time the acetaldehyde is fixed by the sulfur dioxide or polyphenolic compounds and the aerated odor disappears. It occurs after filtration, racking, bottling (where the odor is known as bottle sickness; Flaschenkrank, Ger.) or simply from exposure to air in unfilled containers. It is also produced by film yeasts growing on the surface of wines or in submerged yeast cultures. Longtime contact of red fortified (or high alcohol) wines with air leads to a rancio q.v. odor. Long time exposure to air leads to a frankly oxidized q.v. odor. Heating wines in the presence of air may lead to madeirization q.v., though this odor may have overtones of caramel q.v. Not all the ill effects of exposure to oxygen can be accounted for by the formation of carbonyls (Anon., 20).

Aromatized. It is quite legal to add aromatic material to vermouth and special natural wines. Addition of dried resin to produce retsina wines in Greece is also a normal practice. The turpentine odor is highly undesirable in other types of wines. An off-odor may arise by inadvertent mixing of aromatized and non-aromatized wines through poor cellar operations.

During the phylloxera period 100 years ago when there was a great shortage of grapes in Europe, wines were made from raisins and whatever was available. There was a sale of fruit and flavor extracts to give a special wine-like flavor to these neutral wines at that time. Almonds, coriander, iris, nutmeg, walnuts, etc. were used. The practice has practically disappeared, though I confess to having tested at least one raspberry-flavored sparkling burgundy about 1936 (in California!).

Ascescent. See ethyl acetate and acetic.

Asbestos. Owing to some asbestos fibers passing into the wines filtered through an asbestos filter pad, this type of pad is not now used in the United States and most other countries. The odor was present in the first hundred or so liters of wine passing through the filter. Usually asbestos pad filters were used at the time of bottling so the first wine had the asbestos odor. The solution was, of course, to blend the tainted wine back with the whole lot. There is some evidence that the asbestos odor faded with time. At any rate it was rare and is now mainly of academic and historical interest.

Baked. The caramel q.v. odor of wines that have been heated too long or at too high a temperature. Webb and Noble (21) attribute these to Maillard and carmelization reactions.

Bacterial. The odors due to bacteria or yeasts producing acetic, butyric, geranium-like, mousy or sauerkraut odors are discussed elsewhere. Yeasts (Brettanomyces sp. or Dekkera sp.) may also produce off-odors (Phaff and Amerine, 22).

Biacetyl. The threshold is from 1 to 10 mg/l depending on the individual. At concentrations up to about 4 mg/l it apparently adds complexity to red wines and is not considered undesirable. In the range of 6 to 8 or more mg/l it is disagreeable to some consumers. In beer Hannack (15) reports a threshold of 0.2 mg/l. In the presence of acetaldehyde it is difficult to establish a threshold (Amerine and Roessler, 23).

Bitter Almond. To aromatize wines bitter almonds were once used. The odor is that of hydrogen cyanide. This is now rare (absent in the United States). This compound is also produced when potassium ferrocyanide is used to reduce excess metals (blue fining) and its reaction products with copper and iron are allowed to remain in contact with acid solutions. The same reaction may occur if the proprietary product Cufex (legal in the United States) is used in excessive amounts. The odor threshold for hydrogen cyanide is far below its harmful or lethal dose.

Botrytis. The botrytis odor is expected and appreciated in sweet table wines. In dry table wines it may be undesirable as a moldy q.v. odor.

Burnt. See baked and caramel. Maybe also for an acrid odor from poor spirits in dessert wines.

Butyric. I have personally not found the highly disagreeable rancid butter-like odor in a California wine but Robertson and Rush (17) in New Zealand suggest that it may be a problem and suggest a maximum level of not over 3 mg/l. It is certainly of bacterial origin. Hannack (15) reports a threshold value of 1 mg/l in beer.

Caramel. Except for Marsala and some baked sherries and Madeiras wines the caramel odor is normally not present or desirable. Baking or concentrating musts (or boiling them down in open fire-heated containers) or baking sweet wines will produce hydroxymethylfurfural and caramel-like compounds. Caramel is used in making vermouths and brandies. If a caramel containing brandy is used for fortifying a wine, it will communicate a caramel-like odor to the wine. Probably caramel is undesirable in ports and muscatels and angelicas. Otherwise it may be a matter of taste. Overly high temperatures in pasteurization, particularly for too long a period, may result in this type of odor.

Cooked. This is probably just another manifestation of the preceeding. See overripe for its natural occurrence.

Cresol. A number of compounds are used to impregnate wooden posts for use in vineyards. There are reports of contamination of grapes, and thence of wines, with this odor. See also petroleum.

Corked. (goût de bouchon, Fr.) This unpleasant odor is found in wines that have been aged in bottles with cork closures. The odor is attributed to mold growth in the cork: *Penicillium multicolor*, *P. frequentans*, and *P. velutina*, according to Schaeffer et al. (24) It may occur in only one cork in thousands. However, in one case most of a lot of several thousand corks produced the corked odor in the wines in which they were used. It develops rapidly in some cases and slowly in others. Porous corks and corks which allow leakage (ullage) are more prone to develop the corky smell. Götz (25) notes that bottles often leak (are on ullage) due to cork borers. The wine may acquire the corked odor from mold in the cork -- the cork borer simply exposing a larger surface to the wine -- not from the borers. Many amateurs fail to recognize the corked odor but once recognized it is seldom forgotten. There is a recent report (Tanner and Zanier, 26) that an off-odor may be acquired from corks which have been exposed to microbial agents (methyl thiopyrazine). See mousy. They also find an off-odor from corks which have been improperly stored or over-treated with sulfur dioxide. Broadbent (12) separates corked from corky. He defines the latter as "an 'off'oxidized and thoroughly obnoxious smell."

Earthy. The term is widely used (goût de terroir, Fr.; Erdgeschmack, Ger.). But it is of unknown origin and difficult to recognize. It has been attributed to use of manures and fertilizers, but wines of unfertilized vineyards may develop the odor. The soil itself has been blamed and this may be true in some cases. At Davis washed grapes had as much of the earthy odor as non-washed grapes from the same variety and vineyard. The microflora of the vineyard and of the winery may also be involved. Growth of several fungi, *Actinomycetes* sp. or *Cladothrix odorifera*, may in certain conditions produce the odor, but not, so far as is known, experimentally in the winery. A malt distillate with an earthy odor was found to contain trans-1, 10-dimethyl-9-decalol (19700-21-1) (Bemelmans and Te Loo, 27). In the enological literature it has been associated with wines as diverse as those from the Moselle and Nahe in Germany to those of Argentina and Davis, California. See also flinty. It is likely due to a rather high boiling point compound because it does not become apparent until the wine is well warmed-up in the mouth. A vineyard or regional flavor need not be undesirable if it is not too dominant. The Germans make a distinction between earthiness and a hot climate-soil condition (Bodengeschmack) which implies oxidation and a sherry-like odor.

Ethyl acetate. (ascescent, Fr.) Produced by Acetobacter sp. as acetic acid is. Also produced by various yeasts: Hansen-iaspora sp. and Kloeckera sp. The odor is similar to that of acetic acid and more particularly of wine vinegar. Addition of 44 mg/l caused a noticeable change in the odor of a wine and 176 mg/l gave the wine a completely spoiled vinegary odor (Peynaud, 28). Robertson and Rush (17) suggested 150 mg/l as the upper limit.

Faded. See aerated.

Fermentation. The yeasty-like odor of fermenting musts and wines. Probably some hydrogen sulfide is present q.v. Generally attributed to yeast autolysis. Addition of thiamin may or may not increase the degree of yeasty odor (Höhn et al. 29, Jakob 30).

Filter. See aerated and asbestos. Filtration with kieselguhr or through a paper may also give a filter odor.

Flinty. This term is also widely used, especially for low-alcohol high-acid wines from grapes grown on silicious soils. This may be partially a taste reaction with perhaps some unconscious odor associations of such wines. Laederer (31) specifically calls it a flavor particularly noted in wines from the silicious soils of the Dezaley in Switzerland.

Foxy. Traditionally this odor has been associated with wines made from varieties and hybrids of Vitis labrusca, a native American species. Thus cultivars such as Catawba, Concord, Niagara, etc. were said to have a foxy odor. This has generally been attributed to ethyl and methyl anthranilate although recent research indicates that the composition of the foxy odor is more complex. There is no doubt as to the distinctiveness of the odor. Europeans detest the odor because some American varieties with strong foxy odors were planted in Europe during the phylloxera invasion to provide some grapes for wine. The foxy flavor was considered unnatural, and as reducing the quality of their traditional non-foxy V. vinifera wines. California consumers generally prefer non-foxy flavored wines although they drink plenty of foxy-flavored grape juice! Apparently it is an acquired taste. Only a misguided vinifera zealot would criticize the anthranilate odor. See also raspberry.

Frosted. Certainly this is very rare in California. The Frostgeschmack in Germany seems to apply to wines from grapes frosted during the autumn. Otherwise Eiswein would also have a frosted odor. Perhaps they do because in both cases the wines appear to be somewhat oxidized, even though sulfur dioxide may have been used.

Fusel. The higher alcohols, a by-product of fermentation, that are present in wine have pungent odors: amyl, isoamyl, isobutyl, etc. These are collectively called fusel oils and their odor as fusel. During distillation of brandy the concentration of the higher alcohols in the product may be increased in the product, especially if the still is improperly operated. When such spirits are used to fortify dessert wines the fusel oil odor may be found in the resulting wines. Table wines rarely have a fusel odor, though it has been identified as such in a number of wines, possibly from bacterial action, at about 300 mg/l. Some wines fermented at a low temperature have an amyl odor. People differ in their dislike of the fusel odor -- even preferring a moderate amount in some Portugese ports.

Geranium. Wines or musts to which sorbates have been added may develop a very undesirable geranium-like odor if bacteria develop. Crowell and Guymon (32) have identified the undesirable odor as 2-ethoxy-hexa-3,5-diene. Not garlic-like (Broadbent, 12).

Green. The odor of wines made from immature grapes, possibly due to leaf alcohol and leaf aldehyde (Berg and Webb, 11).

Hail. In many viticultural regions hail is common. Depending on its severity it may denude the vine of fruit and leaves. Often it only bruises the fruit. Unless harvested promptly the fruit may brown and oxidized musts result. If the fruit is not harvested promptly, microbial infection may occur with molds, yeasts, and bacterial all present. The resulting wines are said to have a hail odor (goût de grelé, Fr.). Hail is rare in California viticultural areas and I have not encountered wines here with such an odor.

Heated. There are two possible sources: from an unduly high temperature during fermentations (pomace q.v.) and from holding the wine at too high a temperature during pasteurization. If the wine is sweet a caramel odor q.v. may be produced, or the wine may appear aerated q.v. or oxidized q.v.

Herbaceous. (erbaceo, It.) This is not the odor of aromatized wines q.v. due to added herbs. It is found mainly in certain varieties with strong varietal aromas, such as Cabernet Sauvignon and Sauvignon blanc. It is notably found in wines when the grapes are harvested late in the season in cool regions usually when vegetative growth continues, or when the fruit is unripe. It has also been called the bell-pepper odor in California. The odor is difficult but not impossible to blend-out. It has been noted in the Loire and Bordeaux in France, and in California in grapes of the Santa Cruz, Monterey, and other South Coast areas. Hexanal and hexanol do not seem to be the responsible compounds but 2-methoxy-3-isobutylpyrazine may be. Not objectionable at low concentrations. The French herbace has a connotation of added herbs and green fruit.

Hydrogen sulfide. (Böckser, Ger.) This is the rotten-egg odor. The threshold is not more than 0.001 mg/l. It is produced during alcoholic fermentation, primarily by reduction of elemental sulfur from the grapes. Currently enologists (Vos and Gray, 33) attribute some sulfide formation to nitrogen deficiency and possibly to arise from sulfur-containing amino acids, from sulfate, from elemental sulfur and in cases, from reduction of sulfite. Judicious use of sulfur in the vineyard and addition of ammonium phosphate to low-protein musts is recommended.

To remove hydrogen sulfide was traditionally accomplished by early racking (removal from the lees), aerating, and adding sulfur dioxide. Addition of a trace amount of copper (1 to 2 mg/l is usually sufficient) as copper sulfate. Early removal is recommended before less volatile mercaptans are formed.

Lactic. See sauerkraut.

Leafy. See stemmy.

Lees. Wines that have been allowed to remain on the first lees for several months, and especially at a warm temperature (over 20° C) develop an obnoxious odor. In some cases it seems to be of bacterial origin and resembles the mousy odor q.v. In other cases autolysis of yeasts occurs and a rotten odor (sulfides or mercaptans) appears. Some such wines have a bitter taste.

Maceration carbonic. Wines so fermented have a special odor -- disagreeable to some and not so to others. Certainly an acquired taste if desirable.

Madeirized. (maderisé, Fr.) In Madeira wines this is obviously a desirable odor. But old white wines, particularly sweet wines, may develop a madeirized odor that is inappropriate to the type and which is also often unpleasant. Caramelization q.v. is probably involved.

Medicinal. See cresol and petroleum. Corks and open barrels stored near certain pharmaceuticals, especially phenols and chlorophenols may acquire and transmit to wines a medicinal or "drug-store" (Apothekergeschmack, Ger.) odor. See Andrews (2). Rare.

Mercaptans. These high-boiling point compounds are disastrous to the quality of any wine. They have an onion-like or garlic-like odor. They have very low thresholds so little is required to spoil a wine. Ethyl mercaptan and ethyl sulfide are especially obnoxious. Not all mercaptans are disagreeable. Dimethyl sulfide is present in normal bottled wines. The threshold value, 0.05 mg/l, is exceeded in old white wines during storage (20). For its formation in wines see Marais (34). In sake it is considered an off-flavor above 6.2 to 11.6 mg/l (Takahashi et al. (35).

Mildew. See moldy. Amerine and Roessler (8) made a subtle distinction between the mildew and moldy odor, perhaps too subtle.

Moldy. This odor may arise from molds growing on the grapes or from contact of the must or wine with surfaces on which molds are growing. *Penicillium* sp., *Aspergillus* sp., *Rhizopus* sp., *Uncinula necator* (oidium) etc. may be involved. An ounce of prevention is worth a pound of cure in this case. The odors are persistent and difficult to blend-out. There are reports of trying to extract the foreign odor by mixing the wine with olive oil. Unless the olive oil is very pure, the treatment may create more problems than it cures. See also corky. Sweet wines made from white grapes attacked by *Botrytis cinerea* have a moldy botrytis odor, but not an undesirable one unless too pronounced. It is even possible for wines to pick up a moldy odor from molds on the outside surface of barrels (Ruiz Hernandez, 36). This may be simply an indication of general unsanitary winery practices.

Mousy. (vin tourné, Fr.; girato, It.) This was one of the most common words in the enologists' vocabulary, 75 to 100 years ago. (goût de souris, Fr.; Mauselgeschmack, Ger.). It is produced by bacterial action, often in low acid wines. Various microorganisms are responsible. Sometimes there is a production of gassiness. Often the pH increases and the total acidity decreases. In most cases, the volatile acidity is not unduly high, but the lactic and propionic acid contents may be higher than normal. It is probably due to high boiling point compounds, since it is most easily recognized when some of the suspected wine is placed between the palms of the hands and the hands rubbed together and then smelled. Acetamide and acrolein have been suggested as causal agents. The suggestion of Hallgarten (19) that it may be due to added chemicals is unique and doubtful. It is rare nowadays because of the regular use of sulfur dioxide and pure yeast cultures, and our super-clean cellars.

Musty. This term is used by English connoisseurs for the unpleasant odor of red table wines when first poured into the glass. Possibly the presence of threshold amounts of some moldy constituents may disappear and explain the phenomenon. Amerine and Roessler (8) could not define it. Lachman (37) wrote, "The musty taste (sic) is not noticed until fully half a minute after the mouth has dried." Did he mean earthy q.v.?

Overaged. See oxidized. However, overaged is more than just oxidized. It is a general breakdown of the desirable characteristics of the wine plus the accumulation of acetaldehyde and related compounds. The French suranne or passé and the Italian decrepito or passato express it better.

Overripe. As grapes ripen, they finally begin to shrivel and eventually acquire an overripe odor. With some varieties (Pinot noir, for example) this ripe-grape odor is not considered undesirable. But as the drying continues and the grapes take on a raisin q.v. flavor, then the odor becomes undesirable (except in the case of Malaga wines, which, of course, are made from partially raisined wines). Zinfandel is a variety which is subject to early shriveling of some of the berries, and an overripe odor in its wines is not uncommon. Amerine and Roessler (8) recommended distinguishing overripe from raisin, "if you can." The French figue expresses it well.

Oxidized. (eventé, Fr.) One of the most used terms in the modern enologist's vocabulary. It is a general term for an undesirable oxidation of a wine, primarily of white table wines, and indicates an accumulation of acetaldehyde. The descriptions of the Office International de la Vigne et du Vin (5) are confusing in the different languages. Undesirable contact with air with formation of acetaldehyde is the best definition. The OIV listing confuses it with butyric q.v. spoilage or possibly to the rancid odor of red wines aged in contact with the air for a long period of time. With 1 mg/l of biacetyl, Pisarnitskii et al. (38) found 8 to 20 mg/l of acetaldehyde gave an oxidized odor. See aerated for a less severe form, and madeirized for a different form, rancid for a more severe form.

Peppery. See herbaceous. English connoisseurs use it for

the odor of young red wines of the Rhone and also for young vintage port. Neither is apparently related to the varietal-derived herbaceous odor described above. There is, however, a peppery odor in some young white table wines. Amerine and Roessler (8) ask whether it is due to some sulfur-dioxide related concentration (or compound). Broadbent (12) suggests "probably higher alcohols," though none of the usual higher alcohols have a peppery odor.

Petroleum. Petroleum is used in such a variety of ways in the vineyard and in the winery that it is no surprise that it may occasionally become a contaminant. See phenol for a similar case. Petroleum is used for grape stakes, in smudge pots, and in all sorts of machinery and engines. Compressed air is sometimes used to mix the fortifying spirits with the fermenting must being fortified. If oil from the compressor gets into the compressed air line, it may get into the wine. Railroad cars that have been used for petroleum or other chemicals are sometimes used for moving wine. Obviously, unless they are thoroughly cleaned, they may contaminate the wine. Bottles that have been used for petroleum products are sometimes returned to the winery for re-use (not so much in this country as in Europe). It is very difficult in the usual automatic washing machine to remove all of the petroleum residue. Grapes grown near freshly macademized or tarred roads or near where these are prepared may acquire an off-odor which they transmit to the resulting wines. Murphy's Law again!

Phenol. Pentachlorphenol (PCP) is used as a preservative for wooden stakes in vineyards. Its own odor is not so unpleasant, but it is methylated and dechlorinated by various soil fungi (bacteria?) thus producing obnoxious odors. Haubs (39) believes it is related to the Muffton of some German wines. As long ago as 1950, Brown (40) noted that use of Gamma Hex 100S (Registered trademark). This contains 1% other isomers of hexachlorocychohexane, 50% sulfur, and 47% inert ingredients for spraying vines and could result in an off-odor (phenol?) in the wines. No more recent report appears.

Plastic. Reported in wines stored in plastic containers containing free styrene.

Pomace. Most California enologists identify the odor as being produced from hot fermentations of red wines on the skins. Others identify it with fermentations that have remained on the skins too long (Trestergeschmack, Ger.). See the stemmy taste. It is often associated with a rather high pH (3.8 - 4.0). It is also present in some pomace brandies, apparently from rancidification of seed oils. See rancid.

Raisin. Wines made from raisins have the odor. See overripe for an earlier stage. Except where designed, as in Malaga, it is considered a defect, especially in table wines.

Rancid. I have not been unfortunate enough to come across a rancid wine. It is said to be found in wines that have been held in contact with the seeds for a long while. Williams and Strauss (41) reported unsaturated odoriferous aldehydes in spirits. They were derived from oxidation of fatty acids in grape seeds. The odor was removed by passing through a strong anion-exchanger in the bisulfite form. In beer, Hannack (15) suggested isovaleric acid as a source of a rancid odor (at about 8 mg/l). When olive oil was used to cover the surface of wines, rancidification of the olive oil would be a possible source. That is no longer practiced. Not to be confused with the desirable rancio odor of old red wines.

Raspberry. This odor is reported in some of the new interspecific hybrids tested in Switzerland. See primarily foxy.

Reduced. (reduit, Fr.) A wine that has been insufficiently aerated. See hydrogen sulfide and mercaptan.

Resin. Few, if any, containers for wine in this country are made of pine. Wines stored in such containers will acquire a resin (turpentine) odor. Of course, ground-dried resin is added to wines in Greece in making retsina. If non-retsina wines are placed in containers that have contained retsina, they may pick up the odor. Even filters and hoses that have been used for retsina must be very thoroughly cleaned before passing non-retsina wines through them, lest they become tainted.

Rubber boot. Brown (40) gave cooked cabbage, straw-like, and amyl alcohol-like as synonyms. He postulated two possible sources: use of fortifying brandy with a fusel-like odor, and use of fortifying brandy with a "burnt" and rancid grape seed oil odor. A similar odor, rubbery, he believed to be due to hot wines being passed through poor quality rubber hose. Berg and Webb (11) associated it with high (4.0) pH dessert wines. In Australia usage rubbery has a connection with hydrogen sulfide and organic sulfides (Rankine, 42).

Sauerkraut. Wines infected with lactic acid-producing bacterial will acquire this odor. Wines that have undergone too much malo-lactic fermentation sometimes show it. Laederer (31) noted that bacterially-spoiled wines, tourné, Fr., have a sauerkraut odor. It has been noted in wines fermented under pressure, especially if the original musts have an above-normal pH. Since pure lactic acid has a mild odor, Amerine et al. (7) suggested it to be due to secondary odorous by-products.

Smoky. Roman wines that were smoked in clay amphora were said to acquire a smoky odor. But wines are no longer "smoked", and even if they were, how would the smoky odor come in contact with the wine?

Sophisticated. Addition of a foreign odor. See aromatized.

Stagnant. Cooperage is often filled with water when not required for wine. Unless an adequate level of acidity and sulfur dioxide is maintained, various slime-producing microorganisms may develop and grow in the water, producing a foul odor (some hydrogen sulfide). This stagnant water odor is difficult to remove, and unless removed will contaminate the wine placed in the cooperage. Goût de croupi, Fr., and sapore di muffaticcio, It. perhaps express it better. Except in small poorly-managed cellars (and with home wine-makers) it is rare.

Stale. See moldy or musty or stagnant.

Stalky. See stemmy.

Stemmy. See also the stemmy taste. An odor reminiscent of green stems is believed due to too long a contact of green stems in the fermenting must. The French goût de rafles and the Italian sapere di graspo are similar. Its herbaceous q.v. odor has been noted. Leafy and stalky are probably similar. Berg and Webb (11) suggested that stemmy was a "modified complex of the bitter and green sensations."

Straw. The practice of drying grapes after the harvest on straw to increase the sugar concentration has largely disappeared. Presumably the straw was the source of the odor. A straw-like odor has been noted in a few California wines of high pH. Possibly this is due to some bacterial infection, but no studies are available.

Sulfide. See hydrogen sulfide.

Sulfur dioxide. This is surely the major off-odor of modern white table wines. In any wine it is an off-odor and ipso facto therefore undesirable. But its desirable chemical properties make its use ubiquitous. The enologist must ask himself, which is the least undesirable, a spoiled or oxidized wine or an unspoiled unoxidized wine with some sulfur dioxide odor? Thanks to more careful laboratory control, excess sulfur dioxide is now much less common than it was twenty years ago. Still the free sulfur dioxide odor in some low alcohol, high acid, low pH white table wines is sometimes more than just undesirable, and at best is regrettable. Robertson and Rush (17) suggest 150 mg/l as the limit. The threshold for sulfur dioxide is difficult to establish, probably because of the rapidity of adaptation (Amerine et al., 7; Amerine and Roessler, 23). Some experienced judges are either insensitive to or do not object to the odor.

Sunburned. This is reported to occur when immature grapes are sunburned. I have not observed it, and overripe q.v. and raisin q.v. may be close enough synonyms. However, Amerine and Roessler (8) say it "may not be the same odor as raisin."

Vinegary. See acetic.

Weedy. Various weeds growing in the vineyard may get into the grapes as they are being harvested. They may communicate undesirable odors to the resulting wines. Chenopidium and Aristolochia may be involved among other weeds. A similar situation was noted in wines from a certain California vineyard which had a row of eucalyptus trees along one side of the vineyard. Enough of the leaves of the eucalyptus got mixed with the harvested fruits to contaminate the resulting wines.

Whiskey. Empty charred oak barrels that have contained bourbon whiskey are occasionally used by wineries. Unless thoroughly cleaned and preferably re-coopered, they may acquire a whiskey odor.

Woody. (goût de bois, Fr.) Wines left in the wood for too long a time eventually acquire a distinct and unpleasant woody odor. This occurs more rapidly in small containers, owing to their greater surface to volume ratio. Some white wines need little wood aging. Some red wines need more. It is the enologist's command decision to decide what the proper amount of wood character his wines should have. Different types of wood give slightly differing types of odors to the wine. American oak gives the wine a distinct vanillin odor which some winemakers like and others dislike. Charcoaled barrels, such as are used in the whiskey industry, are seldom used for wines, and if they are, they should be scrapped or re-coopered before use. See whiskey. One of the undesirable compounds seems to be 3-methyl-4-hydroxyoctanoic acid-4-lactone ("oaklactone") in the trans form (Kepner et al. (43). The odor is reminiscent of coconut. Salo et al. (44) reported the threshold for this compound to be 0.05 mg/l (in 9.4% w/w/ethanol). For the cis- and trans- forms Otsuka et al. (45) reported thresholds of 0.8 and 0.07 mg/l only with wines stored too long in a new cask. Not so. With time, even in used barrels the wine may acquire a woody odor. Singleton and Noble (9) note that many phenols of the oak extract contribute to the "woody" odor.

Yeasty. See fermenting. In Australian usage, yeasty is used more for oxidized wines (Rankine (42).

Tactile

Astringent. (rude, Fr.; Herb, Ger.) This puckery sensation is more common in young wines and generally is reduced to not undesirable levels as the wines age. The sensation is more pronounced in the presence of a moderate amount of tartaric acid along with the polyphenolic compounds. The resulting feel is harsh. Singleton and Noble (9) attribute the puckery sensation with precipitation of the proteins, saliva, and the mucous surfaces. This is due mainly to "tannins," the larger (molecular weights of 500 or more) natural polyphenols. Mildly astringent red wines have a total phenol content of 1300 mg/l (as gallic acid equivalent). A wine with 2000 mg/l is excessively astringent. Singleton and Noble (9) indicate that in deference to the public's preference the total phenol content of commercial California wines has been decreasing.

Not to be confused with the bitter taste q.v. Although astringency decreases with age, who can wait 20 years? Andrews (2) also reported that detergents that accidentally get into beer may result in an astringent sensation. Ribéreau-Gayon (46) defined bad tannins as those "which give wine an aggressive astringency."

Burny. See hot.

Gassy. A certain amount of carbon dioxide is present in all wines and sparkling wines require an excess. Still, white and red table and dessert wines should not be perceptibly gassy. The prickly sensation on the tongue is somewhat unpleasant, particularly in dry red wines. It is not difficult to remove a good deal of the excess carbon dioxide by decanting the wine before serving -- one of the justified uses of decanting. Note also that young wines which contain a slight but noticeable odor of hydrogen sulfide or some other undesirable volatile odors may be similarly decanted and thus made palatable, or at least less undesirable. But this is true only when the amounts present are very near the threshold concentration. White Swiss wines may be gassy, and Swiss connoisseurs use a variety of words to express the nature of the gassiness: déluré, dévergondé, eveillé, pimpant, as well as petillant (Laederer, 31). Recently the Bordeaux enologists have shown that there is a low but optimum amount of carbon dioxide for Bordeaux red wines.

Hard. (dur, Fr.) Wines of high acidity and low alcohol and extract often have a hard (dureté, Fr.) tactile sensation which is unpleasant. It may be that some wines with just high acidity and a slight astringency also give this or a similar sensation. In European languages there is also sometimes a connotation of high acetic acid content (Ribéreau-Gayon,(46).

Harsh. (acre, raide, Fr.) This means lots of astringency q.v.

Hot. (chaud, brûlant, capiteux, Fr.) Wines of high alcohol content are almost always hot (goût d'echaud, Fr.). For reasons that are not clear, some wines and brandies of the same alcohol content are more hot than others. The sensation is sometimes at least partially masked by the sweet taste. It is, therefore, more often observed in high alcohol (over 14 percent) red wines and in ports of 20 or more percent alcohol. What the "hot" constituent of the ethanol solution is, I do not know. Burny is a synonym. Some enologists associate it with wines fermented at high temperatures, but I do not know of a controlled experiment. Singleton and Noble (9) discuss a "hot" pungency with an odor parameter. It is not clear whether pungency also has a tactile sensation vis-a-vis "hot."

Metallic. Probably some juxtaposition of taste and tactile sensations. But when to use? The French equivalent is styptique, which indicates it is a tactile sensation.

Rough. (âpre, Fr.; aspro, It.) Perhaps more astringent than hard and less than harsh.

Thin. (faible, Fr.) Wines of low alcohol and extract, particularly if dry, are properly so designated. However, what constitutes the proper alcohol for a non-thin wine seems to vary with the type of wine. Even fairly low-alcohol dry white table wines may not be downgraded by many consumers as being thin. On the other hand, someone who is accustomed to drinking dry red wines of 12 percent ethanol will reject a 10 percent dry red wine as being thin or watery. The French equivalent is mince, but anémique seems more descriptive.

Watery. See thin.

Difficult to Classify

Amerine and Roessler (8) have given a list of words, both positive and negative, which they believe should be used cautiously, because they have not been defined, or are undefinable, or at least there is no agreement on their meaning. They do note that if the judges all agree on their meaning they might be used. Some negative words are listed here with suggestions as to what the problems are.

Some if these words have to do with quality judgements (ordinary, poor, for example). The problem is that a visual-minded judge will use them for off-color or off-appearance, but not for undesirable taste, odor, tactile, or over-all impressions. Other judges will use them for other combinations of sensory response.

Austere. (austère, Fr.) Is it too much or too little taste, odor, or tactile sensation? Or is it astringent and rough, q.v.?

Bacon. Used by some California enologists. Meaning not clear, but undesirable.

Bite or Biting. Maybe related to high acidity or astringency or bitterness, or a combination of these?

Chalky. Earthy?

Coarse. General disagreeableness. Berg and Webb (11) associate it with poor balance due to excessive acidity or astringence. Others associate it with poorly-made wines, while some enologists use it for immature red wines. Take your choice, but tell us what you mean. Not related to corsé, Fr., which has to do with flavor and alcohol content.

Common. Generally lacking character.

Dirty. Off-odors from whatever source.

Dumb. (muet, Fr.) Like neutral? or lacking character? or what? Used commercially in France (muet) for lack of odor in wines following racking. Such wines may develop well later.

Empty. See dumb.

Flabby. Maybe flat?

Goaty. Surely an odor. See butyric.

Heady. (capiteux, Fr.) Too much alcohol. See hot. Also, too much carbon dioxide (Grazzi-Soncini (13).

Heavy. Too much alcohol?

Insipid. Neutral?

Lean. See thin or?

Light. See thin or?

Lithe. Good or bad?

Meager. (maigre, Fr.) See thin.

Meaty. See bacon, unless used figuratively for a wine of full body.

Old. (vieillardé, passé, Fr.) See oxidized.

Ordinary. In what?

Poor. A quality judgement based on what? Lack of charm, one French enologist writes!

Rotten. (pourri, Fr., Faul, Ger.) From what? See hydrogen sulfide, moldy, stagnant, etc.

Severe. See austere. Presumable harsh q.v. from immaturity or from excess tannin and acidity (Broadbent,(12).

Sharp. Used by some for excess acidity or specifically for excess acetic acid, q.v.

Short finish. Lacks after-taste? Does this imply that a bitter-tasting wine with a long after-taste is good?

Sick. Bacterial q.v. or yeasty q.v.

Small. See thin or?

Soapy. Possibly flat q.v., but what is the odorous component due to? Possibly from residual detergents, Andrews suggests (2).

Steely. See metallic, if anything.

Superficial. Without much character. See ordinary.

Tired. Aerated or oxidized.

Unbalanced. Off-odors or -tastes or -tactile sensations or one sensation predominating. The French desequilibre is expressive.

Unclean nose. Specify which odor.

Undesirable odor. Best to specify its chemical nature or origin.

Weak. See thin.

Withered. Tired? or oxidized?

Without charm. Easy to define if you know the meaning of charm.

Unharmonious. See unbalanced. The Italian angoloso expresses it well.

Literature Cited

1. Adlum, J. "A Memoir on the Cultivation of the Vine in America, and the Best Mode of Making Wine" 2d ed., William Greer: Washington, D.C., 1828; p. 179.

2. Andrews, D.A.; Taints: flavour descriptions and origins, Brewer, 1979, 65, 717-726.

3. Amerine, M.A.; Organoleptic examination of wines, Wine Technol. Conf., Univ. Calif., Col. Agr., Davis, Aug. 11-13, 1948, p. 15-26.

4. Harper, R.; Bate Smith, E.C.; Land, D.G.; Griffiths, N.M.; A glossary of odour stimuli and their qualities, Perf. Essent. Oil. Rec., 1968, 59, 22-37.

5. Office International de la Vigne et du Vin. "Lexique de la Vigne et du Vin"; Office International de la Vigne et du Vin; Paris, 1963.

6. Moll, M.; Flayeux, R.; Vinh That; Noel, J.P.; Relations entre les parametres physico-chimiques de la biere et les resultats de degustations, Bios, 1974, 5, 328-333.

7. Amerine, M.A.; Roessler, E.B.; Filipello, F.; Modern sensory methods of evaluating wines, Hilgardia, 1959, 28, 477-567.

8. Amerine, M.A.; Roessler, E.B. "Wines. Their Sensory Evaluation"; W.H. Freeman: San Francisco, 1976; p.230.

9. Singleton, V.L.; Noble, A.C.; Wine flavor and phenolic substances, ACS Symposium Series, 1976, 26, 47-70.

10. Ough, C.S.; Berg, H.W.; Powdery mildew sensory effect on wine, Am. J. Enol. Viticult., 1979, 30, 321.

11. Berg, H.W.; Webb, A.D. "California Wine Types"; Wine Institute: San Francisco, 1955; p. 20.

12. Broadbent, J.M. "Wine Tasting. A Practical Handbook on Tasting and Tastings"; Christie Wine Publications: London, 1973; p. 54.

13. Grazzi-Soncini, G., "Wine. Classification; Wine Tasting; Qualities and Defects"; Translated by F.T. Bioletti. State Printing Office: Sacramento, (Appendix E, Biennial Report, Board of State Viticultural Comm.), 1892; p. 56.

14. Von Sydow, E.; Akesson, C.; Correlating instrumental and sensory flavour data, in "Sensory Properties of Foods"; Applied Science Publishers: London, 1977, p. 113-127.

15. Hannack, H., Schwellenwerte als Kriterium für Qualitäsbiere, Brauwelt, 1973, 113, 699-702, 704.

16. Engen, T.; Kilduff, R.A.; The influence of alcohol on odor detection, Chemical Senses Flavor, 1975, 323-329.

17. Robertson, J.M.; Rush, G.M.; Chemical criteria for the detection of winemaking faults in red wine, Food Technol. N.Z., 1979, 14(1), 3-11.

18. Amerine, M.A.; Ough, C.S. "Methods for Analysis of Musts and Wines"; John Wiley & Sons: New York, 1980; p.x., 341

19. Hallgarten, S.F. "Rhineland Wineland"; Elek Books: London, 1955; p. 119.

20. Anon. "The Australian Wine Research Institute, 25th Annual Report"; Adelaide, 1979; 36 p. (p. 13, 15).

21. Webb, A.D.; Noble, A.C.; Aroma of sherry wines, Biotechnol. Bioengin., 1976, 18, 939-952.

22. Phaff, H.J.; Amerine, M.A.; Wine, Microbial Technology, 1979, II, 131-153.

23. Amerine, M.A.; Roessler, E.B.; Pleasantness and unpleasantness of odors added to white wines, Am. J. Enol. Viticult., 1971, 22, 199-202.

24. Schaeffer, A.; Meyer, J.P.; Guillerm, A.; Étude sur l'origine du 'goût de bouchons', Rev. Franç. Oenol., 1978, 16, 25-29.

25. Götz, B.; Ueber 'Korkwürmer' und ihre Bekämpfung, Deut. Wein-Ztg., 1976, 102, 1222-1226.

26. Tanner, H.; Zanier, C.; Erfahrungen mit Flaschenverschlussen aus Naturkorken, Weinwirtschaft, 1978, 22, 608-613.

27. Bemelmans, J.M.H.; Te Loo, N.L.A., Analysis of earthy taint in malt distillate, Nord. Symp. Liusmedels Sens. Egenskaper, 1976, 4, 129-141.

28. Peynaud, E.; Etudes sur les phenomenes d'esterification, Rev. Viticult., 1937, 1, 86, 209-215, 227-231, 248-253, 299-301, 394-396, 420-423, 440-444, 87, 49-52, 113-116, 185-188, 242-249, 278-295, 297-301, 344-350, 362-364, 383-385.

29. Höhn, E.; Solms, J.; Roth, H.R.; Untersuchung der Geschmacksstoffe der Hefe, Lebensm. Wissen. Technol., 1975, 8, 212-216.

30. Jakob, L.; Neue Weinhandlungsmittel - Eigenschaften und Anwendung, Weinwirtschaft, 1979, 115, 847-850.

31. Laederer, B. "Les Vins Suisses," Éditions Generales, S.A., Génève, 1968.

32. Crowell, E.A.; Guymon, J.F.; Wine constituents arising from sorbic acid addition, and identification of 2-ethoxyhexa-3.5-diene as source of geranium-like odor, Am. J. Enol. Viticult., 1975, 26, 97-102.

33. Vos, P.J.A.; Gray, R.S.; The origin and control of hydrogen sulfide during fermentation of grape must, Am. J. Enol. Viticult., 1979, 30, 187-197.

34. Marais, J., Effect of storage time and temperature on the formation of dimethyl sulphide and on white wine quality, Vitis, 1979, 18, 254-260.

35. Takahashi, K.; Ohba, T.; Takagi, M.; Sato, S.; Namba, Y.; Identification and determination of an off-flavor compound, dimethyl sulfide, in sake brewed from old rice, Hakko Kagaku Kaishi, 1979, 57, 148-157.

36. Ruiz Hernandez, M.; En torno a una degradación de la madera de roble de las barricas de Rioja por effecto de mohos, La Semana Vitivinícola, 1979, 1737, 4439-4443.

37. Lachman, H.; The manufacture of wine in California, U.S.D.A. Bureau Chem. Bull., 1903, 72, 25-40.

38. Pisarnitskii, A.F.; Rodopulo, A.K.; Bezzubov, A.A.; Gregorov, I.A.; Vosprosu ob okislenii vina (Oxidation of wine), Vinodel. Vinograd: USSR, 1969, 29(1), 12-14.

39. Haubs, H., Untersuchungen uber die Ursachen des "Mufftones," Jahresbericht Forschungsanstalt Weinbau, Gartenbau ... Geisenheim am Rhein, 1977, 51-52.

40. Brown, E.M.; A new off-odor in sweet wines, Proc. Am. Soc. Enol., 1950, 1, 110-112.

41. Williams, P.J.; Strauss, C.R.; Spirit recovered from heap-fermented grape marc: nature, origin and removal of off-odor, J. Sci. Food Agric., 1978, 29, 527-533.

42. Rankine, B.C. "Glossary of Wine Tasting Terms"; Roseworthy Agricultural College: Roseworthy, 1979; p. 8.

43. Kepner, R.E.; Webb, A.D.; Muller, C.J.; Identification of 4-hydroxy-3-methyl-octanoic acid lactone 5-butyl-4-methyledihydro-2(3HO-furanone as a volatile component of oak-wood-aged wines of Vitis vinifera var. Cabernet Sauvignon, Am. J. Enol. Viticult., 1972, 23, 103-105.

44. Salo, P.; Nykänen, L.; Suomalainen, H.; Odor thresholds and relative intensities of volatile aromatic components in an artificial beverage imitating whiskey, J. Food Sci., 1972, 37, 394-398.

45. Otsuka, K.; Zenibayashi, Y.; Itoh, M.; Totsuka, A.; Presence and signification of two diastereomes of β-methyl-γ-octalactone in aged distilled spirits, Agric. Biol. Chem., 1974, 38, 485-490.

46. Ribéreau-Gayon, P.; Wine Flavor in "Flavor of Food and Beverages; Chemistry and Technology" (Charalambous, G.; Inglett, G.E., eds.), Academic Press: New York, N.Y., 1978; p. 355-380.

Index

E

F

G